Bioinformatics

Second Edition

BIOS INSTANT NOTES

Series Editor: B.D. Hames, School of Biochemistry and Microbiology, University of Leeds, Leeds, UK

Biology
Animal Biology, Second Edition
Biochemistry, Third Edition
Bioinformatics, Second Edition
Chemistry for Biologists, Second Edition
Developmental Biology
Ecology, Second Edition
Genetics, Third Edition
Immunology, Second Edition
Mathematics & Statistics for Life Scientists
Medical Microbiology
Microbiology, Third Edition
Molecular Biology, Third Edition
Motor Control, Learning & Development
Neuroscience, Second Edition
Plant Biology, Second Edition
Sport & Exercise Biomechanics
Sport & Exercise Physiology

Chemistry
Consulting Editor: Howard Stanbury
Analytical Chemistry
Inorganic Chemistry, Second Edition
Medicinal Chemistry
Organic Chemistry, Second Edition
Physical Chemistry

Psychology
Sub-series Editor: Hugh Wagner, Dept of Psychology, University of Central Lancashire, Preston, UK
Cognitive Psychology
Physiological Psychology
Psychology
Sport & Exercise Psychology

Bioinformatics

Second Edition

T. Charlie Hodgman, Andrew French

School of Biosciences,
University of Nottingham, UK

and

David R. Westhead

Faculty of Biological Sciences,
University of Leeds, UK

Taylor & Francis
Taylor & Francis Group

Published by:
Taylor & Francis Group

In UK: 2 Park Square, Milton Park
        Abingdon, OX14 4RN
In US: 270 Madison Avenue
        New York, N Y 10016

First published 2002; Second edition published 2010

ISBN: 9780415394949

Library of Congress Cataloging-in-Publication data

Hodgman, T. Charlie.
    Bioinformatics / T. Charlie Hodgman, Andrew French, and David R. Westhead. — 2nd ed.
        p. cm. — (BIOS instant notes)
    Includes bibliographical references and index.
    ISBN 978-0-415-39494-9
1.   Bioinformatics.   I. French, Andrew.   II. Westhead, David R.   III. Title.
    QH324.2.H63 2010
    570.285—dc22   2009026566

Editor: Elizabeth Owen
Editorial Assistant: David Borrowdale
Production Editor: Karin Henderson
Typeset by: Phoenix Photosetting, Chatham, Kent, UK
Printed by: T J International Limited, Padstow, Cornwall, UK

Printed on acid-free paper

10 9 8 7 6 5 4 3 2 1

Taylor & Francis Group, an Informa business   Visit our web site at http://www.garlandscience.com

CONTENTS

ABBREVIATIONS

AC	approximate correlation	FAD	flavin adenine dinucleotide
ADME	absorption, disposition, metabolism, and excretion	FE	false exon or finite-element
		FN	false negative
ADR	adverse drug reaction	FP	false positive
AE	annotated exon	GA	genetic algorithm
AI	artificial intelligence	GASP	Gene Annotation aSsessment Project
ALU	arithmetic logic unit	GATE	General Architecture for Text Engineering
AN	actual negative		
ANOVA	analysis of variance	GIGO	garbage in, garbage out
ANSI	American National Standards Institute	GIS	geographic information system
		GUI	graphical user-interface
AP	actual positive	GOLD	Genomes Online Database
ArMeT	architecture for metabolomics	GRAIL	Gene Recognition and Assembly Internet Link
ASCII	American Standard Code for Information Interchange		
		GSS	genome survey sequence
ATP	adenosine triphosphate	GXD	gene-expression database
BE	boundary-element	HCA	hierarchical clustering analysis
BioPAX	Biological Pathways Exchange	HMM	hidden Markov model
BLAST	Basic Local Alignment Search Tool	HSP	high-scoring segment pair
BLOB	binary large object	HT	high-throughput
CASP	Critical Assessment of Structure Prediction	HTG	high-throughput genomic
		HTML	hypertext mark-up language
CCD	charge-coupled device	HTTP	hypertext transfer protocol
cDNA	complementary DNA	IDE	Integrated Development Environment
ChEBI	chemical entities of biological interest		
		IE	information extraction
COSY	correlated spectroscopy	IEEE	Institute of Electrical and Electronic Engineers
CPU	central processing unit		
CT	computed tomography	ILP	inductive logic programming
CVS	concurrent versions system	IP	Internet Protocol
DARPA	Defense Advanced Research Projects Agency	ISO	International Organization for Standardization
DAS	Distributed Annotation System	IT	information technology
DDBJ	DNA Databank of Japan	JRE	Java runtime environment
DIGE	differential gel electrophoresis	KEGG	Kyoto Encyclopedia of Genes and Genomes
DIKW	data, information, knowledge, and wisdom		
		KGML	KEGG Mark-up Language
DNA	deoxyribonucleic acid	LOG	Laplacian of Gaussian spot detection
EBI	European Bioinformatics Institute		
		LOPIT	localization of organelle proteins by isotope tagging
EC	Enzyme Commission		
ECG	electrocardiogram	MAGE	microarray and gene expression
EM	expectation-maximization	MALDI	matrix assisted laser desorption/ ionization
EMBL	European Molecular Biology Laboratory		
		MC	Monte Carlo
ESI	electrospray ionization	MCMC	Markov chain Monte Carlo
EST	expressed sequence tag	MeSH	medical subject headings

MIAME	minimum information about a microarray experiment		PSSM	Position Specific Scoring Matrix
MIAMET	minimal information on a metabolomics experiment		QSAR	quantitative structure-activity relationship
MIAPE	minimal information about a proteomics experiment		RMSD	root mean square deviation
mmCIF	macromolecular crystallographic information file		RNA	ribonucleic acid
			RT-PCR	reverse transcriptase polymerase chain reaction
MMDB	Molecular Modeling Database		SAGE	serial analysis of gene expression
MRC	Medical Research Council		SBML	Systems Biology Mark-up Language
MRI	magnetic resonance imaging		SDEs	stochastic differential equations
mRNA	messenger RNA		SMART	Simple Modular Architecture Research Tool
MS	mass spectrometry			
NAD	nicotinamide adenine dinucleotide		SMARTS	an extension of SMILES
NASA	National Aeronautics and Space Administration		SMILES	Simplified Molecular Input Line Entry System
NCBI	National Center for Biotechnology Information		SMRS	standard metabolic reporting structure
NJ	neighbor-joining		SNOMED	Systematized Nomenclature of Medicine
NLP	natural language processing			
NMR	nuclear magnetic resonance		SNP	single nucleotide polymorphism
NNSSP	Nearest Neighbor Secondary Structure Prediction		SOM	self-organizing map
			SQL	structured query language
NOESY	nuclear Overhauser effect spectroscopy		SRS	sequence retrieval system
			TAP	tandem affinity purification
ODE	ordinary differential equation		TCA	tricarboxylic acid
OMIM	Online Mendelian Inheritance in Man		TCP	transmission control protocol
			TE	true exon
OODB	object-orientated database		TIC	total ion chromatogram
OOP	object-oriented programming		TIFF	tagged image file format
ORF	open reading frame		TN	true negative
PAGE	polyacrylamide gel electrophoresis		TP	true positive
PAM	accepted point mutations		tRNA	transfer RNA
PAUP	phylogenetic analysis using parsimony		UDDI	Universal Description, Discovery and Integration
PCs	personal computers		UML	Unified Modeling Language
PCA	principal components analysis		UMLS	Unified Medical Language System
PDB	protein data bank		UniProt	Universal Protein Resource
PDE	partial differential equation		USB	universal serial bus
PE	predicted exon		UTF	unicode transformation format
PES	potential energy surface		UV	ultraviolet
PHP	personal home page		WE	wrong exon
PHYLIP	Phylogenetic Inference Package		WSDL	Web Services Description Language
PN	predicted negative		WST	watershed transformation
PO	plant ontology		WWW	worldwide web
PP	predicted positive		XML	extensible mark-up language
PRPS	phosphoribosyl pyrophosphate synthetase			

PREFACE

Since the first edition of *Instant Notes – Bioinformatics*, the field has progressed substantially and is well on the way to becoming an established discipline in its own right. We are grateful, therefore, to the publishers for giving us the opportunity to produce a second edition. This has enabled us to restructure the book into a form that has two aims. The first is to provide material for informatics students irrespective of their chosen field – biology, chemistry, medicine, neuroscience, etc. The second is to show how these generic informatics skills are being applied to most aspects of the life sciences, and not simply molecular biology where bioinformatics first flourished.

The Sections have been grouped into three parts. The first (Sections A and B) provides an introduction to the subject. Section A outlines the factors leading to bioinformatics becoming an essential activity. Section B is a brief history of the subject (through a series of definitions of the term *bioinformatics*) from its origins in the 1960s through the exhilarating (if not intoxicating) 1990s to the twenty-first century, where bioinformatics is being applied to all types of biological information.

The second part comprises the building blocks of informatics (Sections C–I): physics, mathematics, and computer science. It contains one major omission, however. Computer programming is an essential skill in bioinformatics, but the space constraints here do not allow us to include adequate training in any useful language. Since this is a particularly practical activity, it is better to leave this to the many other books that are available. However, we have tried to outline the rudiments of good data management and programming practice.

The third part contains the biological application domains (Sections J–R). It comprises three subgroups: molecular biology; metabolism, anatomy, physiology; and complex information sources (specifically image datasets and natural-language texts). The latter remain the hardest places from which to extract accurate and quantitative data. The second and third parts interlock as shown in the figure below. This emphasizes the foundational importance of the building blocks. The consequent reliance on them from all of the application areas is clear from the tight mesh of interconnections linking the two together.

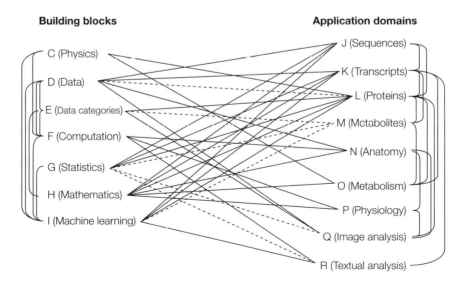

Interconnections between sections. The sections of the book for the building blocks and application areas are shown in two columns. The connections between them are shown by solid lines where they are linked by related topics, and dashed lines where the connections are implicit in the text.

Bioinformatics is now sufficiently broad that even the three main authors of this work felt it appropriate to draw in others to write particular Sections. Thus, we wish to thank several people for contributing: Sections J (Nicola Gold), K (Alex Marshall), L (Nicola Gold and Tom Gallagher), and M (Rob Linforth). Various people also checked the accuracy and clarity of various Sections, so we gratefully acknowledge the help of Alastair Middleton, Leah Band, Tom Gallagher, Kim Kenobi, and particularly Jane Hodgman (who also proofread many of the sections). We are grateful to members of the UK Centre for Plant Integrative Biology for providing microscope images ahead of publication. Readers are liable to find some duplication, but this is retained for the purposes of clarity. Finally, we hope that students and teachers alike will appreciate the breadth of the subject and enjoy reading this work.

A THE CHANGING FACE OF RESEARCH BIOLOGY

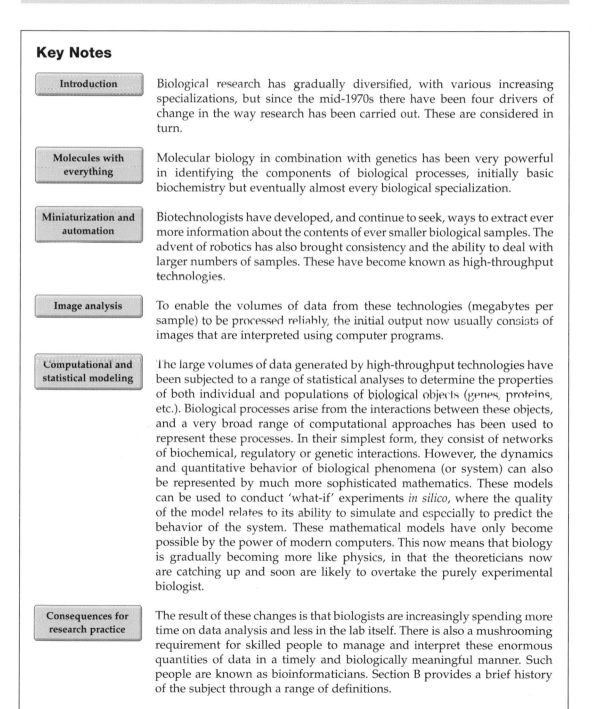

Key Notes

Introduction

Biological research has gradually diversified, with various increasing specializations, but since the mid-1970s there have been four drivers of change in the way research has been carried out. These are considered in turn.

Molecules with everything

Molecular biology in combination with genetics has been very powerful in identifying the components of biological processes, initially basic biochemistry but eventually almost every biological specialization.

Miniaturization and automation

Biotechnologists have developed, and continue to seek, ways to extract ever more information about the contents of ever smaller biological samples. The advent of robotics has also brought consistency and the ability to deal with larger numbers of samples. These have become known as high-throughput technologies.

Image analysis

To enable the volumes of data from these technologies (megabytes per sample) to be processed reliably, the initial output now usually consists of images that are interpreted using computer programs.

Computational and statistical modeling

The large volumes of data generated by high-throughput technologies have been subjected to a range of statistical analyses to determine the properties of both individual and populations of biological objects (genes, proteins, etc.). Biological processes arise from the interactions between these objects, and a very broad range of computational approaches has been used to represent these processes. In their simplest form, they consist of networks of biochemical, regulatory or genetic interactions. However, the dynamics and quantitative behavior of biological phenomena (or system) can also be represented by much more sophisticated mathematics. These models can be used to conduct 'what-if' experiments *in silico*, where the quality of the model relates to its ability to simulate and especially to predict the behavior of the system. These mathematical models have only become possible by the power of modern computers. This now means that biology is gradually becoming more like physics, in that the theoreticians now are catching up and soon are likely to overtake the purely experimental biologist.

Consequences for research practice

The result of these changes is that biologists are increasingly spending more time on data analysis and less in the lab itself. There is also a mushrooming requirement for skilled people to manage and interpret these enormous quantities of data in a timely and biologically meaningful manner. Such people are known as bioinformaticians. Section B provides a brief history of the subject through a range of definitions.

Related sections	Definitions of bioinformatics (B)	Transcriptomics (K)
	Probability and statistics (G)	Proteins and proteomics (L)
	Modeling and mathemetical techniques (H)	Image analysis (Q)

Introduction

Up until the final quarter of the twentieth century, biological research consisted of observing and making descriptions of the form of biological 'objects' and their behavior or function in response to different environmental and genetic contexts. Initially, these 'objects' were individual plants or animals that were visible to the naked eye (biology), or groups of organisms and the way they interact with each other and the environment (population biology, ecology). Surgeons and botanists began to study the organs within animals and plants: their physical distribution/connectivity and their role in the organisms' life (anatomy, physiology). Microscopy paved the way to study smaller objects (cell biology and microbiology). Then from the end of the nineteenth century, chemists took an increasing interest in the molecules found in living things (biochemistry), and these efforts were spurred on by the influx of physicists, especially after the Second World War, who developed the technologies to purify and investigate the structure and function of macromolecules and study the metabolic fate of radio-labeled biochemicals. However, four developments are driving biomedical research away from the descriptive to a more hypothesis-driven approach. These drivers will be examined in turn.

Molecules with everything

Gene cloning and sequencing techniques were developed in the 1970s and this resulted in a change to the way people studied their biology during the 1980s. Instead of trying to identify and characterize the protein involved in some biological mechanism, molecular genetic techniques led to characterization of its gene or transcript, from whose sequence the protein sequence was determined. These biological mechanisms moved from microbiology to aspects of the physiology and pathology of higher plants and animals. In *Drosophila*, major inroads into embryonic development of body pattern formation took place. Others turned to looking at the same gene in different organisms of the same species – this became even easier to do with the development of polymerase chain reaction techniques in the late 1980s. Hence, population genetics and population dynamics began to look at specific sequence variants within and between populations. For humans, this lead to the proposal of a 'mitochondrial Eve,' who represented the small number of women who migrated out of Africa very early in human population dispersal. Many more such migration studies have taken place, taking molecular biology into the realm of anthropology, and some studies have tried to link such molecular data to the development of languages. In a different vein, certain genomic sequences evolve more rapidly so that some sequences are unique to a small number of closely related family members. Apart from their use in legal cases of paternity or maternity, they have drastically changed the way that serious crime is investigated. It is routine now in forensic science and scene-of-crime investigation to collect samples for DNA testing. Similar DNA testing has been used also in archaeology to determine the origins of unearthed skeletons.

Miniaturization and automation

Most people are familiar with the progressive use of smaller and smaller containers in laboratories: test tubes gave way to 1.5 ml plug-cap tubes to multi-well plates. To handle the ever smaller volumes, glass pipettes gave way to micro-pipettes with disposable tips (dispensing down to ~0.5 microliters), micro-syringes (down to nanoliters), and glass-drawn capillaries (down to picoliters). This smaller scale makes expensive reagents (e.g., purified enzymes) go further; one batch can be used for more experiments, resulting in service companies and organizations (such as hospital laboratories) carrying out multiple experiments simultaneously. However, carrying out experiments in parallel is very tedious laboratory work for whoever carries out the work. The development of machines to carry out these repetitive tasks (robots) has, on the one hand, saved the sanity of many laboratory technicians, and on the other, enabled the throughput of tests to increase ever further. Hence, we have a range of techniques referred to as **high-throughput** (often abbreviated to 'HT') technologies. Other laboratory techniques have been transformed also into HT technologies.

Polyacrylamide gel electrophoresis has been used to separate biological macromolecules since the 1960s. Throughput has been increased in several ways. Smaller sample volumes could be handled by using thinner gels and narrower lanes. From the late 1980s, some DNA sequencers have reduced the gaps between lanes by the use of a 'shark's tooth comb.' Some gel electrophoresis apparatus allows multiple gels to be run simultaneously.

A further key development was the ability to immobilize macromolecules on filters that are then probed using labeled reagents. Ed Southern developed this approach to identify homologous DNA fragments; thus were 'Southern blots' invented. After not very many months, the ability to immobilize and probe RNA (ribonucleic acid) and proteins was developed (known as northern and western blots, respectively). Eastern blots have never caught on, however. The filters were made originally of nitrocellulose, which was physically strong enough to be handled while chemically capable of irreversibly bonding to polynucleotides. They were also, however, somewhat brittle and highly flammable when dry, so it is no surprise that plastic-based alternatives were developed. Miniaturization of these filter approaches resulted in a move to using photolithography on glass slides to pack spots far closer together. Thus were created the 'microarrays,' on which an area of 1 cm^2 carries ~50 000 spots (see Section K for further details).

Spectroscopic methods have been consistently refined over the last 50 years with particular emphasis being placed on sensitivity. Mass spectroscopy originally could only be used for separating metabolites. However, procedural and technical developments made it possible to detect the masses of oligopeptides with accuracies of <1 Dalton. The masses of an oligopeptide mixture whose lengths differ by successive amino-acids can be captured, and the peptide sequence inferred with some possible ambiguities (for example, leucine and isoleucine have the same mass) from the mass differences. The values coming out of the spectrum also can reveal data on post-translational modification and the distribution of isotopes that might have been used as part of the laboratory procedure. There will be more on this in Section L. The proteins that constitute the source material for these spectroscopic analyses often come from one- or two-dimensional polyacrylamide gel electrophoresis. However, recent developments have turned to combinations of ion exchange and reversed phase chromatography to avoid losing low levels of protein during the gel extraction phase. The disadvantage of this approach is that larger numbers of aliquots need to be analyzed as part of a single experiment.

The increasing strength of magnets has enabled biomedical research developments in several directions with respect to nuclear magnetic resonance

(NMR). On the one hand, it is now possible to tease apart and determine quantitatively the fine interactions between metabolites and proteins. Metabolite mixtures can be resolved sometimes, though in complex mixtures one can still obtain information about compound classes. On a different scale completely, magnetic resonance is now used routinely in hospitals and medical research settings for non-invasive imaging of what is happening inside tissues (http://www.medicinenet.com/mri_scan/article.htm). On this scale, there are also positron emission tomography and computerized tomography.

Finally, with the above robotics procedures and advances in laboratory techniques, the time required to produce proteins in large quantities and crystallize them has decreased considerably. For soluble proteins, what took years now often only requires months. Novel protein structures derived by both X-ray crystallography and NMR are now being determined at ever faster rates. Where one protein sequence is similar to that of another, whose structure has been determined experimentally, then substantial time can be saved using isomorphous replacement, rather than having to produce extra crystals containing heavy metal derivatives to solve the X-ray phases that lead to electron density maps.

Image analysis

All the above are leading to a situation where the volume of data that can be generated on some biological process is growing exponentially. This situation is facilitated, some would say exacerbated, by the growing trend of capturing data in the form of images for subsequent computer processing. The first images were, of course, drawn by hand, but photography has been the mainstay for publishing research results for many decades. The advent of computer capture and analysis of images made the step change in the way that automated techniques generate, interpret, and output biological data. The techniques for image analysis will be outlined in more detail in Section Q, but at this stage the reader should note just how ubiquitous image generation and analysis have become. A non-exhaustive list includes gel electrophoresis (including DNA sequencing) and probing of blotted filters, array profiling (of multi-well plate experiments and immuno-assays, as well as microarrays), microscopy, video footage, and the various types of scans mentioned in the previous section. A single image file can range in size from a few kilobytes to over a gigabyte. If a dataset consists of a series of images, either as a time-series dataset or for a 3D-image reconstruction, the dataset can extend easily into the tens of hundreds of gigabytes. That is a lot of data to store, retrieve, and analyze.

Computational and statistical modeling

A **model** is a simplified representation of something. This 'something' could be either some biological process or behavior, or a body of data relating to biological objects. Experimental observations of one or a population of some object may be imprecise. Where the imprecision is large, the resulting experimental data are said to be **noisy**. Statistics (see Section G) are required to extract meaningful results from such data, to define properties of a population as a whole, and to provide quantitative measures of how similar (or different) one item is to another when compared with the rest of the population (for example, the similarity between two protein sequences or between leaf shapes).

A **biological system** is the collection of processes that take place within and between objects, which may be molecules, cells, organisms or populations of different species. System behavior, in terms of changes over time or the response of the system to some externally induced change, can also be modeled by statistical techniques. However, such models tend to be used to give an overall impression of the system's behavior and say little about the actual reactions taking place within the system. These approaches underpin many attempts to predict the effects of external factors on a system. Examples include carcinogenicity of compounds, and binding affinities of ligands to proteins (often called quantitative structure-activity relationships, QSAR).

Models that include representations of the internal mechanisms of the system require other types of mathematics. These mathematical techniques are summarized in Section H. Such models have tremendous potential, and have come into their own only in recent decades as computers have become powerful enough to run simulations of biological system behavior in realistic timescales. A model is only as good as its ability to simulate the behavior of the system or, better still, to predict the behavior of the system in response to perturbations. Statistical approaches can provide quantitative measures of how good a model is. However, mechanistic models can also be used, in conjunction with actual data on the behavior of the system, to determine system parameters that might not be measurable directly. This is known as **parameter fitting**. Perhaps the archetypal example of this is the representation of enzyme activity by the Michaelis-Menten equation. This is a model of such long standing that many biochemists treat it as reality, rather than as just a representation of how enzymes actually behave. Most undergraduate biochemists have measured then plotted initial reaction velocities of an enzyme at a range of different substrate concentrations to determine the kinetic parameters intrinsic to the enzyme: V_{max} (maximum velocity) and K_m (Michaelis constant).

Consequences for research practice

These automated or high-throughput techniques, coupled with large volumes of data to manage through ever more powerful computer systems, have changed the way research biologists work. Increasingly, an experiment might only take a few days to carry out in the laboratory, but then several months in the office, to process the resulting data. Where software solutions have been found to streamline the data analysis, the whole research process is speeded up, and biologists are free to choose the path of finding new ways to make use of the data they already have, or develop new techniques to investigate their research area in a novel way. Paradoxically, the IT (information technology) specialist and computer scientist are coming out of their offices and into the laboratories to interface new laboratory equipment to computers. As a result, they are discovering the technical details relating to the quality of the data, how it might best be managed and analyzed, and how they might be integrated with the rest of the biological information space.

There is a particular role for people working at this interface between the laboratory and the office. Such people are now known as **bioinformaticians.** Their world and work will be outlined in the next section.

B DEFINITIONS OF BIOINFORMATICS

Key Notes

Introduction	Bioinformatics is an emerging discipline whose scope has broadened and importance increased over time. A series of definitions is given to reflect these changes and the continuing excitement in the subject.
Antediluvian origins	The term *Bioinformatics* was first used in 1968 and the content of a bioinformatics course was first defined in 1978. These predate the techniques leading to the flood of biological data in which we now swim.
Royal definition	In the mid-1990s, there was great excitement about the subject, which generated a spectrum of strong opinions both for and against it. On occasions, there was more money than sense spent on the area.
Canonical definition	Bioinformatics is the discipline at the interface between biology, information science, and mathematics. This definition broadens the subject from just the analysis of biomolecular sequences and structures to include other types of biological data (biomolecular profiles, interactions, population and cell biology). The multi-disciplinary nature of the subject and the role of a bioinformatician as an interpreter are implicitly acknowledged.
Functional definition	This specifies what people endeavor to find out using bioinformatics, and shows how the subject's building blocks fit into this scheme.
Public services definition	This wordy definition attempts to capture all the aspects of the subject, including its associations with other flavors of informatics.

Related sections

Essentials of Physics (C)	Genomes and other sequences (J)
Data and databases (D)	Transcriptomics (K)
Data categories (E)	Proteins and proteomics (L)
Computation (F)	Metabolomics (M)
Probability and statistics (G)	Supramolecular organization (N)
Modeling and mathematical techniques (H)	Biochemical dynamics (O)
	Physiology (P)
Artificial intelligence and machine learning (I)	Image analysis (Q)
	Textual analysis (R)

Introduction   As with every emerging discipline, bioinformatics has had more definitions than practitioners, and these definitions themselves have evolved over time as people's understanding has increased. Rather than being prescriptive, we have opted to give a selection of definitions that highlight this evolution.

Antediluvian origins

Most bioinformatics textbooks define the subject in terms of biological data (often macromolecular sequences) and computers, and the emergence of bioinformatics being a result of the flood of biological data to be managed and analyzed. The few that mention the origins of the term almost certainly have got it wrong by a couple of decades. Bio-informatics, or more strictly the French translation, *bio-informatique*, first appeared in a textbook by Rybak in 1968. To put this into context, the longest nucleotide sequence at that time was for a tRNA (76 bases), the genetic code had only recently been cracked, and the computer on the *Apollo* spacecraft to the moon had both a vastly slower central processor unit and about a million-fold less data storage capacity than a mobile telephone. The entire computer power of NASA was less than an average desktop computer today.

Over three chapters, Rybak outlined the ways in which biological molecules, cells, tissues and organisms encode information, and how, in response to the laws of thermodynamics, this information is relayed. This encoding includes:

- the chiral forms of amino acids, sugars, and other metabolites;
- the order (sequence) of bases, amino acids and sugars, respectively, in nucleic acids, proteins, and polysaccharides;
- the spectrum of small and macromolecules in different complexes, cellular compartments, cell types, and organisms; the anatomy of different tissues and organs;
- the density and distribution of populations of organisms. Information relay included macromolecular transcription, translation, and replication;
- hormonal signaling within organisms through a circulatory system, and between them through diffusible metabolites (now referred to as pheromones or quorum sensing); nerve impulses;
- epidermal coloring, sounds, and touch.

Following on from this, Rybak then outlined the content of a bioinformatics course in 1978. This predated other university postgraduate courses in the subject by over a decade. Again to put this into context, the longest sequence was that of the bacteriophage *phi*X174 (5.3 kbases), and any microcomputers were still only prototypes. The flood of sequence data came in the late 1980s, and the deluge in the mid-1990s when automated sequencers first appeared on the market.

Royal definition

There is an apocryphal story that in 1995, when Her Majesty Queen Elizabeth II was introduced to a head of the Bioinformatics Unit, she said 'Bioinformatics is a horrible word.' Her Majesty's economical definition encapsulated the following opinions.

1. The term has a clumsy spelling, which gave ammunition to the Europhobes and Francophobes in the UK and elsewhere.
2. It is very easy to mistype, and a person must be a specialist in the subject if they are able to spell it correctly or pronounce it at speed or without singing the rising and falling tones.
3. Given the opinion in major pharmaceutical companies and research funding agencies of the importance of the subject, people with bioinformatics skills were mopped up by industry, even those who knew little more than how to spell the word. The effect of this was to stifle the ability of the academic sector to train more people in the necessary skills. This also resulted in people with Master's degrees being recruited on salaries normally reserved for postdoctoral scientists.

4. Long-standing computer scientists and mathematicians were aghast at the way bioinformaticians were being courted and paid ludicrously large salaries for applying old algorithms and mathematical techniques to biological data. Other than the 'application domain,' where was the novelty to justify such a fuss?

5. Successful bioinformatics projects often did not follow good IT development practice, of fully scoping the problem, developing a phased project plan, and then implementing a solution as the final step. Instead they provided quick, partial solutions to data analysis problems as rapid technological developments resulted in different data content and different biological questions arose. For the bioinformatics developer, the work more closely resembles that of an automobile mechanic trying to redesign a car while driving it.

6. Bioinformatics involves biological data, and it is such data that are truly horrible! The output from experiments is usually imprecise (noisy), variable, and incomplete. Biological research practice over decades has resulted in the inconsistent use of terms such that a single gene, protein or metabolite may be given many different names depending on who identified it first and what was found out about it subsequently. To make matters worse the same term has been applied to different molecules by groups from different branches of biology. For example, there was a time when new proteins were defined by their apparent molecular weight in polyacrylamide gels – 'p40' was applied to a range of different proteins, because they happened to have an observed molecular weight of 40 kDaltons. We can be thankful that muscle molecular biologists dropped the acronym for de-naturalized actomyosin (DNA) many years ago.

Canonical definition

Bioinformatics is the discipline at the interface of *Bio*logy, *inform*ation science and mathem*atics*. This definition tacitly implies a range of issues.

The discipline is intrinsically multidisciplinary. This means that:

a. Nobody can be competent in all aspects. In the short period when some people set themselves up as bioinformatics gurus, they fairly quickly suffered embarrassment and lost credibility through not knowing some obvious part of one of the disciplines. This was exacerbated by the fact that biological research technologies and information science/technology were and still are moving on quickly.

b. Bioinformaticians aim to remain current in one discipline, but are sufficiently conversant with the others to act as interpreters between specialists in one domain and another. This also led to the catchphrase that a bioinformatician is a 'jack of all trades and master of one.'

c. Practitioners must work as part of a team who bring different expertise to bear on a given problem. This is less of a problem in the industrial sector (where multidisciplinary groups can be created at the whim of the company president) than for academic groups (which often are able to raise the needed funds for individuals only rather than the teams). It took universities longer to develop the administrative structures to create bioinformatics groups that serve multiple departments.

d. Novelty in bioinformatics comes from at least one of three directions. First, a well-recognized set of mathematics or computing algorithms are applied to a novel category of biological data. Computer science may develop some new hardware or interface to make some aspect of biological research more tractable by informatics methods. Lastly, novel algorithms or mathematical techniques are developed to crack some bottleneck in an existing biological problem. For example, given the exponentially increasing size of sequence databases, what novel approach to database searching can be found to ensure that searches run in an acceptable timescale?

e. Biologists address the question of **why** a particular project is important enough to be done; while information/computer scientists are most adept at defining **how** a particular piece of software should be developed and implemented; and mathematicians can best define **which** is the best algorithm or technique for a given data analysis problem. In any given project there will always be alternative ways forward, and the best from one perspective is often not the best (or possibly even feasible!) in another. At each stage in the development, decisions need to be reached jointly.

f. The balance between the disciplines must be maintained. Otherwise there is a risk of poor software development or, worse still, effort wasted on a biological issue that has already been solved. This was perhaps best highlighted at the world's major bioinformatics conference (*Intelligent Systems in Molecular Biology*) in the mid-1990s, where a computer scientist gave a long description of an elegant and major piece of software development. This was shown to be worthless by the first biologist to speak during questions, who pointed out how the biological problem had been solved through new laboratory techniques some years previously.

The canonical definition does not specify which aspect of biology is the focus of bioinformatics, which is perhaps just as well, as its focus has shifted over time. In the mid-1960s, these three disciplines came together for the first time to apply informatics techniques to the growing quantities of protein X-ray crystallographic data. This work shifted more to protein-structure prediction and was joined in the late 1970s by sequence capture and analysis, when the volume of such data increased by new laboratory techniques. The 1990s saw the appearance of '-omics' (genomics, transcriptomics, proteomics, metabolomics, and many more '–omics' than one cares to consider). It also saw the development of population profiling through small nucleotide sequence variations (initially the distribution of certain sequence repeats, but now down to single nucleotide polymorphisms – SNPs). These techniques have shifted the focus from single entities to populations. With software pipelines (programs that chain together software executables to carry out a set of repetitive tasks) and high-performance computing clusters, why study each member of the population in turn when you can work on all of them simultaneously?

Building on earlier developments, the twenty-first century is seeing bioinformatics turn its attention to the dynamic processes occurring between biological objects, especially those at different physical scales. For example, the physiology of one cell in a multi-cellular organism depends as much on its location in the body, the organism's feeding status, and environment in which the organism finds itself, as on its developmental origin and current gene expression profile. Section P gives examples of this. Data for biological systems at these higher scales are determined by either analysis of images of various kinds (see Section Q), or quantitative extraction from other biophysical techniques (e.g., pressure sensors of electrical conductivity).

Functional definition

(Bio)informatics seeks to generate knowledge of the properties, populations, and processes of (biological) entities.

This definition grew out of a more generic definition of informatics, whose domain specificity is defined by the prefix and the adjective in front of the word **entities**. In this context, the biological entities could be on any physical scale, ranging from molecules through cells to organisms and food-webs. They might also be virtual objects, such as databases or software that pertain to physical biological entities. The application areas in Sections J–R will look at different classes of entities in turn. The following table sets the three different facets of entities in the context of what is the interest, and the building blocks (Sections C–I) used to investigate them.

Table 1.   Linking book sections to facets of biological entities

Category	Interest	Building blocks
Properties	Composition, Structure, Activities	Sections C, G, I
Populations	Design and mining of databases	Sections D, G, I
Processes	Interactions Networks, Paths Rates and efficiencies	Section H

Sections E and F are common to all three categories. There is an interplay between them also. For example, the biological function of a protein is a property of both its structure and the context in which it is found, and this function is part of some larger physiological process. However, when presented with a new protein sequence, the way to elucidate its biological function is by comparison with a population of other protein sequences whose functions have been determined experimentally.

Public services definition

Bioinformatics is the application of computing and mathematics to the management, analysis, and understanding of data to solve biological questions, and involves links to medical, chemo-, neuro-, etc. informatics.

In 2000, the UK Department of Trade and Industry gathered together the leading representatives of the bioinformatics arena to define the subject and how it should be developed and applied for commercial benefit. In true public-services fashion, the definition contains multiple clauses each of which carries its own special significance.

Bioinformatics is the application of computing and mathematics...

See canonical definition above.

...to the management, analysis, and understanding of data...

Bioinformatics is concerned with the entire process from initial capture of the data, managing it in appropriate databases, analyzing this data and formulating the results into a context that results in genuine new understanding. The subject is driven by the quantity and quality of data.

...to solve biological questions,...

Bioinformatics is applied to biological problems and not purely computer science issues.

...and involves links to medical, chemo-, neuro-, etc. informatics.

Mathematics and computer science developments that have been applied to data in other application domains may also be useful in bioinformatics. In another vein, data in one application domain might inform studies in biology. For example, chemo-informatics has developed tools to define structural features of organic molecules, such as how similar one molecule is to another. This can provide indications of how these molecules might interact with enzymes or other proteins, including their kinetics. In other words, chemo-informatics can link to bio-informatics to give clues to the medical impact of organic molecules as potential medicines, say in neurological disease.

Conclusion   Bioinformatics has become an essential part of biological research activity, and is rapidly becoming as ordinary in the biosciences as molecular biology has been since the 1980s. We now turn to the building blocks upon which bioinformatics is based.

C ESSENTIALS OF PHYSICS

Key Notes

Conservation of mass

This Law states that the overall mass of a system remains constant. This means that in all biological reaction equations, the mass of its substrates must equal that of its products.

Thermodynamics

The first Law of Thermodynamics states that the energy of a closed system remains constant. This is akin to the above law, but means that energy released by a chemical reaction will appear as heat and vice versa. The second Law states that the entropy of a system not at equilibrium will tend to increase until equilibrium is reached. This law has very major implications in biology as macromolecules will tend to degrade into smaller fragments (increasing the entropy), and energy must be expended to overcome the loss of entropy in creating them. The third law states that entropy will tend to decrease as temperature decreases (reaching a minimum at absolute zero). Hence, macromolecules are generally more stable at lower temperatures. This principle has also been applied to computer optimization techniques, the most common of which is simulated annealing.

Applications of physical principles in computing

Physical principles apply to computing in two main ways. First, the laws of physics limit the computational power of computers by affecting the components as their temperature increases. Second, physical principles can be built into algorithms to improve their effectiveness, such as the simulated annealing search, which simulates the physical process of cooling, to help prevent the search becoming trapped in local minima.

Approach to physics research

Over centuries, physics research has followed the approach sometimes known as Occam's razor. This effectively states that the simplest description in terms of the number of components or concepts and their relationships is kept to a minimum and must only be made more complex when the simpler descriptions are no longer adequate.

Related sections   Computation (F)

Introduction

Physics underpins biology. A physical system is a set of objects possessing inherent energies and modes of interaction, often in some defined volume. Our quantitative understanding of changes to the state or behavior of these systems is represented in a simplified form by mathematical equations. These equations, which also expand into laws and theories expressed in human language, are sometimes referred to as **models**. Since biological processes are physical systems operating in benign conditions, they must obey these physical laws as well. Hence, an understanding of the essential concepts of physics provides a basis for understanding how biological systems and processes behave. As well as looking at a few important physical laws, we will outline the general approach taken by physicists to make discoveries, and how this can be applied to bioinformatics research.

Conservation of mass

The Law of Conservation of Mass states that the total mass of a system remains constant. Mass is neither created nor destroyed. When considering processes in biological systems, the descriptions (or models) that result in loss or gain of mass have overlooked some detail, and technically are flawed. This most commonly arises because biologists use some form of shorthand to emphasize a point they are making. For example, the equation:

$$\text{sucrose} = \text{fructose} + \text{glucose}$$

shows how sucrose hydrolyses to form the two monosaccharides, but overlooks the fact that a molecule of water is required to generate the products. Similar liberties might be taken with oxygen, carbon dioxide, reducing hydrogens on NAD (nicotinamide adenine dinucleotide), etc., and electrical charges on molecules. For individual equations this precision usually does not matter, but if sets of equations are being grouped together to discover their net effect, then this become critical. Such failures can result in gain or loss of organic carbon, or errors in pH changes caused by a biochemical system.

Thermodynamics

The first Law of Thermodynamics is analogous to the conservation of mass but refers to energy instead, that is, the energy of a closed system remains constant. Thus if some biochemical reaction or process is going to take place, it must obtain its energy from somewhere, and quantitative understanding must include where the energy comes from as well as how it is used. Some energy is present simply from the fact that the biological system is at a given temperature. Hence some reactions will produce heat that others might then consume.

The second Law sheds a better light on this. It states that the entropy of a system not at equilibrium will tend to increase until equilibrium is reached. **Entropy** can be thought of as a measure of the disorder of a system. If something is highly chaotic or random, it has a high entropy.

A series of important equations can be derived from the second Law, one of them being the equation for the Gibbs energy, also called the free energy:

$$\Delta G = \Delta H + T\,\Delta S$$

This equation describes the overall change in energy of a system because of some process as a function of both the intrinsic energy, or **enthalpy** (ΔH) produced by the process, the change in entropy (ΔS) the process produces, and the absolute temperature (T). Examples of processes that this equation can be applied to are as diverse as the hydrolysis of ATP (adenosine triphosphate) or protein folding. For a given process, the sign of the value of ΔG indicates whether the process will be spontaneous at a given temperature. Processes with a negative ΔG will be spontaneous; with a positive ΔG they will not. Viewed another way, as a system tends towards equilibrium, the overall free energy will tend to zero. Processes that have a positive ΔH term can proceed spontaneously, when they are offset by a greater negative value for $T\Delta S$.

This has major implications for the processes sustaining life. The absolute temperature where living things thrive does not vary very much, so increases in entropy are achieved by two mechanisms. Perhaps the most obvious is a change from an ordered to a disordered structure. The second is by breaking molecules down into a larger number of smaller molecules – a larger number of molecules can diffuse around in more ways than can a macromolecule. Energy is required to generate macromolecules; the equivalent of one ATP molecule for every sugar and nucleotide added, respectively, to polysaccharides and nucleic acids. The formation of fatty acids requires roughly equivalent amounts of energy per acetyl group, though amino acids need more to join a polypeptide chain.

The formation of stable conformations of macromolecular structures also complies with this second Law. Any ordered tertiary structure in polysaccharides (for example, the high tensile strength of cellulose) is mediated via a combination of the covalent bonds, hydrogen bonds, and to a lesser degree by van der Waals interactions. Duplex DNA is the same. Protein tertiary structures are stabilized by these but (entropy-driven) hydrophobic interactions also play a major role. The latter involve the hydrophobic amino acid side chains moving to bury themselves inside the center of the protein away from the water-filled surface. In so doing, water molecules otherwise trapped along the hydrophobic surface are released to diffuse at random. In other words, the loss of entropy by the formation of a folded protein structure is compensated for by an increase in the entropy of the water molecules released from interacting with the protein chain.

Protein folding was confirmed as a purely physical process in the 1950s, when Anfinsen inactivated the enzyme ribonuclease by harsh chemical treatment until its biological activity was lost, then by slowly returning the solution to mild conditions, the activity gradually reappeared as the 3-D structure of the enzyme was restored. This was later shown to be the case for other proteins (sometimes by heating), though some only regained their activity if the solution was cooled slowly. The implication is that rapid freezing resulted in stable 3-D structures that were not the functionally active conformation. Similar changes in the structure of metals have been known for centuries, where the process of slow cooling is known as **annealing** (see *Fig. 1*).

Biological membranes consist of phospholipids that are hydrophilic at one end and hydrophobic at the other. When mixed together in aqueous solution, they will tend to form bi-layers in which the hydrophobic parts of the molecules are buried in the middle of the membrane and the hydrophilic portion on the surface. Hydrophilic molecules will only diffuse very slowly, if at all, across membranes because the hydrophobic layer repels them. Often special (protein) transporters are required to enable molecules to cross membranes. This separation of compartments by membranes then leads into cellular level functions. Thus in mitochondrial membranes there are proteins which force the hydrogen ions on certain molecules (especially reduced forms of NAD and FAD) to be pushed specifically to one side of the membrane. This imbalance will cause the ions to diffuse in one particular direction through protein pores, and this directional flow is used by cells to generate ATP or move metabolites across membranes.

The third Law of Thermodynamics states that entropy will tend to decrease as temperature decreases. Hence at higher temperatures, macromolecules will writhe with greater force and are more likely to break down into larger numbers of smaller molecules. Conversely, motion is reduced at lower temperatures leading to greater macromolecular stability. This is one reason why food and medicines are stored in refrigerators or freezers.

Application of physical principles in computing

The physical principles described apply to all walks of life, nothing can escape them – obviously then, computer hardware must also comply with these laws of physics. Computer memory and functional central processor units (see Section F) will get scrambled if they get so hot that the binary data or instructions are no longer stable in the silicon components. Thus the temperature they run at is one governing factor in their processing capability. Generally, the more powerful the processor, the hotter it will run. This is why all computers include mechanisms for cooling. Over time, chip technology is improving, allowing them to generate less heat for any given computation speed.

However, the physical principles can also be encoded in computer algorithms. If we have a situation where a variety of items need to be arranged in some optimal order (for example, a university lecture timetable in which modules and lectures in different departments are taught for a broad range of different degrees), then systematic algorithms that exhaustively search all the alternatives will take too long to find a solution. Alternative approximation strategies are required, and one, known as **simulated annealing**, is analogous to the process

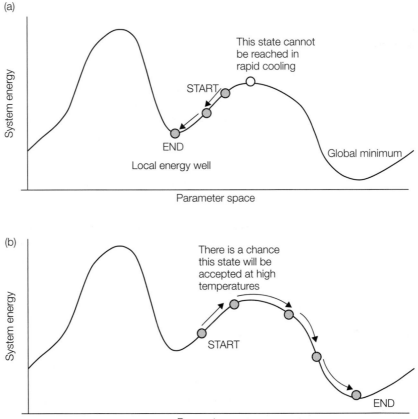

Fig. 1.   *Examples of annealing with fast and slow cooling schedules. (a) Fast cooling annealing is effectively equivalent to finding the steepest descent down a hill. However, this can lead to a local, rather than global, minimum being found. This is equivalent to the molecules in a physical structure rapidly losing energy and so finding the nearest energy minimum, rather than the optimal global minimum. (b) In the early stages of slow cooling annealing (at a high temperature), the state is able to move to 'worse guesses' allowing the algorithm to jump out of local energy wells. As time progresses the temperature falls, slowly changing to the steepest descent approach seen in (a). In reality, these slow cooling processes allow molecules to accumulate into the globally best energy minimum, rather than moving to a local minimum.*

of annealing in physical materials. In this process, how quickly a substance is cooled affects how the molecules are arranged. Slow cooling allows the molecules to order themselves, finding a minimum energy state; faster cooling leads to more disorganization. In the algorithm, this 'cooling' process is represented by the algorithm converging on a solution (defined by some **fitness function**); the application of 'heat' can prevent the algorithm from becoming stuck in a local minimum – results which are locally good but are not the global minimum. At high 'temperatures' the algorithm is able to jump out of local minima by randomly accepting worse solutions with a certain probability; this is analogous to the particles having more heat energy. As the system 'cools', worse solutions are accepted with decreasing chance, causing the algorithm to tend towards a downhill search at the end of the process (see *Fig. 1*). Of course, a further option is to carry out the simulated annealing a number of times to discover the most optimal solution.

Approach to physics research

For many decades, physics research has almost always followed the philosophy that the simplest explanation is the best. Models to describe quantitatively the key parts of the system are kept as simple as possible, using the simplest mathematics. Simple models may be replaced later by more complex ones, when more detail is required or some circumstance is found when the model ceases to explain observations. It is possible also that models have to be expanded when their purpose changes. We may want to look at smaller scale effects, for example, which might require a more detailed understanding of larger-scale processes (e.g., atoms as small marbles, led to the proposition of subatomic particles, led to quantum physics). One common example of this is with the modeling of gravity. In the seventeeth century, Newton proposed a model for gravity that explained the motion of the planets. Indeed, the existence of Neptune was predicted from the motions of other planets using this model. However, detailed observations of Mercury's orbit centuries later showed perturbations that could not be explained by Newton's models. An alternative model was needed which eventually appeared in Einstein's General Theory of Relativity.

When this extra complexity is not needed, it is often best to use a simpler model. This principle of the simplest yet still-functional explanation being the best is commonly referred to as Occam's razor – named after the fourteenth century English Friar, William of Ockham, who first developed the approach. There are several definitions, a common one being 'Of two equivalent theories or explanations, all other things being equal, the simpler one is to be preferred.' The 'razor' refers to the concept of shaving away any unnecessary components. Of course, the simplest model may *not* be the best in technical terms, but the simplest that still captures the system at an adequate level prevents over complication of the model. One explanation of this is that every component of a model has the possibility of introducing error; while that component does not increase the accuracy of the model output, it may still introduce errors, therefore it is best to discard it. Additionally, simpler models tend to reduce the required number of parameters needed to specify the model, thereby reducing the amount of laboratory work needed.

D   DATA AND DATABASES

Key Notes

Digital data	Digital data are at their lowest level stored in a binary form, comprising bits (0s and 1s), bytes (8 consecutive bits), and words (the number of bytes concurrently examined by a central processor unit (CPU).
Representing numbers	These take the form of whole numbers (integers) or decimal numbers (usually referred to as floating-point numbers). The former are far more economical on memory space and can be used in arithmetic more efficiently and accurately. As part of the definition of the number, you might also have to reserve a bit for the sign (+/-) of the number. Floating point numbers commonly require 32 or 64 bits depending on the precision you require.
Representing text	This is achieved by arranging single (or, more often nowadays, multiple) bytes to represent defined keyboard characters, which include the nonprintable characters such as blank spaces, carriage returns, and tabs. The early single-byte sets include ASCII, where the up-to-four-byte unicode transformation format (UTF)-encoding allows the printing of many other alphabets (Russian, Hebrew, etc.), and oriental and other characters.
Representing groups of data	Where data of a similar type can be grouped together for a common purpose, they can be stored in arrays or hashes or combinations thereof. The individual elements of these are labeled by numbers and character strings, respectively, and modern programming languages allow these to be embedded in each other resulting in quite complex data structures.
Data models	The efficient storage and querying of complex sets of data requires an understanding of how items in the dataset are or need to be related. A data model is a definition of the entities, attributes, and their relationships within datasets. They are usually represented initially in UML, from which the context of the data can be stored using extensible mark-up language (XML). The latter is now a very widely used standard, within which various subsets are extremely common, for example hypertext mark-up language (HTML – the language of the worldwide web [WWW]).
Databases	Where large quantities of information need to be stored and retrieved in an efficient manner, then databases are required. Database management systems come in two main types. Relational databases consist of a set of connected tables, analogous to a set of connected spreadsheets, while object-oriented databases come into their own when the number of entities is large, the attributes are few, and the entities/attributes are big. Developments to enable relational databases to contain large objects have resulted in so-called object-relational databases.
Ontologies	These are a hierarchical way of capturing concepts and knowledge as a set of related objects having associated definitions. Probably the most widely used ontology in bioinformatics is the gene ontology.

Knowledge hierarchy	This is the hierarchy in computer science that links data, information, knowledge, and wisdom.
Related sections	Computation (F) Image analysis (Q)

Digital data

All electronic data stored in a computer are reduced to a series of ones and zeros that encode the original data. A bit is the smallest unit of data, being represented by either a one or a zero. The system used to represent data using bits is called binary. Combinations of bits are read together to represent numbers. A common grouping in binary systems is a byte, which most commonly consists of a group of 8 bits that are read together to interpret their meaning. How many different values can an 8-bit byte represent? Each bit is capable of having two different states, a one or a zero. Therefore a single bit can encode two data values. Two bits together have four combinations (00, 01, 10, 11), and can therefore encode four different values. Combining three bits produces more possibilities. In fact, to work out how many different values a set of bits can store, you can raise two to the power of the number of bits you have. For example, one bit has $2^1 = 2$ possible values, and two bits have $2^2 = 4$ different values. The Table below lists the number of values that different numbers of bits can represent.

Table 1.   The possible values stored by increasing the numbers of bits

Number of bits (n)	Possible number of values (2^n)
1	2
2	4
3	8
4	16
5	32
6	64
7	128
8	256
32	4294967296
64	18446744073709551616

You should notice from *Table 1* that with each additional bit, the possible number of values stored is doubled. So, an 8-bit byte can represent 256 different values. What these values actually are depends on how the byte is defined; for example, the 256 values may represent the numbers 0 to 255 or the numbers –128 to +127. These are the kind of options which must be decided when setting up programs, and we'll look at this in more detail next. The table also shows the number of values that can be stored by 32 and 64 bits (i.e., 4 and 8 bytes). Central processing units (CPUs) process data in 4-byte or 8-byte chunks, which are referred to as words.

Representing numbers

Various packages and software programming environments present the user with a number of options about how to store numbers. It is worth highlighting here the options, as making the wrong choices can complicate further data processing. There are two common sets of options here, storing a number: (a) as a positive-only value or allowing +/– values, and (b) either with a decimal point or as whole number. Choosing to store a whole number only (an integer) makes representing the number as a series of binary bits a simpler task, and computers handle arithmetic with integer numbers more efficiently than decimal numbers.

Choosing between being able to store negative numbers or not is often made using the signed or unsigned option. Signed numbers can represent negative numbers; the name 'signed' relates to the ability to represent the negative sign, which indicates a number less than zero. The disadvantage with storing a number as a signed version is that while the range of the number remains the same, the maximum positive or negative size is half that of the unsigned version. This is because one of the bits previously used to represent the number is now used to represent the sign. Looking again at the example of an 8-bit integer number, the unsigned version can store numbers between 0 and 255, a range of $2^8 = 256$. However, the signed version can store numbers from -128 to $+127$; the range is still 2^8; however, the maximum and minimum values that can be stored have changed.

These constraints are worth planning for when building systems, as choosing the wrong format to begin with can be costly in the future, when you cannot store the numbers you need. The year-2000-problem was an example of this; one issue was that programmers had only used two digits to store dates, causing a potential problem when the date changed from 99 to 00, as some systems believed this represented 1900 rather than 2000. A lot of money was spent to fund software engineers to correct the problem before the millennium. Choose your numerical types carefully!

A number with a decimal point is referred to as a floating-point number. The term 'floating-point' refers to the presence of a decimal point, and the fact that its position can be moved or 'floated' relative to the numerals in the number. A 'fixed-point'number is limited, e.g., a 4-digit decimal with a two-place fixed-point representation could store 00.55, 12.34, 15.00, 00.01, but not 0.001 or 122.5. All six numbers are possible only in floating-point numbers, which allow you to decide where to place the point. Floating-point numbers are able to store a much wider range of numbers as a consequence. It is usual for floating-point numbers to be represented by 32 or 64 bits, corresponding to single and double numbers. The former comprises 1 bit for the sign (+ or –), 8 for the exponent, and 23 for the mantissa (i.e., the significant digits), respectively. Double floating-point numbers have 1, 11 and 52 bits, respectively, for the sign, exponent, and mantissa. This standard is known as IEEE 754 and full details of how this is done can be found at http://www.psc.edu/general/software/packages/ieee/ieee.html

It is worth noting that the actual range a number type can store can vary from system to system. Normally this is related to the 'word size' – the number of bits handled together by design for a particular machine. Today, most desktop personal computers (PCs) have a 32- or 64-bit architecture. An integer represented by 64 bits has a range of 2^{64}, in other words it can represent the whole numbers from 0 to 18 446 744 073 709 551 615. 64-bit systems are able therefore to address more memory locations than previously, as the address is stored as a 64-bit rather than 32-bit number.

Representing text Textual characters are represented also by integers. Originally, the main standard was the US-ASCII character set, in which the value of one byte (8 bits) defined the particular alphabetical character of interest (see *Fig. 1*). Note that these character sets include codes for keyboard commands, such as carriage return, line feed, and tab. These nonprintable characters are handled in different ways

ASCII value	Character	Control character	ASCII value	Character	ASCII value	Character	ASCII value	Character
000	(null)	NUL	032	(space)	064	@	096	`
001	☺	SOH	033	!	065	A	097	a
002	☻	STX	034	"	066	B	098	b
003	♥	ETX	035	#	067	C	099	c
004	♦	EOT	036	$	068	D	100	d
005	♣	ENQ	037	%	069	E	101	e
006	♠	ACK	038	&	070	F	102	f
007	(beep)	BEL	039	'	071	G	103	g
008	■	BS	040	(	072	H	104	h
009	(tab)	HT	041	)	073	I	105	i
010	(line feed)	LF	042	*	074	J	106	j
011	(home)	VT	043	+	075	K	107	k
012	(form feed)	FF	044	,	076	L	108	l
013	(carriage return)	CR	045	-	077	M	109	m
014	♫	SO	046	.	078	N	110	n
015	○	SI	047	/	079	O	111	o
016	►	DLE	048	0	080	P	112	p
017	◄	DC1	049	1	081	Q	113	q
018	↕	DC2	050	2	082	R	114	r
019	‼	DC3	051	3	083	S	115	s
020	π	DC4	052	4	084	T	116	t
021	§	NAK	053	5	085	U	117	u
022	▬	SYN	054	6	086	V	118	v
023	↨	ETB	055	7	087	W	119	w
024	↑	CAN	056	8	088	X	120	x
025	↓	EM	057	9	089	Y	121	y
026	→	SUB	058	:	090	Z	122	z
027	←	ESC	059	;	091	[	123	{
028	(cursor right)	FS	060	<	092	\	124	|
029	(cursor left)	GS	061	=	093	]	125	}
030	(cursor up)	RS	062	>	094	^	126	~
031	(cursor down)	US	063	?	095	_	127	⌂

Fig. 1.   US-ASCII character set.

depending on the computer operating system, and can cause problems when compiling software from different sources. Now, the use of unicode characters is encouraged, as they allow for multilingual support. This is because using more than one byte allows a greater range of different characters to be represented. One example of these new formats, UTF-8, uses from one to four bytes to represent a character. One byte can be used if the character is from the US-ASCII set. Text is, thus, represented by a set of consecutive characters: `'"HELLO!" he said.'` becomes:

```
034,072,069,076,076,079,033,034,032,104,101,032,115,097,105,100, 046
  "   H   E   L   L   O   !   "       h   e       s   a   i   d   .
```

Representing groups of data

Just as text strings can be represented by a set of consecutive bytes, data can be linked as a set of consecutive elements into what is known as an array. In strict programing languages, the data in all the elements of an array have to be of the same type (integer, floating-point number, character string) and a maximum size has to be declared. This is to enable it to be stored in the most efficient way, with the array starting at one memory location (l) and the position of the nth array element occurring at $l + n *$ *datatype size* (in bits). To the irritation of non-programmers the first array element is always given the number 0 so that the computer does not have to look anywhere other than the first position (see *Fig. 2(a)*).

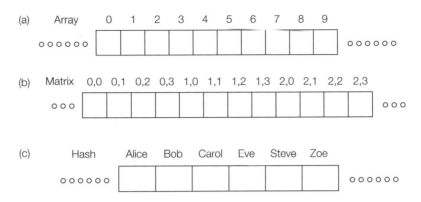

Fig. 2.   The layout in computer memory of an array (a), matrix (b), and hash (c).

An array can have two or more dimensions, in which case it is known as a matrix, and the layout of a 3×4 matrix is shown in *Fig. 2(b)*. An image file is an example of a 2-D matrix (see Section Q). It is not surprising that multidimensional arrays can quite quickly consume large amounts of memory. But why remain restricted to numbers? *Fig. 2(c)* shows a hash. This is like an array but the label for each element is a string, usually of bytes corresponding to meaningful character strings. The labels themselves reside in an array, in which the values of the elements point to the location in memory of the hash element.

The descriptions have so far indicated only that numbers or text reside in the elements of arrays and hashes, but the elements can themselves be arrays or hashes. This can lead to some otherwise quite complex data structures that are not uncommon in bioinformatics. For example, a hash of hashes can be a convenient way to store annotations associated with genes, proteins or other biological entities.

Data models

When a project requires a range of connected data to be stored in a consistent way, then the connections between these different items of data need to be defined precisely, so that data deposition and retrieval are simplified as far as possible. The process of reaching this definition (known as a data model) is as much an art form as a science. One must ask a range of sometimes conflicting questions, such as what are the entities in the data? What are the attributes (i.e., features) of these entities? How are the entities and attributes connected? What are the types and frequency of queries of the data?

It is not always obvious what should be an entity and what should be an attribute, and the final data model can end up being dictated by the question to be asked rather than the structure of the data themselves. However, there are standards for data modeling that help this process. Unified modeling language (UML) is the way of capturing and defining data models, and often results in an entity-relationship diagram. UML can define software processes/pipelines also.

UML data models can be converted readily into a data standard known as XML. Initially used for enabling the transfer of data from one type of database to another, XML is often the *de facto* data standard in bioinformatics because the tools are being developed alongside the laboratory techniques and biological questions being asked. The data model is subject to such frequent amendment that storage in typical database management systems would be too time consuming to revise. HTML (the data standard for WWW pages) is one specific example of XML.

Databases   Databases are stores of structured information. There are several ways to structure the information and therefore several types of databases. Two common examples are relational databases and object-oriented databases. A single database entry is called a record. Each record is composed of a number of fields. A field can be one of a number of different types, including numerical, text, etc.

In relational databases, records are related to other records by sharing common fields. Relational databases can be thought of as being organized into tables. Each table (also called a tuple) contains groups of the records with the same fields (also called attributes). Below is a simple example.

EXPERIMENT

Experiment No.	Scientist No.	Lab	Start date
1	22	A14	10–6–08
2	19	A14	3–5–08
3	18	LabC	1–6–08

SCIENTIST

Scientist No.	First Name	Last Name	Office
18	Dave	Smith	B12
19	Louis	Bergman	B13
20	Jon	Grey	C22
21	Claire	Greening	C14
22	Susie	Ardman	C15

Fig. 3.   Example of relational database tables.

Indeed, a spreadsheet can be thought of as a single table in which the rows are the records and the columns are the attributes. One advantage of storing data in this form is that new representations can be created without changing the original format of the table. For example, by relating the tables using the 'Scientist No.' attribute we can easily display information about experiments and scientists together.

Experiment No.	Scientist No.	Lab	Start date	First name	Last name	Office
1	22	A14	10–6–08	Susie	Ardman	C15
2	19	A14	3–5–08	Louis	Bergman	B13
3	18	LabC	1–6–08	Dave	Smith	B12

Fig. 4.   Example of combining tables.

If Dave Smith were to change offices, the data can just be amended in his original **scientist** record, and the information in *Fig. 4* would be updated as it links to the original data. A primary key is a column in a table that has a unique value for every row/record. So in Table 3, **Experiment No**. and **Scientist No**. are the primary keys fields, respectively. Using these keys allows data to be related among tables.

The standard language for querying such relational databases is SQL (structured query language). This language is covered by both the ANSI (American National Standards Institute) and ISO (International Organization for Standardization) standards organizations, although most versions of the language have extensions specific to the particular vendor, which are not covered by these standards. A simple example might be:

```
SELECT *
FROM EXPERIMENT
WHERE Lab='A14'
```

This will return the records from the experiment table for which the laboratory location is A14. SQL allows us to query data across tables; variations on the **join** keyword are often used to produce results that involve multiple tables.

Where relational databases are organized into tables, object-oriented databases (OODBs) organize data into objects. These can be defined in a hierarchical manner, just as they can in object-oriented programming languages like C++ and Java. Objects inherit properties from classes further up the hierarchy. These properties tend to be more general further up the hierarchy, and more specialized at the lower levels of the hierarchy. An example of a class might be **Animal**, and an example of a subclass might be **Dog**. The **Animal** class might have attributes such as Latin name, height, age. The **Dog** subclass would inherit all these general animal properties, and additionally have some more specific ones, such as tail length, coat color, etc.

The advantages of OODBs include the flexibility of data organization, easy incorporation of data from XML files, and the wide range and size of objects that can be stored. They can be the best choice when the number of relational tables would be large, with only a few records in each table, or when data objects (e.g., pieces of executable code) must be included also. Using a hierarchical data model is advantageous as much real-world data naturally adopts this organization. Disadvantages include unnecessary overheads when data and relationships are simple.

Some database-management systems are a hybrid of the two and are referred to as object-relational databases. These are usually relational databases that have extra functionality for holding binary large objects (BLOBs) often in one specific table, and/or cartridges, which are executable programs that reside within the database management system that carry out specific analyses or computations on particular subsets of data. While BLOBs are increasingly common in bioinformatics, for example in storing image data repositories, cartridges are not because the focus of bioinformatics is moving too quickly. A final category of management system is the logical database, in which entities are connected by logical operators (e.g., AND, OR, NOT, etc.). They have found little use so far in bioinformatics.

Ontologies

An **ontology** is a way of representing an existing knowledge domain, in such a way that it is intelligible to an informed user, and also formal enough that the data can be automatically processed. Perhaps the easiest way to understand the structure and role of an ontology is to consider a science textbook. After the title is a list of contents structured hierarchically from general to ever more specific concepts. This hierarchy is reflected in the main text, which comprises a set of ever more precise definitions and explanations as one gets deeper into the different parts of the hierarchy. Of course, some concepts may be pertinent in more than one part of the book and these parts are linked either within the text (e.g., the Related Sections found in this book) or, more often, in the Index.

The concepts in an ontology should each have a definition and are linked via a series of *is-a* relationships: *is-a* synonym of, *is-a* type of, and *is-a* sub-topic of (can also be *is-an* example of). One can think of them as an extreme form of object-oriented database. *Fig. 5* is a simple example of how alanine is linked to more generic names for biochemical molecules. The most widely used ontology in bioinformatics is probably the gene ontology (http://geneontology.org/), which captures our knowledge of the cellular compartment, biological process, and biochemical function of gene products. Each concept has a GO-ID, and for many genes and proteins, the most specific GO-IDs have been defined in association tables (analogous to the index of a textbook). Another widely used ontology in biology and medicine is the medical subject headings (MeSH) terms found in the journal indexing resource known as Medline.

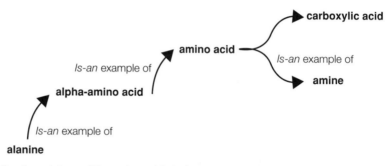

Fig. 5.   An ontology of the amino acid alanine.

An ontology is usually a high-level data model, rather than a low-level method of storing data. Ontological representations can be persisted by saving to a low level data format, which might commonly be an XML file (see below) and/or to a database. A good open-source ontology editor is Protégé (http://protege. stanford.edu/). There are also some useful background information and tutorials about ontologies in the documentation section of this site.

Attempts to develop an **ontology of everything** have repeatedly stalled because terms in one domain of knowledge can mean very different things in another and synonyms are rife. A pharmaceutical company researcher would refer to a drug target, but a clinician would call this a drug receptor, while the precise definition of a drug target varies in each company. When something is constitutive, this means one thing to a mathematician but something entirely different to a molecular biologist. The meaning depends upon the context. Hence, it is now widely accepted that ontologies should be restricted to specific domains of knowledge, so that at least the context is consistent. However, we may wish to link concepts between these domains, and a mechanism to do this is called description logic. The primary source of further information on this is: http://dl.kr.org/

Knowledge hierarchy

In computer science, the knowledge hierarchy is a ranked list of all knowledge and intellectual skills. It condenses down to data, information, knowledge, and wisdom, sometimes abbreviated to DIKW.

Data are the specific values found in variables, database elements or cells of a spreadsheet. On their own they are meaningless. However, other data (e.g., the headings on each column or row of a spreadsheet) can provide some context for the data to someone who knows what the headings mean. These explanatory data are referred to as metadata, and convert the data into information, as they give the data some meaning. It is increasingly common for the metadata to take up more space than the datum itself. For example, the activity of an enzyme may be given by a floating-point number (data) but the metadata must include the units, and often includes the assay conditions, reaction equation, and source of the enzyme. The collation of information into a coherent whole (say in an ontology) is referred to as knowledge, while wisdom is the accumulated experience that links areas of knowledge into contexts, causes, and effects. This enables decisions to be made that are likely to have the best outcome in the future. While knowledge working may be a hot topic in computer science, wisdom is still a very abstract activity.

E1 DATA CATEGORIES

Key Notes

Dimensions	Data can be thought of as having a dimension, which represents either some spatial element to the data, or where the data sit in relation to other data.
0-D data	A data point existing in isolation from other data has zero dimensions, e.g., integers, letters, Booleans, etc.
1-D data	Data that consist of a 'string' of data points, e.g., sequences representing nucleotide bases and amino acids, SMILES, etc.
2-D data	Data that have a spatial component, such as images, NMR spectra, etc.
2.5-D data	Data can be stored in the computer as a 2-D matrix, but can represent biological entities in three or more dimensions, e.g., PDB records.
3-D data	Data with a 3-D spatial component, e.g., image voxels, e-density maps, etc.
Geographic Information Systems	GIS analysis and visualization usually revolve around data relating to the earth. However, such methods and tools have also been applied to large, multidimensional biological datasets.
Related sections	Data and databases (D)

Dimensions

Biological data consist of a large variety of different data types, from numerical results, through to descriptions of molecules and cells or tissues, to image data, etc. Such varied sources often have a particular category of data that can represent the information they create the best. These categories can be thought of as having dimensions; for example, an isolated integer can be thought of as 0-dimensional (0-D).

A sequence on the other hand is 1-D as it consists of a series of 0-D components strung together; the one dimension represents the position in the string. An image is an example of a two-dimensional category, as the information at a particular point in the image is located using a 2-D coordinate specifying how far along the width- and height-axes of an image the data point belongs. However, we shall see that the number of dimensions used to store data does not have to correspond to the number of dimensions in physical reality. There are semantic tricks that allow us to capture higher dimensional data in arrays and 2-D matrices.

Having looked at how data are encoded in computers in the previous chapter, this chapter will present some examples of the more common and interesting data structures relevant to bioinformatics. The chapter is broken down into categories on the basis of the different dimensions involved.

Zero-dimensional data (0-D)

A 0-D data type consists of data in isolation; that is, data that is not read in the context of other data. A number is an example of a 0-D data category, as it exists as an individual data point. Boolean data, namely that which represents either a true or false value, is also 0-dimensional – essentially this is a special case of numerical data that can only represent 0 or 1 and has a particular semantic meaning attached, that of the concept of 'true' or 'false'.

In bioinformatics, there is very heavy emphasis on the use of textual terms. At the simplest level, we have single letters of the alphabet representing individual nucleotide bases or amino acids, each of which have further meaning in physical reality. It is very common practice to represent biological entities (e.g., genes, RNA, proteins, metabolites, organelles, cell/tissue types, and organisms) by either a standardized term, or some form of database identifier. Taken alone, these identifiers are zero-dimensional. As we shall see, however, they can also be combined into 1-D strings.

One-dimensional data (1-D)

1-D data consist of a sequence or 'string' of data points. The relative order is important in relation to the information it conveys. An example might be a telephone number; while still technically a number, it is generally considered to be composed of separate digits strung together to form a unique combination of numbers. When we read a telephone number, we read it as 'zero-one-six-two-two' rather than 'one thousand-six hundred-and-twenty-two,' for example. The sequence of the digits is important.

Essentially, data in this category consist of sequences of subunits of data strung together, hence the term 'strings'. This simple data structure can represent complex, higher-dimensional data in a simple, accessible manner. Combining the 0-D data units described above, such as those representing nucleotide bases and amino acids, into sequences represents more complex information, which mirrors the biological context in which these simple units occur, coding regulatory or structural DNA sequences and protein structure, function, and co-localization. Much has already been published on sequences and their meaning, so for further details the reader should refer to any of the many textbooks on biochemistry or molecular biology.

SMILES (simplified molecular input line entry system) is a technique used to represent the structure of organic molecules as strings of textual characters. The approach was developed by David Weininger in the 1980s, and is based on graph theory, which is a formal method of describing relationships between pairs of objects. The method simplifies the structure description where possible.

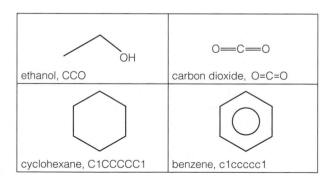

Fig. 1.   Some examples of SMILES. For many more, see http://www.daylight.com/dayhtml_tutorials/languages/smiles/index.html

For example, the proper number for the number of implicit hydrogen bonds is assumed; water is represented simply as O. Double-bonded carbon dioxide is represented as O=C=O. Triple bonds are represented with a # symbol. Aromatic atoms are specified using lower case letters; for example, cyclohexane is represented by C1CCCCC1 whereas benzene is represented by c1ccccc1. Many other combinations of symbols provide the ability to store a wide variety of information about structure within a single-character string.

Some extensions to SMILES are in use. One example is SMARTS. This extension allows the inclusion of wildcard terms and other operators, which permit the searching of databases in a flexible manner. Another variant is isometric-SMILES, which can be thought of as higher-dimensional as it can convey limited information about 3-D structure. The extra characters aid the differentiation of asymmetric carbon atoms; for example, D- or L-conformations of amino acids and sugars, which is the difference between molecules being readily incorporated into biological macromolecules or requiring specialized enzymes to be metabolized. These strings are only one-dimensional but capture the details of higher dimensions and could be thought of as 1.5-D.

Two-dimensional data (2-D)

In this category we look at data that have a spatial component. An example is a coordinate marking a location on some sort of map, or information about a particular pixel within a digital image. A digital image is a good example of a 2-D dataset. An image is composed of pixels (picture elements), each of which stores information about the color and brightness of the light recorded at that particular location. The resolution of an image determines the number of pixels that make up a row and a column. The information in a digital image is represented usually in a computer as a 2-D matrix, with the data at each pixel being referenced by an (x, y) coordinate that describes the spatial location of that pixel. Referencing pixels in this way is analogous to the way we reference points on a 2D map, which normally consist of a location and information about what is at that location.

Another example of a 2-D dataset is the 2-D spectra generated by NMR techniques, such as NOESY and COSY 1-D protein NMR spectra can be very complex and the signals hard to interpret because of the cluttered nature of the individual peaks in the signal. The aim of introducing multidimensional spectra to the analysis of NMR data is to tease apart the complex data in a 1-D spectrum, making it easier to comprehend and also to reveal new information. The development of the 2-D NMR technique was considered so important it won a Nobel prize for chemistry (http://nobelprize.org/nobel_prizes/chemistry/laureates/1991/illpres/two.html).

These 2-D techniques rely on nuclei exchanging magnetism during a mixing phase; these signals are called cross signals and they indicate an interaction between the nuclei. In the COSY technique – which stands for correlation spectroscopy – magnetism is transferred by scalar coupling. Only signals from protons two or three bonds apart appear in a COSY spectrum. Another technique is NOESY (nuclear Overhauser effect spectroscopy). NOESY experiments are important for determining protein structure; the information produced relates to space not bonds (as with COSY). Both techniques produce data in the form of a 2-D plot or map, rather than a line graph.

While we are describing 2-D data structures, it is worth noting that the size of memory required to store a data structure increases exponentially according to the number of dimensions. For example, a 1-D sequence consisting of 4000 single byte numbers requires 4000 bytes to store the data. However, a

2-D image with a resolution of 4000 rows and 4000 columns, with each pixel storing one byte of information, would need $4000 \times 4000 = 16\ 000\ 000$ bytes or 16 megabytes. Adding a third dimension compounds the problem of required memory; this is something to keep in mind when working with higher dimension data structures.

Two-and-a-half-dimensional (2.5-D) data

The two-and-a-half dimensions in this category refer to the way the data are stored in the computer as a 2-D matrix, but which can in fact represent biological entities in three or more dimensions. The archetypal example of this can be found in the Protein Data Bank (PDB: http://www.wwpdb.org/). The PDB is a large repository containing the three-dimensional structural information of large biological molecules. Established in 1971, the original text-file-based data format for the PDB files has had to be adapted over time.

The 2-D arrays found in PDB records comprise a set of rows corresponding to each atom. Each row has multiple columns, many of which provide data about the atom's identity and context (parent residue, sequence position, and chain numbers). Other columns refer to the atom's x, y, and z co-ordinates, the so-called occupancy and temperature factors, and the charge on the atom. Thus, this 2-D array captures six structural dimensions on top of the contextual dimensions mentioned above. A full description of the format can be found at: http://www.wwpdb.org/documentation/format23/sect9.html

The main standard used today is the mmCIF (macromolecular crystallographic information file) format. The PDB was recently re-structured to become a relational database, although the success of the conversion has been debated. The challenge of managing data formats over such long periods of time is highlighted by the mottled history of PDB formats. The failure to follow ontological standards when converting PDB records to relational databases has been cited as one reason for the shortcomings of the final database.

Software used, for example, to visualize macromolecular structures or to analyze functional sites can import PDB files. Discrepancies in the format as implemented by different organizations causes problems here, as the files cannot always be parsed directly. This illustrates the importance of modern data standards: automated processing by software packages requires the data to be in a form that a computer can make sense of. Poorly specified input files can produce erroneous results or simply cause the software not to function at all.

Three-dimensional (3-D) data

Three-dimensional sets of data are commonly used to represent data with a 3-D spatial component. A 3-D image representation has just such a spatial component, with its position within the world defined by an (x, y, z) coordinate. You should recall that 2-D images were composed of pixels. 3-D images are composed instead of voxels (volume elements). 3-D representations can be found in atlases, like the mouse 3-D atlas. These three coordinates allow you to specify a point on a plane, just as in a 2-D image, and a depth of this plane as well. Often this bears a direct relationship with how the 3-D images are generated. For example, a confocal microscope allows images to be taken from inside a translucent sample. A plane through the sample is imaged at one depth, giving a 2-D image, then the depth of the plane is incremented by a small amount. Another 2-D image is formed at this depth, then the depth is incremented again. Therefore, the resultant 3-D data really are composed of a series of 2-D images or 'slices' through the sample. The final resolution of the 3-D data depends both on the resolution within the imaging plane (as in a traditional 2-D digital image) and also on the size of the steps between the planes. 3-D rendering software can be used to display the resulting data, and view the reconstruction from any angle.

Other 3-D data examples include electron density maps, which are created by looking at the scattering pattern of X-rays passed through crystal structures. The diffraction pattern formed by the X-rays can be used to reconstruct 3-D information about the electron density within the material. These data are normally viewed either as 2-D images of contours of constant electron density, viewed from each of the three axes of the model, or rendered with iso-surfaces producing what looks like a solid 3-D shape.

Geographic Information Systems (GIS)

GIS software is commonly used to store, analyze, and display information related to geospatial information; that is, information relating to the earth. Common use includes building maps with multiple data overlay levels; for example, one layer might show elevation information, another may show roads, another houses, another pollution levels. Often these layers can be switched on and off independently. You may have encountered GIS data displayed in popular earth data-viewers such as Google Earth (http://earth.google.com/)

Being aware of techniques from other domains such as GIS systems is important for bioinformaticians as sometimes it is useful to analyze and interpret biological data in novel ways. Displaying genomes spatially, for example, has been implemented in a system named GenoSIS built on an existing GIS software package. This new system supports spatial queries in addition to basic keyword searches. It also utilizes 'data layers' that can be toggled on or off according to user's needs. This is analogous to activating layers on maps to display features such as roads, buildings or elevation information. Being able to build just such interdisciplinary tools can be a good way to achieve a new level of understanding of data sets, as it allows them to be approached from new perspectives.

E2 BEST PRACTICE FOR DATA REPRESENTATION IN BIOINFORMATICS

Key Notes

Good informatics practice in data representation	Storing your data in a sensible way makes it easier for others to understand and have programmatic access to it, and also helps it to remain useful into future years. Using common data standards and being careful about how data are actually stored in the system are critical concepts for successful management of data.

Related sections	Data and databases (D)	Computation (F)

Good informatics practice in data representation

Below are some suggestions for tips and things to watch out for when deciding how to represent data.

Data standards

- In all long-term or multi-user projects, it is imperative that data conform to open, community-agreed standards (data formats). Adherence to a proprietary standard (i.e., one defined by a private company) puts the software developer at the mercy of that company's priorities. Software can be rendered useless by unscheduled changes in such standards. Adherence to particular software standards rather than data standards is even worse because it makes the linking of software tools considerably harder.

- Develop data standards, conventions, and models as early as possible and stick to them. If datasets do not conform to a specific standard, then new software will always generate erroneous output. In computer science, this has been known for decades as GIGO (garbage in, garbage out). Along these lines, if data are being stored in a filing system, then even the directory and file names must conform to the standard. The consequence of not doing this is that software may either fail to find, or exclude or lose data without informing the user. Some data standards can be found in *Table 1*, and throughout Chapters J–R.

- Databases can be an efficient way of storing and retrieving data. However, substantial care is needed when designing the structure of a database. On one hand, the data model should reflect the structure of the data as closely as possible, but this should not outweigh the need for rapid retrieval of data for the most common types of query. It is not unusual for projects to begin generating data that must be stored in a structured way long before anyone can define what the final data model should be. In these circumstances, it is often better to store the data in an XML format, whose schema or document-type definition can evolve as the project proceeds.

Table 1.   A selection of some common data standards

Acronym	Name	Subject
OGSA-DAI	Open Grid Service Architecture-Data Access and Integration	Assist with data access and integration via the grid
RDF	Resource Description Framework	A semantic web framework to assist the encoding of knowledge, such as information about and relationships between real world things
MIAME	Minimum Information About a Microarray Experiment	A standard for describing microarray experiments and results

These and others will be evolving all the time, so it is always worth checking for up-to-date information. Such information can be found usually by searching the internet. A good place to start would be the website of an organization or governing body that oversees the area you are interested in.

Data types
- Be careful when specifying data 'types' For example, be aware that specifying a number as a double type in one piece of software and converting to an integer type in another piece of software will cause the number to lose all information beyond the decimal point. Also be aware that in this situation the number may be rounded up or down, depending on the software.
- Never mix data types, unless your software can handle the ambiguity. Probably the simplest case of this is the occurrence of ambiguity codes in DNA sequences; for example, R refers to a puRine (i.e., A or G), and Y refers to a pYrimidine (i.e., C or T). Software that does not handle such codes cannot be blamed for generating erroneous results, if the user inputs sequences for which it was never intended.
- Be aware of data range limitations. For example, 8-bit numbers only have a range of 256 values (2^8). Data overflow can occur if you try and store numbers larger than the maximum permitted. For example, in the Java language, a signed long can store a maximum value of 9 223 372 036 854 775 807 (2^{63}-1). This may seem like a large number, but beware that doing multiplicative calculations with array data, such as that derived from images, can easily exceed this amount. In this case, the system may not return an error message, and may instead store a nonsense number in the variable. This is called data overflow, and is tricky to trace when debugging software. Therefore it is a good idea to take this into account when deciding how to represent your numbers in the first place!
- Storing continuous numbers as discrete numbers. A lot of real-world data exists that are continuous; that is, the data do not naturally break up into discrete chunks. When these data are digitized using analog to digital converters, the numbers have to be made discrete so that they can be stored in a computer. The 'real' numbers stored by types such as Java's double may look precise, but actually they are only an approximation of the infinite-precision real data. However, this storage precision is usually greater than the precision of data capture, so often it can be treated as a good approximation to the actual data.
- In a related problem, some real decimal numbers cannot be represented exactly using binary. Converting 0.1 to binary leads to an infinite repeating binary sequence; therefore it cannot be represented exactly. This error is comparable to the problem of representing the rational fraction 1/3 as the decimal number 0.33333… The number is infinitely repeating and therefore can only be approximated by any significant number of digits. Such issues can lead to

unexpected results in calculations, for example in decimal $3 \times 0.333 = 0.999$, not 1.0. The same is true of the endlessly repeating binary representations; for example, 10×0.1 does not exactly equal 1.0 in binary.

Longevity of data
- Backing up data. It is always good practice to back up your data regularly in a safe and secure way. Normally this involves copying data to a remote network drive, although if you are working in the laboratory or field, backing up to a USB hard disk or memory stick is better than nothing (though bear in mind such portable media devices are prone to failure!).
- Meta data. It is often helpful to store information about your data, as well as the actual raw data. For example, details such as where and when the data were produced, how they have been processed so far, and who has processed the data allow us to understand the data more easily when we come to look at them again in the future. It also allows us to see where errors may have crept in, and can aid the repetition of experiments. Often, meta-data tags or notes can be stored in data files, although the exact implementation can vary between different standards and software implementations.

F1 COMPUTATION

Key Notes

Central processing unit

The central processing unit (CPU) is primarily composed of two components, the control unit and the arithmetic logic unit (ALU). The former retrieves and executes instructions comprising basic operations such as addition and subtraction. The latter performs arithmetic and logical tests on numbers.

Internal memory

The CPU's internal memory is organized like an array, the individual elements of which are called registers and allow the processor to store small amounts of data for calculations. These registers can be accessed much more quickly than the main memory store. There are often several kinds of registers used for different purposes, such as data registers for storing data, registers for storing memory addresses, stack registers for maintaining stack data structures, and so on.

Machine code

The control unit in the CPU executes machine code instructions. Machine code itself is essentially a set of binary codes that is made more legible by writing in assembly language, which replaces raw binary with mnemonics. Machine code is often specific to particular types of processor, and is therefore referred to as native code.

High-level languages

Such languages move away from machine code towards a style that is much more like natural language. As they are separate from machine code, they are often more portable as well, as they can be converted to native machine code on a particular machine. Java, C++, Perl, Python, and PHP are introduced as examples.

Object-oriented programming

Java and C++ are examples of object-oriented languages, which means that both the data and processing commands are encapsulated inside 'objects', which can communicate with each other. An object is defined by a class, which contains information about what kind of data can be stored and the processing which can be done.

Pointers

Pointers refer to the location of data. Passing a pointer to a subroutine or object, rather than the actual data, can save a considerable amount of processing time.

Regular expressions

Regular expressions are a way of describing search patterns to find matching text strings. They use a combination of text, wildcards, and special characters to allow the user to specify exactly which set of strings they are searching for.

Software versioning

As software is developed, it is useful to think of the development as happening in finite steps. There are ways of documenting these steps, using major and minor version numbers, and 'alpha', 'beta' and 'release candidate' labels. The CVS (concurrent versions system) software is introduced as a solution for managing software versions.

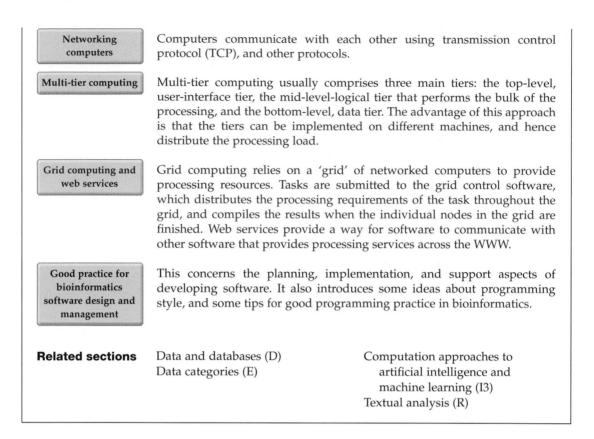

| Networking computers | Computers communicate with each other using transmission control protocol (TCP), and other protocols. |

Multi-tier computing usually comprises three main tiers: the top-level, user-interface tier, the mid-level-logical tier that performs the bulk of the processing, and the bottom-level, data tier. The advantage of this approach is that the tiers can be implemented on different machines, and hence distribute the processing load.

Grid computing and web services Grid computing relies on a 'grid' of networked computers to provide processing resources. Tasks are submitted to the grid control software, which distributes the processing requirements of the task throughout the grid, and compiles the results when the individual nodes in the grid are finished. Web services provide a way for software to communicate with other software that provides processing services across the WWW.

Good practice for bioinformatics software design and management This concerns the planning, implementation, and support aspects of developing software. It also introduces some ideas about programming style, and some tips for good programming practice in bioinformatics.

Related sections   Data and databases (D)   Computation approaches to
                     Data categories (E)   artificial intelligence and
                                                              machine learning (I3)
                                                          Textual analysis (R)

The previous two Sections focused on how data are represented in computer systems. This Section outlines how computers process these data, which also requires an understanding of computer architectures. An introduction to good informatics practice in software design will then be presented.

The core components of a computer

The main components of a computer consist of storage devices, comprising internal memory and disks, and components with processing ability; principally, the **central processing unit** (CPU) and specialist peripherals such as graphics acceleration cards.

The CPU is primarily composed of two components, the control unit and the arithmetic logic unit. The first of these retrieves and executes instructions, and the latter performs arithmetic and logical tests on numbers. Modern CPUs are composed also of a number of other components, such as an internal memory, multimedia extensions, high-speed cache-memory, etc. Modern chips are likely to contain more than one **core;** that is, they are able to operate on two sets of instructions at once (if programmed to do so). The internal memory itself is organized like an array, the individual elements of which are called **registers** and allow the processor to store small amounts of data for calculations. These registers can be accessed much more quickly than the main memory store. There are often several kinds of registers used for different purposes, such as data registers for storing data, registers for storing memory addresses, stack registers for maintaining stack data structures, and so on.

The **control unit** is the part of the CPU responsible for retrieving and executing instructions. These instructions are written in **machine code,** which allows the CPU to perform elemental steps. Machine code itself is essentially a set of binary codes, which is made more legible by writing in **assembly language** that replaces raw binary with mnemonics. Machine code is often specific to particular types of processor, and is therefore referred to as **native code**.

The **arithmetic logic unit** (ALU) performs simple operations on data in the registers, and stores the results back to the registers. The simple arithmetic operations include basic calculations such as addition and subtraction, and possibly multiplication and division. The logical tests would include bitwise logical operations such as *and*, *or*, and *not*. Some ALUs may implement more complex processing capabilities: it is a trade-off between the cost and power requirements (and hence heat generation) of producing ALUs with more powerful processing abilities, versus the extra time required to replicate the process using the simpler arithmetic building blocks. It is worth noting though that *all* the processing the computer does must be built from these basic processing steps.

An example of calculating a simple sum in the CPU is presented below to illustrate the steps involved, where the task is to calculate 45 + 105. The machine code operations are

1. Retrieve value '45' from main memory address X and place in register 1.
   45 in binary: `00101101` [remember the place values are: 128 64 32 16 8 4 2 1]
2. Retrieve value '105' from main memory address Y and place in register 2.
   105 in binary: `01101001`
3. Add register 1 and register 2.

   ```
     00101101      [rules: 0+0 = 0; 1+0 = 1; 0+1 = 1, 1+1 = 0 carry 1]
   + 01101001
     --------
     10010110      [= 150]
   ```
4. Store the result in register 3.
5. Copy the value of register 3 to a main memory address location.

In a computer, all of these instructions would be given to the computer as machine code. As mentioned earlier, assembly language is a more human-readable form of machine language. For example, `mov eax, 10` means 'move the value 10 into the eax register of the CPU.' Likewise, `add eax, 16` will add the value 16 to eax. Assembly language is not commonly used, even if you do a lot of computer programming. Its main advantage, however, is speed, and there are times when it might be worth writing a section of a program in assembly. It is possible sometimes to write assembly language within another programming environment, such as C++, for those occasions when a speed increase is really necessary.

Higher-level languages

This leads us on to higher-level languages. These tend to be easier to understand, as they move away from machine code towards a style that is much more like natural language, and hence easier for programming. Because they are separate from machine code, they are often more portable as well, as they can be converted to native machine code on a particular machine. Often, programming environments that support high-level languages may have a visual toolkit for designing **graphical user-interfaces** (GUIs) as well. Such environments, which may also include debugging tools, code formatting tools, and source-code versioning (more later), are referred to as **integrated development environments** (IDEs).

Two of the most popular higher-level languages are **C++** and **Java**. C++ is an advancement of the C language. The '++' is a reference to the command in C++ to increment a variable; i++ adds one to the variable i. Therefore, the name C++ is a kind of joke, meaning 'increment C' or ' plus one.' Both C++ and Java have freely available compilers as well as commercially available alternatives. A **compiler** is a program used to convert the source code into machine code that the computer can understand. Rather than compiling directly to machine code, Java compiles to **bytecode**. This is 'virtual machine code', which is designed to be interpreted by a **virtual machine** – a piece of software designed to emulate a physical machine. Despite being slower than machine code, this process allows the bytecode to be distributed to different platforms, each of which has its own version of the virtual machine software required to interpret it (the Java runtime environment, or JRE). In theory, this makes Java easier to distribute across different computer platforms.

There are other languages encountered in bioinformatics, and it will be useful to give a brief introduction to some of the most common ones here.

'Perl' is a freely available, high-level, scripting language, which is particularly good at text processing, and is arguably the most used language in bioinformatics. It has found useful roles in systems administration, programming for the web, and for 'gluing' different software packages and scripts together. It is particularly adept at handling 'regular expressions' (see below), and so lends itself to parsing data files.

'Python' is another free, high-level, interpreted language commonly encountered within bioinformatics. It comes with a large library of functions that can handle text processing, networking protocols, mathematical functions, operating system interaction, etc., and it supports object-oriented programming (see below). Unusually, white-space (space and tab characters) is significant in Python and so cannot be altered freely. Python has found common use in web-based applications, as well as being embedded as a scripting language in larger software packages, among other varied uses.

'PHP' is an open-source, general-purpose scripting language that is particularly useful for binding together web-based applications. It is embedded often into the HTML of web pages, which causes code to be run on the web server and results in a web page being generated for the user. PHP features support for a wide range of databases, simplifying the process of creating a database-reliant web site. However, it can be at risk from some security issues and needs to be implemented with care.

FORTRAN is one of the oldest high-level languages and is designed primarily for efficient arithmetic. Certain other languages specialize in logical operations. Examples include lisp, prolog and its variant progol. Their use has been restricted primarily to Inductive Logic Programming (see Section I3) and natural language processing (Section R).

Object-oriented programming (OOP)

Both Java and C++ are **object oriented**, which means that both the data and processing commands are encapsulated inside 'objects' that can communicate with each other. An object is described by a **class**, which contains information about what kind of data can be stored and the processing which can be done (in the form of **methods**, which are akin to programming 'recipes' that, once defined, can be called many times to process data in some way). Objects are created by creating **instances** of classes. For example, suppose there is a class called *Experiment*, which is able to store as an array the name of the experiment as a string (see Section E), an experiment number as an integer, and a list of apparatus. This class might also contain a number of methods, one of which might be called *PrintInfo*, which describes how to write out to screen the experiment name, number, and apparatus items. To create an object from this class, it is necessary to create a specific instance of the class. That is, we fill in the name, number and apparatus information, and give the object a name, such as *MyExperiment*. We can create many instances of this class with different specific information, such as one representing *YourExperiment* and another called *DavesExperiment*.

Object-oriented programming has several advantages over traditional procedural approaches. The modularity of the class/object system allows large programs to be broken down into separate modules. These can be worked on independently, so long as the **interface** between them is specified and adhered to; we must define what data can be sent into and out of the object, and the method and format for this process. Also, this modular approach involving classes is often an intuitive way for the original problem to be broken down in the first place. It is convenient often to be able to represent many different instances of the same blueprint, such as different car models specified by a *Car* class. There are several other tricks which object-oriented programming allows us to use, such as abstraction, inheritance, and encapsulation, which, while too in-depth to describe here, can be of great value when writing programs. Disadvantages of OOP include the speed overhead in managing the classes.

Pointers

It is common in modern programming to use large data structures, which need to be accessed in different parts of the program. This might mean passing a large record to a function for processing, or passing an image from one object to another. Copying the whole data structure in its entirety can be very time consuming. However, a better alternative is to pass an address to where the data reside rather than copy the whole set of actual data. Clearly, a memory address of a few bytes is much smaller and hence quicker to pass around than the several kilobytes (or megabytes) of the data.

A useful analogy here might be to think of a web site you want to email to a friend. Is it easier to copy the whole site, attach all the data to an email and send that, or just to send an email containing the URL to the page? The hyperlink points the reader to the source of the data, and this is exactly what a **pointer variable** does in a program: it represents the memory address of the data rather than the actual data. Passing this pointer to a function is called passing 'by reference', rather than 'by value'. Pointers can greatly increase the speed at which a program runs.

While using pointers is very powerful, they can also be problematic. For example, a pointer might contain an address to a memory location that has been de-allocated, and re-used with different data; the pointer has no knowledge of this change in data and the results can be very erratic. It is important therefore to keep track of what is happening to the data to which your pointers are referring. Some newer languages can help to take care of this for the programmer by using 'managed pointers' and 'automatic garbage collection'.

Regular expressions

Regular expressions are a way of describing search patterns to find matching text strings. They use a combination of text, wildcards, and special characters to allow the user to specify exactly what sort of text they are searching for. This works in a similar way to the common use of the '*' when listing files; '*.doc' might tell the computer to find all files ending in '.doc.' Regular expressions allow the specification of much more complex and powerful searches. Some common expressions are

- Matching one of several characters
  h[aou]t matches hat, hot, and or hut.
  h[a-z]t matches any character between a and z, eg. hat, hbt, etc.
- Match any character
  h.t matches hat, hqt, h%t, h2t etc.
- Match preceding token 0 or more times
  [abcd]* matches d, abb, bbddca, bbdcdadcdaaa, ccccd, etc.

The start and end of a line of text are defined sometimes by specific characters (usually '^' and '$'), and brackets can be used to define more complex expressions. Certain other characters also have a special role, for example '.' and '+'. To search for these characters a stop-character is needed, usually '\'. Thus, '\$a\+b' will search for '$a+b.'

There are many possible ways to define matches using a library of special tokens like the ones above, which can be used to validate email address, parse HTML tags, etc. Many modern programming languages support some form of regular expression handling.

Software versioning

Writing a large program often involves working with many different source-code files and resource files, the latter containing things like parameter lists, or icons for user interfaces. As the software is developed, it is useful to think of the development as happening in finite steps. Some of these steps may be small (such as small bug fixes), or large (such as switching to a whole new method of communication with a database). For future reference, it is useful to document all these changes as much as possible, and keep track of these development steps using version numbers. Keeping such a record allows you to know exactly what features exist in a particular version of the software a user might be running, and can be very helpful in tracing when a bug was introduced into the code.

Version numbers often contain major and minor numbers. If a piece of software is on version 1.4.2, and a major new feature is added, this might increase the version to 1.5.0, or even 2.0.0, depending how large the software change is. A minor change might just change the version to 1.4.3. The decision on how to increment the numbers is fairly flexible and is left to the developers to decide. However, there are some generally adhered-to guidelines. Version numbers starting with a zero, such as 0.7.1, generate indicated pre-release code. This might be referred to as a 'beta release', meaning it is released to the users for testing, and might be indicated by a 'b' in the version number. There are also 'alpha' releases, which come before beta releases, and are usually released only for internal testing. Software version 1.0.0 tends to be reserved to show the first version of the software considered suitable for general release. This might be initially referred to as 1.0rc1, the 'rc1' meaning Release Candidate 1. If further bugs are found, these might be fixed and 1.0rc2 might be released; if no more bugs are found, the latest release candidate would be renamed version 1.0.0.

There are several software tools available that help the programmer to manage software versions, one of the most notable being the freely available **CVS** (concurrent versions system). This software manages source code and associated files by initially copying all associated files into a central store called a **repository**. This is the initial version. The programmer works on the code locally, and then 'checks in' the code to the CVS tool along with some notes about what has been added or changed. CVS then stores all the changes made to the code since the last version, and stores the changes and notes under a new version number. Using the system, the programmer can 'check out' the code at any stage in the development by asking CVS for the appropriate version number or supplying a date for the required snapshot. CVS will provide the user with all the files in the state they were at that particular version or time.

Networking computers

A network of computers is a set of interconnected computers which can share data and communicate with each other. The internet is an example of an enormous computer network. Managing the transfer of data is not as simple as just sending the bits down a wire and forgetting about them. The transmission is likely to be noisy, leading to data corruption, and data may be lost completely if intermediate computers or connections are lost. The main management system to handle this is known as the **transmission control protocol** (TCP).

TCP begins by breaking down the data into smaller parts, known as **packets**, which also contain metadata about the data being transmitted. Think of this as breaking a large machine down into parts, and packaging them ready to post separately. This is then passed to the **internet protocol** (IP) whose job is to carry out the actual transfer between machines. The machine receiving the packet then passes it back to TCP, which checks that the packet has not been corrupted en route. If it has then the packet is sent again. Once all the packets have been correctly received, TCP reassembles the data in the correct order. Together, these protocols are referred to as TCP/IP, one of the foundation protocols on which the Internet is built.

In practice, there is actually a whole suite of 'layered protocols' that handle network communication, from the top-level application layer down to the layer which handles the physical network itself. TCP/IP just happen to be the protocols most often mentioned. Data are passed from applications down through the layers, across the network, and back up through the levels to the application at the receiving end. At each layer, there may be several protocols available, and particular ones might be better suited for particular kinds of data or network.

Multi-tier computing

Applications can be thought of often as comprising three separate functional modules, referred to as 'tiers' because of the layered communication between them. The top-level tier is the user interface. This is the part of the program the user sees, and is the part that allows the user to interact with the program. It may be a command-line console, a graphical user-interface, or a web page. The middle-tier handles the actual computation and logic of the program, and for this reason is commonly referred to as the logical tier. The user input from the top tier is fed into the middle tier, and results are fed back to the top-level tier for presentation to the user. The middle tier reads data from and to the bottom tier, called the data tier as it handles data storage. This is what is meant by **multi-tier computing** (see *Fig. 1*).

These three tiers might all be physically located on the same machine, but for heavily used resources it is often sensible to split the tiers across separate computers. The user interface might be viewed as a web page on a user's personal computer. This would feed input into the logical tier, which could reside on a remote server with plenty of processing power. In turn, the data may be held on a networked system capable of storing and backing up large amounts of data. The advantage of dividing the application up in this manner is that the processing

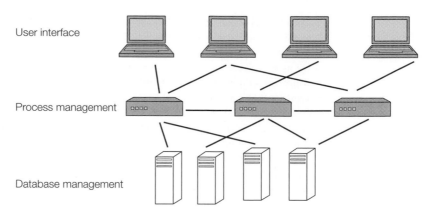

User interface

Process management

Database management

Fig. 1.   Multi-tier architectures. This figure shows a three-tier architecture involving the user-interface, processing, and data management layers. A given web site might receive heavy use. So if one of the processing servers receives many calls from user-interface machines, it will pass the query to a less busy server. The same applies to the data layer. The big advantage of this architecture is that individual machines can be removed or replaced without disturbing the whole system.

requirements are also divided between the computers; this is referred to as **load balancing**. Tiers can be replicated; it may be advantageous to have many middle-tier processors if there are many requests that require large amounts of processing.

A web site could be designed with the multi-tier approach in mind, and would typically require a web server to serve the web pages, an application server to perform the data processing, and a database management system and server to store the actual data. Other advantages of multi-tier systems include the ability to switch tier equipment for replacements without affecting the rest of the architecture, the ability to provide specialist equipment for each tier, and the ability to manage security for each tier machine separately.

Grid computing and web services

From the basic idea behind multi-tier computing, that is dividing up the processing and data requirements of an application and assigning them to separate computers, we arrive at the concept of **grid computing**. The name *grid* alludes to the power grid for distributing electricity, and the idea that you can distribute processing power the same way and simply plug an application into the grid to make use of that processing capability. The physical structure of a grid is composed of many separate computers connected, potentially over large distances, by computer networks such as the internet. The individual computers might only have the power of standard desktop computers, but because a great number can work together on solving a problem, the processing time can be greatly reduced. The concept is similar to that of a super-computer with many CPUs, but has the advantage of being much cheaper. The computers which form the grid (called 'nodes') may be fully dedicated, or they may perform calculations only when they are not being used for other tasks.

Grid computing methods require software to coordinate the use of the computing resources. Usually, a task is submitted to the control software, which then allocates the grid resources to the task when they are available, and returns the results to the user when the processing is finished.

So using multi-tier computing and grid computing, it is possible to access remote computing power and resources. This can be done manually, but there are occasions when it can be more useful to let software applications communicate directly with remote resources. This is what **web services** allow. This term really relates to a collection of tools and methods that allow software to communicate over the internet using the hypertext transfer protocol (HTTP), on which the WWW is built. A web service provider offers processing services which are made available over the internet. These are like methods or function in programs, except that they are available over the WWW.

Web services can be listed in central registries like Universal Description, Discovery and Integration (UDDI) registries, which act like a phone book allowing clients to find particular services. The functionality of a service is often described using the machine-readable web services description language (WSDL). Client computers who wish to use a web service can look at the WSDL definitions to see which services are available and how to communicate with the services. The applications communicate using XML-based (see Section D) messages. Because of the standard format of these messages, a computer will be able to interpret the message so long as web service standards are complied with, irrespective of the computer architecture. Thus web services can be used to let machines and systems communicate, which they would otherwise struggle to do. Additionally, as the messages are sent using standard HTTP, there is less risk of the communication being blocked by firewalls or security systems, as it uses the same method of communication as the WWW.

One of the powerful features of web services is the ability to string multiple services together, feeding the output of one into the input of another. These are known as **workflows**. Software packages exist to help construct these workflows, such as Taverna Workbench (www.mygrid.org.uk/tools/taverna).

Good practice for bioinformatics software design and management

The process of writing reliable, reusable software can be long and complex. The actual writing of source code for the final software release usually makes up a small percentage of the total time spent developing an application. Large software development projects involve many stages and skills, from talking to potential users and designing how user interfaces will look, to

- defining solutions to complex logical problems that form the backbone of the application;
- writing the code in a way that is understandable to future programmers;
- debugging and testing the code;
- writing the manuals that show users how to use the software;
- dealing with support requests after the software is released.

In a software house, each of these aspects will often be covered by a particular specialist or team of specialists. However, all these tasks can fall to one person or, at best, a small number of people working on a project, which is especially the case in bioinformatics.

While it would be impossible to give a comprehensive set of advice here, this section looks at some of the main concerns to keep an eye out for when writing bioinformatics software. There are some excellent books available (e.g., Weston, 2004) which provide very useful advice in this area, and would be valuable reading for anyone who is likely to do large amounts of bioinformatics software development.

Planning

Before writing any code, it is sensible to plan as much of the project as you can in advance. Break the task down into functions that you know the project must achieve (such as produce the database, or design the web form). Clearly, when writing software for a research project, many of the possibilities will not be known at the start, and much problem solving by trial and error will be required. Nevertheless, it is important to keep in mind from the beginning what your users will want from the system, and what the overall aim is. Talking to potential users is very useful, especially when designing things like the user interface, as things which seem obvious to you as a programmer may well be anything but obvious to a real-world user.

Try to envisage a time-line for development, breaking the software down into functional units if you can. This can be very challenging though, and may be near impossible to determine with some open-ended projects. Invariably, there will be unforeseen challenges along the way that will slow development, sometimes dramatically. It is impossible to factor these in when planning development time, but be aware that these delays can, and invariably will, happen. Often, in a research setting, you will be developing code in an iterative fashion, producing subsequent versions of code for fellow scientists to use.

This is particularly true in bioinformatics where biological research practice may change faster than software development. Hence, effective bioinformaticians maintain a close dialogue with laboratory researchers. Pieces of software are developed to address the most urgent needs first while keeping in mind that these must fit easily into a 'final' system later. Planning needs to be an ongoing process, as the requirements of the system are very likely to evolve over time.

Some developers adopt what is referred to as the **80–20 rule**. This refers to the rule of thumb that states you can develop 80% of the requirements in 20% of the time. So, you develop this 80% then review the users' requirements. As this has only taken 20% of the time, users have been optimally supported, and the review of requirements helps the developer to focus on the new priorities, aiming again to produce only 80% of what is required. The closer the dialogue between users and developers, the more effective the latter can be.

Implementation

When you start a new project, it is important to consider what the most suitable programming language and development tools are. Programming languages each have their own advantages and disadvantages, and spending the time to choose the right one will pay you back later. Will the program break down neatly into objects to implement in an object-oriented language? Will the language need to handle public web-access to the data? Is speed of execution a priority? Other considerations include:

- which languages you know and how long it would take to learn a new one;
- which languages and development tools other programmers in the project are using;
- which existing libraries you can use to save you having to re-write code;
- which computer system the final software needs to run on;
- what kind of network communication is required by your software, etc.

Some programming environments offer suites of bug-testing tools, such as the ability to watch the value of a variable as the program runs, the capability to 'catch' errors such as invalid memory access, or the ability to step through the code line by line. Drawing flow-diagrams of how a process is expected to work, or of the flow of data through a program can help you visualize what is supposed to be happening. It will be useful to catch and log as many errors as you can, as it is nearly impossible to produce bug free code right off the bat; any information you can get your program to give you about how it is failing will be immensely useful in determining the fault. You should test the system during implementation, both at the level of small units of functionality, such as manipulating the value of a variable, through testing larger modules, such as making sure all the database side of the code is working as expected, right up to the overall functionality of the whole program. Testing can be very time consuming, but having an effective testing plan can save much time on fixing errors later on. Indeed, in so-called **extreme programming**, automated bug-testing software is developed alongside the main software (Jeffries *et al.*, 2000), often in the form of **unit testing**, in which code is broken down into small units that are tested throughout development.

Style

Many modern languages allow a fairly free-form production of source code. Variable names can, within a few constraints, be anything the programmer desires, as can function and class names. White space (e.g., spaces, tabs, blank lines) is often (but not always, in some languages) ignored when the program is run or compiled. Most languages define a way to let programmers add free text as comments. Using features such as this to make the code more understandable to humans is referred to as **programming style**. Writing well-styled code can mean the difference between someone understanding it or not. This someone might be yourself when you come back to the code years later, to fix a bug in otherwise forgotten code! Below is a list of some suggestions to help write well-styled code.

- Use white space where you can to help introduce physical distance between sections of code that do different things. Just as breaking text into paragraphs makes books more readable, white space can be used by the reader to group functional lines together to show at-a-glance the structure of the code.
- Where you can, use indentation and tabs to make structures such as nested loops clear.
- Keep code as simple and elegant as possible. Writing over-convoluted code can make it difficult to comprehend, just as writing a lengthy and complex sentence can cause confusion.
- Try to write 'self-documenting' code – that is, code which makes sense without having to be overly explained using many comments. This can be done by using sensible variable and function names, which describe what the variable or function does. Unless the code is very simple, such as a small loop structure, the variable x says little about what it represents. Calling the variable *expressionLevel* may be much more meaningful.
- When writing logical conditions, it is best to keep the logic as simple as possible to avoid confusing readers. For example, writing '*if (!error)*', which is read 'if there is not an error…', is more convoluted than saying *if (normalExecution)*, which means the same thing. While this is a simple example, code quickly can become hard to understand when many combinations of negatives and double negatives are used together.
- Version your software so you can keep track of what was added when, and by whom. Keep an up-to-date list of features that have been added either in your source code or stored with it. It is useful also to have a quick explanation of what the code does, and a list of any known bugs. CVS software can be useful here (see above).

Support   Hopefully there will come a stage when your software will be used by someone not involved in the development process. Packaging the software into an easy-to-install format will make the process of physically distributing the code much easier. While you may think of your software as easy-to-use, invariably naïve users will not think in the same way. This difficulty in use increases exponentially with the complexity of the software. The process can be made easier by writing good-quality manuals, or at least simple 'how to' documents which lead a user through the process. It may be useful to set up discussion forums to report bugs or to allow users to discuss issues and suggestions with yourself and other users. One popular web site for organizing this for open source software is Sourceforge (http://sourceforge.net). Once a community base is built up, it may be that the problems can be resolved between the users themselves. This community may be happy to do formal beta testing of the code. Expect there to be bug reports, as it is rare for software to be released completely bug-free the first time around. Users will be likely also to suggest improvements and changes to your system that benefit them.

Quantifying how much support is to be expected for a project is very challenging, and the amount of time and resources needed to support software should not be underestimated.

G1 PROBABILITY AND PROBABILITY DISTRIBUTIONS

Key Notes

Probability	This is a key concept in bioinformatics, essential to the interpretation of biomolecular data.
Probability for combinations of events	The product rule, P(A and B) = P(A)P(B), for independent events is important in many applications of probability theory. Other important formulae are: P(A or B) = P(A) + P(B) for mutually exclusive events, and P(not A) = 1 – P(A).
Probability distributions and density functions	A probability distribution is a mathematical function giving the probability that a variable will take a particular value or fall within a range of values. Perhaps the simplest is the binomial distribution, which gives the probability of any number (r) of successes in n trials, where each trial can have only two outcomes, success (probability = p) and failure (probability = 1 – p). It applies to the number of heads generated in a sequence of tosses of a coin.
The Poisson distribution	Like the binomial, this is a function that gives probabilities for numbers of events, but this time expressed in terms of a parameter E equal to the expected number of occurrences. It approximates the binomial distribution when the number of trials is very large and the probability of success in any one trial is very small.
The normal (Gaussian) distribution	This is a mathematical function that gives probabilities for a variable that can take continuous (real) values. It is usually expressed as a bell-shaped probability density function with two parameters, the mean and the standard deviation. Probabilities are calculated as areas under the curve of the density function. It is very important in much statistical theory.

Related sections	Conditional probability and Bayes' rule (G2)	Databases and data sources (J1)
	Elementary statistical testing (G3)	Transcript profiling (K1)
	Statistical approaches to artificial intelligence and machine learning (I2)	Interaction proteomics (L2)

Statistics is concerned with the treatment of variability and uncertainty. It has traditionally been seen as an important feature of biological science, because the variability between individual organisms is fundamental to the theory of evolution. But it is equally central in the physical sciences, where the accepted **quantum theory** of mechanics is fundamentally probabilistic, and **statistical mechanics** is used to treat the thermodynamic properties of large systems. On a more mundane level, most scientists use statistical methods to handle experimental errors and precision, but unfortunately the subject is often viewed as rather dull and, for some, difficult. In modern biology, however, statistics has gained new importance, and the questions it seeks to answer should whet the appetite of the most uninterested and apprehensive biologist.

The revolution in experimental techniques in molecular biology since the end of the last century has seen the development of many techniques that generate data in large quantities; for example, whole genome sequences (Section J1), expression measurements from DNA microarrays (Section K1), and protein interaction data from yeast two hybrid screens (Section L2). This quantity of data, along with the noisy and uncertain nature of some of it, means that statistical interpretation is vital. A simple example comes from the analysis of DNA sequences. If two sequences from different species' genomes are seen to be 'similar' (perhaps they can be aligned to show that they share a percentage of identical nucleotides), then this similarity might indicate that the sequences have evolved from a common ancestral sequence. But, all DNA sequences are similar to some extent, and the similarity might just have arisen by chance. A statistical method aiming to distinguish these possibilities would be vital to the generation of an important biological hypothesis.

This book is not intended as a textbook on statistics, but it aims to give an introduction to methods of direct applicability in bioinformatics, with appropriate examples. Readers will probably find it useful to consult a specialist statistics textbook for more detail.

At the core of statistics is the concept of **probability**, and a good understanding of this is vital to any study of bioinformatics.

Probability

Probability is a number between 0 and 1 reflecting how likely something ('an event') is to happen. If we call this event A, then its probability $P(A) = 1$ if it is certain to happen and $P(A) = 0$ if it is certain not to happen. Lesser degrees of certainty are expressed as fractional values. For instance, if an unbiased coin is tossed then the probability of 'heads,' P(heads) = 0.5, and is equal to the P(tails).

Probability can be interpreted in a **frequentist** way, where a probability of 0.5 is taken to mean that if the same 'trial' were carried out many times the event would happen in approximately half of the trials. With increasingly larger numbers of trials this approximation would become better and better, so that while 7 heads is not improbable in 10 tosses of a coin, 700 in 1000 tosses is much less likely, and a number nearer to 500 would be expected. Probability is thus interpreted as the relative frequency of an event in a very large number of independent trials. An alternative to this frequentist view is the **subjective** interpretation, where the probability is viewed simply as a measure of how likely an event is to happen. This interpretation is inherent in **Bayesian** statistical ideas (see Section G2).

Probabilities for combinations of events

It is often useful to be able to calculate the probabilities of compound events. For instance, if A and B are independent events (have no influence over each other) then the product rule applies

$$P(A \text{ and } B) = P(A)\,P(B)$$

So, the probability of getting a six on each die when two dice are thrown is P(six and six) = $1/6 \times 1/6 = 1/36$. This illustration shows why the formula is true: on each die there are six possible outcomes, but any of the outcomes on die 1 can occur with any on die 2, so the total number of possible outcomes when two are thrown is $6 \times 6 = 36$, only one of which is the double six. This is a very important formula, and will be used in several later sections of the book, particularly in the section on Markov models (Section I2). We shall see later how the formula changes when A and B are not independent (i.e., the occurrence of one affects the probability of the other).

When A and B are mutually exclusive (cannot both occur) then

$$P(A \text{ or } B) = P(A) + P(B)$$

In the case of throwing a single die, the probability of getting a 1 or a 6 is P(1 or 6) = 1/6 + 1/6 = 1/3, because throwing a single die cannot have both outcomes. An important special case of this formula is

$$P(A) + P(\text{not } A) = 1$$

Where P(not A) is the probability that event A does not happen. This formula means that since event A must either occur or not, the sum of the two possibilities must be 1. In general P(not A) = 1 – P(A).

Probability distributions and density functions

A probability distribution is a mathematical function giving the probability that a variable will take a particular value or fall within a range of values. Perhaps the simplest is the **binomial distribution**. This concerns binomial trials, which can have only two possible outcomes, like the tossing of a coin. The formula for the distribution is

$$P(r) - \frac{n!}{r! \, (n-r)!} \, p^r \, (1-p)^{n-r}$$

This calculates the probability $P(r)$ of getting r heads in n tosses of a coin[1], or more generally the probability of r successes in n trials with two outcomes, conventionally labeled 'success' and 'failure'. The parameter p is the probability of getting heads (success) in a single toss of the coin, so the formula can cope with the situation of a biased coin where this value would not be equal to 0.5. To calculate the probability of getting 7 heads in 10 tosses of a fair coin you would substitute the values $r = 7, n = 10$ and $p - 0.5$ in the right hand side of the equation (the answer is P(7) = 0.12).

If you want a little understanding of this formula, you can divide it into two parts, $n!/r!(n-r)!$ and $p^r(1-p)^{n-r}$. The second of these uses the product rule (above) and corresponds to the probability of a sequence of coin tosses containing r heads and $n-r$ tails (if p is the probability of heads then $1-p$ is the probability of tails). The first part corresponds to the number of different sequences that have the same net outcome of r heads and $n-r$ tails. For instance, the outcome of 5 heads in 10 tosses can be achieved in 10!/5!5! = 252 ways, and, for example, just two of these are HTHTHTHTHT, and HHHHHTTTTT. The binomial distribution is illustrated in *Fig. 1*.

The Poisson distribution

A probability distribution related to the binomial is called the Poisson distribution. The formula for this is

$$P(r) = \frac{E^r \exp(-E)}{r!}$$

Like the binomial, this gives a probability for a number of occurrences r but this time the parameter is E, which is the expected or average number of occurrences. For example, the Poisson distribution applies to the number of cherries in each

---

1 The notation a! denotes the factorial of the number a, a!=a(a-1)(a-2)....2.1. 3!=3x2x1=6, 4!=4x3x2x1=24, 1!=0!=1.

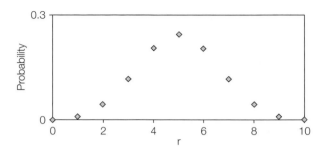

Fig. 1.   The binomial distribution showing the probability of r successes in n = 10 trials where the probability of success in any trial is p = 0.5. This could represent the probability of getting any number of heads in 10 tosses of a fair coin.

slice of a fruit cake. Suppose we put 100 cherries in a well-mixed cake and then cut it into 20 slices. This gives an expected value of $E = 5$ cherries per slice, but of course when you choose a particular slice you might be lucky and get more, or you might be unlucky. The formula above can be used to calculate the probability of any number of cherries in a particular slice, by substituting 5 for the parameter E. The probability of getting 7 cherries is $P(7) = 5^7 \exp(-5) / 7! = 0.1$.

The Poisson distribution is a useful approximation to the binomial distribution when the number of trials (n in the binomial formula) is very large and the probability p (in the same formula) is very small. In this case the binomial and the Poisson are the same distribution so long as the Poisson parameter $E = np$. The approximation is useful because calculating the large factorials in the binomial formula can be difficult (e.g., for 1000 trials you have to calculate 1000!). The Poisson distribution and its relationship to the binomial are illustrated in *Fig. 2*.

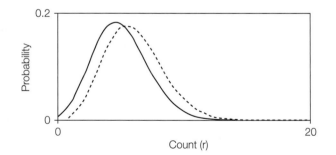

Fig. 2.   The similarity of the binomial (continuous line) and Poisson (dashed line) distributions for n = 100, p = 0.05 and E = np = 5. The approximation becomes increasingly better as n becomes larger and p smaller, with E = np fixed.

The normal distribution

The distributions considered above give probabilities for a discrete variable: a number of events must be a whole number. The normal or Gaussian distribution is different in that it gives probabilities for a continuous variable (a real number that can take values like 1.5, 8.68, etc.). The formula for the normal is

$$N(x) = \frac{1}{\sqrt{2\pi}\sigma} \exp \left(\frac{(x - \mu)^2}{2\sigma^2} \right)$$

The normal distribution is illustrated in *Fig. 3(a)*; you might recognize the characteristic bell-shaped curve. There are two parameters that can be varied to give different normal distributions, the mean (μ) and standard deviation (σ). The mean is the position of the center of the distribution, and the standard deviation controls the width of the bell. An example of data that are normally distributed is the heights of people in some populations; in this case the bell-shaped curve would be centered on the mean (average) height in the population (say $\mu = 1.77$ m), and the standard deviation (say $\sigma = 0.1$ m) would reflect the degree of spread of individual heights around this mean value.

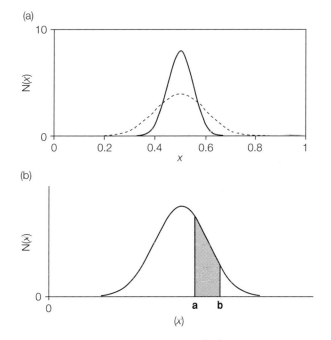

Fig. 3. *The probability density function for the normal distribution. N (x) refers to the number of objects having value x. (a) Distributions both with mean $\mu = 0.5$ and standard deviations $\sigma = 0.1$ (dashed) and $\sigma = 0.05$ (continuous). (b) The calculation of the probability of a value of x in the range a < x < b as an area under the curve.*

Because x is a continuous variable we cannot assign a probability to a particular value. This would not make sense: I might say that my height is 1.83 m, but I do not mean that it is precisely 1.83 m, because I cannot measure it to very high precision. What I really mean is that I estimate my height to be in a range of values, say $1.825 < x < 1.835$. Similarly we assign probabilities to ranges of values of the real number x. The probability that x lies in the range $a < x < b$ is equal to the area under the $N(x)$ curve between these values, as illustrated in *Fig. 3(b)*. The function $N(x)$ given above is known as a **probability density function**; its value is not a probability but a probability density, and probabilities are derived from areas underneath the curve.

In terms of integral calculus, areas under the curve are associated with definite integrals. For instance the probability of a value of x less than x_0 is

$$P(x_0) = \int_{-\infty}^{x_0} N(x)dx$$

This function is knows as a **cumulative distribution function** (or sometimes just a **distribution function**), because it gives the cumulative probability of any value of x up to a certain value.

It is very common to use a standard normal distribution, defined to have a mean of zero and standard deviation of one. Any variable (x) that has a normal distribution can be converted to one with a standard distribution by subtracting the mean and dividing by the standard deviation, $z = (x - \mu)/\sigma$. The variable z is sometimes called a **Z-score**. It measures how far an individual data point differs from the mean, in units of the standard deviation. It is a scale-independent measure of how different an individual is from the average of the population from which he/she originates. For instance, for any normal distribution, the probability of a value of x being more than 2 standard deviations from the mean ($|z| > 2$) is 0.05. Unlike the binomial, the normal does not have a clear interpretation like the number of successes in a set of trials. Nevertheless, it is very important and arises in many areas of statistics. Many data sets (like the heights mentioned above) turn out to be approximately normally distributed, and because a great deal of statistical theory is applicable to normally distributed variables, this can make analysis of data much easier. Another reason for the importance of the normal distribution is the **central limit theorem**, which states that if a sample of n data points is taken from (almost) *any* probability distribution, then the mean of that sample has a distribution that is increasingly close to the normal distribution as the size n of the sample increases.

Example: An application in genome analysis
MAN1 is a hypothetical pattern found in gene sequences to which the transcription factor MANTF binds in order to influence the transcription of nearby genes. The MAN1 element is unusually long, 12 base pairs.

MAN1 = ATAGGCGTAGTC

MAN1 appears 250 times in the human genome (a sequence of 3.1×10^9 bases/letters), but only 10 times in the genome of the fruit fly (3.1×10^8 bases). (a) What is the probability of this element occurring by chance at any position in random DNA? (b) How many times would it be expected to occur by chance in the genomes of fly and man? (c) Is the number of occurrences in man significantly different from this expected rate of chance occurrence? (d) Is there any biological significance to these observations?

a. Making the assumption that each of the four possible bases {A,T,G,C} are equally likely in the genome (not quite true but a reasonable approximation for this example), the probability that any one of them occurs at a particular position in the genome is 0.25. For the 12 base pair pattern we use the product rule to calculate the probability, assuming that the identity of a base at any position is independent of the identities of the surrounding bases (again an approximation). The probability of seeing the first two nucleotides of the pattern is P(A first AND T second) = P(A)P(T) = 0.25×0.25, and the probability of the whole 12 base sequence is $0.25^{12} = 6 \times 10^{-8}$.

b. Using the probability above, we would expect $(3.1 \times 10^9) \times (6 \times 10^{-8}) = 186$ occurrences in man, and $(1.37 \times 10^8) \times (6 \times 10^{-8}) = 8$ occurrences in fly. (Making the assumption that the element can occur independently with equal probability at any point in the genome.)

c. While the actual number of occurrences in fly is approximately equal to that expected by chance (10 compared with 8), in man the actual number is much greater (250 compared with 186). But, is there really a difference? Tossing a fair coin 10 times is quite likely to produce 7 heads, which is greater than the expected number of that under the assumption of a fair coin (5). The extra two heads could just be a chance fluctuation, and they don't constitute evidence of unfairness in the coin. Equally the differences in MAN1 occurrences might just be chance fluctuations.

One way to answer this question is to calculate the probability of the actual number of occurrences in each case, under the assumption that they are occurring by chance with probability $p = 6 \times 10^{-8}$ at each position in the genome. In the human genome, we could use the binomial formula with $n = 3.1 \times 10^9$, $r = 250$ and $p = 6 \times 10^{-8}$. However, because large numbers are involved, it is easier in this case to use the Poisson distribution (which will give the same answer because n is very large and p is very small), in this case with $E = 186$. This gives a probability of $P(250) = 1 \times 10^{-6}$. Since this probability is very small (1 in a million) we might be tempted to conclude that occurrence of the MAN1 element in the human genome is not a simple chance phenomenon, and that there is some underlying reason why there are so many of them. In the case of the fly, we have $E = 8$, and $P(10) = 0.1$. So, this time we might make the opposite conclusion; the observation of 10 occurrences in the fly genome is quite likely to be a chance phenomenon, since under this assumption it has a probability of 1 in 10.

This is an elementary test of **statistical significance** (such tests are discussed further in Section G3). In fact the argument above has one small flaw. Instead of calculating the probability of seeing exactly 250 occurrences in the human genome, we should have calculated the probability of 250 *or more*. This accounts for the fact that precisely 250 might be quite improbable, while the probability of *any* outcome more different from the expected chance rate than 250 could be a lot larger. In this case, however, it makes little difference, for man $P(250 \text{ or more}) = 5 \times 10^{-6}$, and for fly $P(10 \text{ or more}) = 0.3$.

(If you are interested in repeating these calculations, you might use a statistical package like that available in Microsoft Excel. In this program, the above numbers can be calculated as $P(250) = \text{POISSON}(250,186,\text{FALSE})$ and $P(250 \text{ or more} = 1 - \text{POISSON}(249,186,\text{FALSE})$. You will find that the Excel function for the equivalent binomial calculation will not cope with the large numbers involved.)

d. This example is based on a real case of a transcription factor known to be used in man but not fly. The much greater occurrence of the associated sequence pattern by comparison with chance levels in man reflects the fact that the elements are not simply occurring randomly, but have evolved in the genome close to the genes regulated by the transcription factor. In fly there is no such mechanism (this transcription factor is not present in fly), and the elements that occur are simply there by chance.

G2 CONDITIONAL PROBABILITY AND BAYES' RULE

Key Notes

Conditional probability

This expresses the idea that events may not be independent. The known occurrence of one event (A) can affect the probability of another (B), and this is expressed with the notation, P(B|A) meaning the probability that B occurs given that A is known to occur. It leads to the modified version of the probability product rule: P(A and B) = P(A)P(B|A).

Bayes' rule

This is obtained by re-arranging the product rule above:
P(A|B) = P(A)P(B|A)/P(B)

Bayesian statistics

This uses a subjective interpretation of probability. It makes extensive use of Bayes' rule above in the form where A is identified with a hypothesis, H, and B with some data or observations, D, giving P(H|D) = P(H) P(D|H)/P(D). This relates the **prior** probability that H is true (P(H)) to the **posterior** probability (P(H|D)) that the hypothesis is true as influenced by the observed data.

Markov chain models

These are a convenient statistical tool for modeling sequences of states; for instance, the sequences of letters associated with DNA sequences. Within the sequence the identity of a letter depends only on the identity of the letter immediately preceding it, thus modeling some of the simple correlations that are found in biomolecular sequences.

Related sections

Probability and probability
   distributions (G1)
Elementary statistical testing (G3)
Sequence analysis (J3)

Sequence families, alignment, and
   phylogeny (J4)
Multivariate techniques and
   network inference (K4)

Conditional probability

In Section G1 we introduced the product rule for the probabilities of independent events

$$P(A \text{ and } B) = P(A)P(B)$$

If the two events concerned are not independent (i.e., they have influence over each other and the occurrence of one changes the probability of the other), then this equation changes to

$$P(A \text{ and } B) = P(A)P(B|A) = P(B)P(A|B)$$

In this equation, P(B|A) means the probability that event B occurs, given that A definitely happens. Similarly P(A|B) means the probability that A occurs given that B definitely happens. It is very important to realize that P(A|B) and P(B|A) mean different things and have different values. For instance, if A is 'the sky is cloudy' and B is 'it is raining,' then $P(A \mid B)$ is the probability of a cloudy sky when it is raining (probably quite high, say 0.99); $P(B \mid A)$ is the probability of rain when the sky is cloudy (significant, but perhaps not so high, say 0.3). The quantity P(A and B) is different again, and it refers to the probability of it being both cloudy and raining at any given time. To calculate P(A and B) we also need to know either the probability of A or B. Suppose that we are in quite a wet area with P(B) = 0.01 (a chance of 1 in 100 that it will be raining at any given time), then

$$P(A \text{ and } B) = P(B)P(A \mid B) = 0.01 \times 0.99 = 0.0099$$

Bayes' rule

When the second equation above is written

$$P(A \mid B) = P(A)P(B \mid A)/P(B)$$

in order to show the relationship between $P(B \mid A)$ and $P(A \mid B)$, it is known as Bayes' rule (after the man who first derived it).

Bayesian statistics

Bayesian statistics is almost a philosophy about the meaning of probability, and how it should be employed in order to learn things from observations or other data (i.e., to perform statistical inference). It is increasingly popular in bioinformatics. The subjective interpretation of probability (see Section G1) considers the probability of an event to be simply a measure of how likely someone thinks it is to occur. Bayesian statistics shows how such probabilities change when observations are made or data are gathered. Suppose, for instance, that you have a hypothesis (we will call it H), and that P(H) is your initial degree of belief that it is true. You then gather some data (we will call them D), and the influence of these data on the probability that H is true is expressed as $P(H \mid D)$ (the probability that H is true *given* the observed data). Using Bayes' rule above with A = H and B = D gives

$$P(H \mid D) = P(H)P(D \mid H)P(D)$$

This is useful because it is often quite easy to calculate the quantities $P(D \mid H)$ and P(D) on the right-hand side of the equation. Given your initial degree of belief in your hypothesis P(H), these lead to $P(H \mid D)$. P(H) is usually called the **prior**, and the way the data have influenced this prior degree of belief, $P(H \mid D)$ is called the **posterior**.

There is a simple example of the use of Bayesian inference at the end of this section. A more complex biological example is to be found in Section J4 where the problem is to infer the most likely phylogenetic tree for a set of sequences thought to be derived by divergent evolution from a common ancestor.

Markov chain models

A very simple application of the idea of conditional probability is called a Markov chain. These have proved very useful in modeling biomolecular sequences. A very simple statistical model of a DNA sequence might consider it as just a random sequence of letters from the set (alphabet) {A,T,G,C}. DNA sequences are of course non-random for many good biological reasons; for example, where they code protein or RNA molecules, the identities of the letters at different places in the sequence are related by the need to code for stable three-dimensional structures, where elements distal in the sequence can be in spatial contact. Another often quoted reason for non-randomness is that the letter G only infrequently follows the letter C (known as a CpG pair) in some organisms. This is because Cs in CpG pairs are preferentially methylated and this in turn leads to a high chance of mutation from C to T. While modeling all the correlations in letter identities in DNA sequences is a very challenging problem, a very simple modification that is effective in some cases is to consider only correlations between letters that are adjacent in the sequence. Such a model is called a Markov chain. In a Markov chain, a sequence of states (letters from ATCG in this case) is generated in such a way that the next letter depends only on the identity of the current letter. It is defined by a set of conditional **transition probabilities** shown in the following matrix.

Table 1.　Conditional probabilities defining a Markov chain model of DNA sequences

		Next letter			
		A	T	G	C
	A	$P(A|A)$	$P(A|T)$	$P(A|G)$	$P(A|C)$
Current	T	$P(T|A)$	$P(T|T)$	$P(T|G)$	$P(T|C)$
letter	G	$P(G|A)$	$P(G|T)$	$P(G|G)$	$P(G|C)$
	C	$P(C|A)$	$P(C|T)$	$P(C|T)$	$P(C|C)$

For instance, $P(C|T)$ is the probability that the next letter will be C given that the current letter is T. Different values for these probabilities define different models for DNA sequences. In sequences where CpG pairs are under-represented, $P(G|C)$ would be small. We will return to Markov chains in the section on hidden Markov models in sequence analysis.

Example 1. Conditional probability and Bayes' rule applied to medical screening tests
To understand the meaning of conditional probability, it is useful to consider an example. Suppose that a particular disease is relatively uncommon, and only present in 1 in 10 000 people in the population. If A is the event of having the disease then $P(A) = 0.0001$. Suppose also that there is a screening test for this disease, and we call B the event of testing positive. The test is cheap but, however, not perfect, and it has both false positive errors (cases of positive tests in people who are healthy) and false negative errors (cases of negative tests in people with the disease). An independent evaluation has shown that for people known to have the disease, 2 in 10 000 will test negative. In terms of conditional probability this means that

$$P(B|A) = 0.9998$$

which should be read 'the probability of testing positive (B), given that you have the disease (A) is 0.9998.' Similarly the evaluation has shown that 1 in 100 healthy people will test positive for the disease, meaning that

$$P(B|\text{not }A) = 0.01$$

While these figures are interesting, they are not the most useful figures for people taking the test. This is because they only tell you how reliable the test is for people who already know whether or not they have the disease. Much more useful conditional probabilities would be $P(A|B)$ (the probability that you have the disease given that you have tested positive), and $P(A|\text{not }B)$ (the probability that you have the disease given that you have tested negative). To work these out, we first use Bayes rule above

$$P(A \mid B) = P(A)P(B \mid A)/P(B)$$

and then note that

$$P(B) = P(B \text{ and } A) + P(B \text{ and not } A)$$

using the rule for adding probabilities of events that are mutually exclusive. Finally, substituting in this equation from $P(A \text{ and } B) = P(A)P(B \mid A)$ we get

$$P(B) = P(A)P(B \mid A) + P(\text{not } A)P(B \mid \text{not } A)$$
$$= (0.0001 \times 0.9998) + (0.9999 \times 0.01)$$
$$= 0.0100989$$

and

$$P(A \mid B) = 0.0001 \times 0.9998/0.01009898$$
$$= 0.01$$

So even if you test positive, there is still only a one-in-one-hundred chance that you are ill.
Similarly

$$P(A \mid \text{not } B) = P(A)P(\text{not } B \mid A)/P(\text{not } B)$$
$$= 0.0001 \times (1 - 0.9998)/1 - 0.01009898$$
$$= 2 \times 10^{-8}$$

So, if you test negative you are extremely unlikely to be ill. This is the principle of a screening test: a moderate false positive rate can be tolerated so long as the false negative rate is tiny.

Example 2. Which group of bacteria did the GC-rich sequence come from?
A complex sample contains DNA from many species of bacteria. These species can be divided into two broad categories: (a) those with high GC content genomes, and (b) those with low GC content. In category (a) the probability that a GC-rich sequence be obtained by randomly sequencing part of the genome is 0.8, and in category (b) it is 0.1. Suppose that a sequence is obtained by random selection from the sample, and observed to be GC-rich. What is the probability that it came from each of the categories (a) and (b)?

First, we assume that we have no knowledge of the sample except that it contains both bacterial types in the proportion of 1:3 (a:b). In this case our best prior knowledge is

$$P(\text{Sequence came from category (a)}) = 0.25$$
$$P(\text{Sequence came from category (b)}) = 0.75$$

We have the observation that the sequence is GC-rich, and we know that

$$P(\text{GC-rich} \mid \text{category (a)}) = 0.8$$
$$P(\text{GC-rich} \mid \text{category (b)}) = 0.1$$

Using Bayes' rule in the form

$$P(H \mid D) = P(H)P(D \mid H)/P(D)$$

We can identify 'category (a)' as the hypothesis (H) that the sequence came from category (a), and the observation that the sequence is GC-rich with the data (D). We want to calculate the probability that the hypothesis is true given the observed data. For the quantities on the right-hand side, we have

$$P(H) = \text{prior} = P(\text{Sequence came from category (a)}) = 0.25$$
$$P(D \mid H) = P(\text{GC-rich} \mid \text{category (a)}) = 0.8$$

and all we need to complete the calculation is $P(D)$. This may seem a bit confusing, but really this is just a normalization constant to make the total probability equal to 1.

$$P(D) = P(D \text{ and } H) + P(D \text{ and not } H) = P(H)P(D \mid H) + P(\text{not } H)P(D \mid \text{not } H)$$
$$P(D) = (0.25 \times 0.8) + (0.75 \times 0.1) = 0.275$$

Finally

$$P(H \mid D) = P(\text{category (a)} \mid \text{GC-rich sequence}) = 0.25 \times (0.8/0.275) = 0.73$$
$$P(\text{not } H \mid D) = P(\text{category (b)} \mid \text{GC-rich sequence}) = 1 - 0.73 = 0.27$$

So the observed data, in this case a GC-rich sequence, means that the posterior probability of the sequence coming from a bacterial species in category (a) is 0.73, compared with the prior of 0.25, based simply on the known content of the sample.

G3 ELEMENTARY STATISTICAL TESTING

Key Notes

Statistics and variability
Fundamentally, statistics is about dealing with variability and attributing it to different sources. It important to quantify variability from many sources in biology, including experimental imprecision, variation between different organisms of the same species, and variation in environments.

Testing statistical significance
Statistical significance tests aim to distinguish effects attributable to random chance variation from those with real biological significance.

The t-test and alternatives
The t-test aims to test the significance of differences between mean values of normally distributed data. When data are not normally distributed, alternatives are the Wilcoxon and Mann Whitney tests.

ANOVA
Analysis of variance aims to quantify the statistical significance of differences in group means, when there are more than two groups in the experiment (t-tests can only deal with two groups).

Chi-squared and the Fisher exact test
These tests are used with discrete 'count' type data, often expressed in contingency tables. They aim to assess the statistical significance of differences between the counts expected if the null hypothesis were true, and the actual observed counts in each category.

Resampling-based testing
This is an alternative to most statistical tests and usually involves randomized sampling from the data, assuming the null hypothesis is true, in order to calculate p values.

Multiple testing
When more than one statistical test is performed it is very important to correct for multiple sampling, which can give erroneously small p values. Common correction schemes are the Bonferroni and Benjamini-Hochberg corrections.

Related sections

Probability and probability distributions (G1)

Conditional probability and Bayes' rule (G2)

Sequence analysis (J3)

Statistical issues for transcriptome analysis (K2)

Analyzing differential gene expression (K3)

Multivariate techniques and network inference (K4)

This section is a brief account of some statistical tests and techniques commonly used in bioinformatics. It is not an exhaustive account, neither is it sufficient to provide detailed understanding of statistical issues. A reader interested in more detail should consult a textbook devoted to the subject of statistics.

Statistics and variability

Fundamentally, almost all scientific measurements or observations are subject to variability, and typically the variability in any particular measurement can have more than one source. For a very simple biological example, a set of measurements of the level of blood glucose in a number of patients in a hospital would be quite variable. The sources of this variability might include the intrinsic physiological differences between healthy individuals, factors such as the closeness of the patient's last meal, whether the patient had recently eaten a bar of chocolate, effects of treatments the patients had received, factors associated with disease (e.g., diabetes), and also the intrinsic imprecision of the technique by which glucose concentrations were measured. Much of statistical analysis is about quantifying variability and attributing it to different sources.

The most elementary statistical methods are simply concerned with the quantification and visualization of variability in data. Such methods go under the title of **descriptive statistics**. Perhaps the simplest visualization tool is the histogram. Histograms are illustrated in *Fig. 1*, showing a simple dataset where the data clearly show a degree of variability around a single average value (uni-modal data), and more complex (multi-modal) data where there appear to be two distinct populations in the dataset each with a different average value.

In the case of uni-modal data, it is useful to quantify the average value and the variability (spread) of data around this average. There are two common measures of the average of a set of data. The **arithmetic mean** is calculated by adding up all the data and dividing by the number of data. The **median** is calculated by

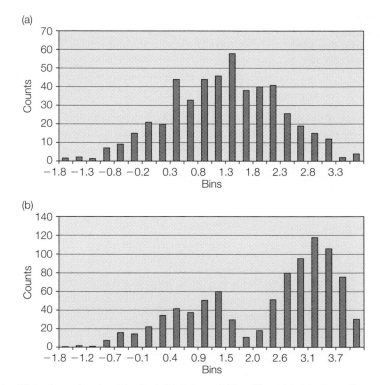

Fig. 1.   Histograms showing (a) uni-modal data, and (b) multi-modal data. In each case the bar heights represent counts of data within the bin ranges shown on the horizontal axis (for instance the largest bar in (a) represents a count of 58 in the range of 1.3 < x < 1.55).

writing down the data in a ranked list from the highest value to the lowest and taking the value in the middle. For instance, with the data on the top row of *Table 1*, the mean is 1.18 and the median is 1.2 (from the ranking 1.4, 1.3, 1.2, 1.1, 0.9). The **standard deviation** is a measure of variability around the average value: it is the square root of the average of the squared difference between each data point and the mean. The mean and standard deviation are appropriate measures of average and variability when data approximately follow a normal distribution (Section G1), when they correspond directly with the parameters μ and σ of that distribution. When data are not normally distributed, and particularly when there are significant **outliers** (outliers are single data points that differ substantially from the remainder of the data), the median is a better measure of the average. It is less sensitive to outliers. Equally, the variability of such data is better measured by the **interquartile range**, rather than the standard deviation. The interquartile range is calculated like the median from a ranked list of the data. In the case of the data below

$$100, 88, 77, 72, 65, 55, 50, 44, 37, 33, 29, 22, 18, 12$$

the median is 47 (half way between 50 and 44 since there are an even number of values). The top quartile is the median of the top half of the data (72 in this case), and the lower quartile is the median of the lower half (29). The interquartile range in this case is $72 - 29 = 43$. Summary statistics like the median and interquartile range are often illustrated on box and whisker plots.

Testing statistical significance

A very common question is whether a significant amount of variability can be attributed to a particular factor. For instance, do (untreated) diabetics generally have different levels of blood glucose compared to patients without this disease? This question can be put differently: is the variability in blood glucose measurements between these two patient groups distinguishable from the variability from other sources?

To be a bit more concrete, let us suppose that measurements are taken as in *Table 1* below.

Table 1. Statistics of patient glucose levels

	Patient glucose levels (arbitrary units)					Mean	Standard deviation
Healthy	1.1	1.3	1.4	0.9	1.2	1.18	0.192354
Diabetic	1.3	1.4	1.5	1.6	1.5	1.46	0.114018

With these data, the mean glucose level is higher in diabetics, but it is clear that there is quite a bit of variation between individuals in both groups. Given that variability exists, with a small sample (five patients) it might well be possible that, simply by random chance, one group of five could be selected with a higher mean glucose level than a second group. What is needed here is a test of **statistical significance**, to see if the data give real evidence of a difference in glucose levels, distinguishable from what might be expected in random chance variations.

All tests of statistical significance are based around a **null hypothesis**: this is the hypothesis that the effect observed is due to random chance. The next step then is to calculate the probability that the null hypothesis is true: this is called the **p value**. If the p value is small enough, then it is assumed that the null hypothesis is not true and the effect is not due to random chance. It is the case often that effects are considered significant if the p value for the null hypothesis is less than 0.05 (95% significance level) or, more stringently, 0.01 (99% significance level). The p value taken to indicate significance is chosen by the user of the test; the smaller it is the lower the chance of a false positive error (attributing a random chance effect to be real), but the greater the chance of a false negative error (attributing a real effect to random chance).

The t-test and alternatives

The t-test (sometimes called Student's t-test) is a significance test that is applied to sample means. It can test if the mean of a sample is significantly different from some fixed value (one-sample t-test), or if the means of two samples are significantly different from each other. In the case of the data in *Table 1*, a two sample t-test is the appropriate choice, with diabetics and healthy patients forming the two samples. It is beyond our scope to give technical details, but essentially what the t-test does in this case is to take the standard deviation as a measure of the variability of the data, and to use this along with some theory about the normal distribution, to calculate the probability that two samples, with means at least as different as those in the table would be produced from a single distribution with a single mean glucose level.

A fundamental assumption of the t-test is that the data are normally distributed, and if this is not true it cannot be applied. There are tests to see if data are normally distributed, but in practice the most important constraint is that t-tests should be avoided if the data contain significant outliers, since these are a strong indication of deviation from the normal distribution.

Fortunately it is very easy to carry out a two-sample t-test in most statistical software packages and spreadsheets, and all that is necessary is to identify the data for the two samples. In the case of the data in *Table 1*, the p value (probability that the null hypothesis is true) is 0.03, and we can assume that the null hypothesis is false with more than 95% confidence (but not 99% confidence). Rejecting the null hypothesis in this way is to provide evidence that some alternative hypothesis is true, in this case that the groups of diabetics and healthy patients have glucose levels drawn from separate normal distributions with different means.

The one-sample t-test is less commonly used than the two-sample version above, and it tests if the mean of a single sample is significantly different from some known value. An example use of such a test might be to test out a new experimental assay. If the new assay was to test the protein concentration of a solution, then it would be sensible to try it out on a standard sample of known concentration. The one-sample t-test could be used to see if a set of repeated measurements using the same assay had a mean answer significantly different from the known concentration.

Another type of t-test arises in the case of paired data. In this case imagine a group of five volunteers having blood glucose measured before and after a meal. It would be tempting to treat the measurements again as two samples, one sample being the before-meal measurements and the other the after-meal measurements. This is not the best thing to do in this case because the data are not **independent**. Each 'before' measurement is linked to a single 'after' measurement by virtue of being taken from the same person. The data are **paired**. Applying a paired t-test (again possible in most common software packages) is the same as converting the ten measurements to a single dataset of five differences in glucose levels for each person before and after the meal, and then using a one-sample t-test to see if the mean difference is significantly different from zero.

When data are not normally distributed the t-test cannot be applied. The alternatives that are applicable are called Wilcoxon tests (the two-sample version is known also as the Mann-Whitney test). Because these tests do not depend on the assumption of an underlying probability distribution, they are called **non-parametric**, and again they are available in most software packages.

Analysis of variance (ANOVA)

Like the t-test, ANOVA is about assigning statistical significance to differences in means between groups of data. The difference is that it can be applied to more than two groups. For instance, it would be applicable if one were trying to assess the effects of 10 different drugs on blood glucose level, using groups of 20 patients with each drug (and probably also a control group). It can be applied also to more complex groupings, such as might arise if one were interested in sex differences as well, and each drug group were divided into 10-person male and female subgroups.

Variance is a measure of variability, and it is actually just the square of the standard deviation. Analysis of variance first calculates the total variance and overall mean of the data by pooling the groups of data into a single group. This total variance is then split into components, the variability within the defined groups (around the group mean) and the variability between them (between the means of the different groups). Intuitively, if the between-groups variance is much larger than the within-groups variance then this indicates a significant effect, perhaps that there are differences in glucose levels caused by drug treatment. Conversely, if the within-groups variance is larger than the between-groups the effect is probably significant. Like the t-test, ANOVA provides p values for the null hypothesis that the groups are all drawn from a single normal distribution (with the same mean in each group).

You might ask why not just perform t-tests between all possible pair-wise group combinations? In the case of 10 groups, to compare each group to every other involves making 45 tests, which might be tedious. But the more serious objection to this is associated with **multiple testing**. If you choose to treat as significant any effect with a p value less than 0.05, then you are accepting a 5% chance of getting it wrong (false positive) in a single test. However, if you then perform 45 such tests, you would expect 5% of them (at least 2) to come out with a p value of less than 0.05, even if there were no significant effects in the data. ANOVA is an elegant way around this problem.

Chi-squared and the Fisher exact test

Blood glucose levels are examples of continuous variables that can take any value in principle. **Categorical variables** on the other hand can only take values from a discrete set. Examples of continuous variables are mass, length, and concentration, while categorical variables could be sex (two possible values), or treatment with different drugs (number of values equal to the number of different drugs studied), or any variable that involves an integer count of a quantity. Categorical variables can be used to define groups for t-tests and ANOVA, but sometimes we need to treat them as dependent variables and test the statistical significance of their values. A commonly occurring case is that of χ^2 tests.

For example, consider the data in *Table 2* below, for three variants of a plant species subjected to drought stress. These data are in the form of a contingency table, where the experimentally observed gene counts are given in normal type. The question here is whether the variants respond differently to the stress. The null hypothesis is that they do not respond differently, and the numbers in italics give the **expected** counts of genes in each category, assuming that the null hypothesis is true. It is easiest to understand how these expected values are calculated with an example. For instance, the value in the top left box (shown in parenthesis) is calculated as $(130/178) \times 60$. Here, $130/178$ is an estimate of the probability that a gene is up-regulated, assuming the null hypothesis of no difference between the variants. The expected number is then just the total number of stress genes in variant A multiplied by this probability. Another way of thinking about it that gives the same answer is as $(60/178) \times 130$, the probability a stress gene is from variant A, multiplied by the total number of up-regulated genes. Similarly, for unchanged genes in variant B

$$17.0 = (48/178) \times 63 = (63/178) \times 48$$

Table 2. Stress gene expression in plant variants

	Variant A	Variant B	Variant C	Total
Number of stress related genes up-regulated	55 *(43.8)*	30 *(46.0)*	45 *(40.2)*	130
Number with expression unchanged	5 *(16.2)*	33 *(17.0)*	10 *(14.8)*	48
Total	60	63	55	178

The statistical test is now whether the observed and expected counts in the table above are significantly different from each other. This is done by calculating the χ^2 statistic, equal to the sum over all 6 boxes above, of $(O-E)^2/E$. The larger this value the less likely it is that the null hypothesis is true, and a p value can be obtained using a special probability distribution for the χ^2 statistic. As usual, this is easily done in many software packages (sometimes requiring you to calculate the expected values and sometimes not). In this case $p = 6 \times 10^{-8}$, giving clear evidence that there is a difference between the variants in their stress responses. χ^2 tests cannot be applied if any of the counts are five or less, and require special modifications for 2×2 contingency tables. The Fisher exact test is an alternative that is not subject to these restrictions.

Resampling-based testing

Nearly all statistical tests have an alternative using the ideas of re-sampling or permutation of the data. For example, suppose we had a sample of blood glucose measurements in 50 untreated diabetics (mean $= m_d$) and 50 otherwise comparable healthy patients (mean $= m_h$). The null hypothesis is that there is no difference on average between the glucose levels in these two groups. With this assumption, what is the probability that we would obtain a difference in glucose levels of $m_{diff} = m_d - m_h$, simply by chance? To calculate this by re-sampling we would simply pool all the data, mixing diabetics with healthy individuals randomly, and then repeatedly divide this group into two equal, randomly chosen groups many times. For each random division the difference in mean glucose levels in each group would be calculated. To get the p value, the number of such differences in mean glucose levels greater than or equal to m_{diff} would be counted. The p value would simply be this number, divided by the number of random divisions. This works as long as enough random divisions are used, and for this reason it can take a lot of computer time to calculate. However, computers these days are very powerful, and re-sampling tests like this can be a very attractive alternative to some of the more standard tests above because they make no assumptions about the data (for instance, that they obey the normal distribution).

Multiple testing   This has already been mentioned above, but it is so important in many areas of bioinformatics that we mention it here again. Currently, biological measurements are often made on a genomic scale. Microarrays, for example, can be used to measure the expression of almost all genes in a genome, often numbering tens of thousands. Thus it is often the case that multiple statistical tests are performed, one for each gene in the case of microarrays. If, for instance, an experiment involves 10 000 genes, and each is subjected to a hypothesis test, then if p values of less than 0.01 in any one test are considered significant, it should be expected that 100 of the tests would show up significant just by chance with no real effect present in the data. Thus, when carried out on this scale many tests that look significant on their own, really are not significant at all. A commonly used correction for this effect is called the **Bonferroni correction**, and involves multiplying all p values by the number of tests (limited so that $p \leq 1$), but there are other more sophisticated alternatives, for instance the **Benjamini-Hochberg** correction.

H1 SYSTEM FEATURES

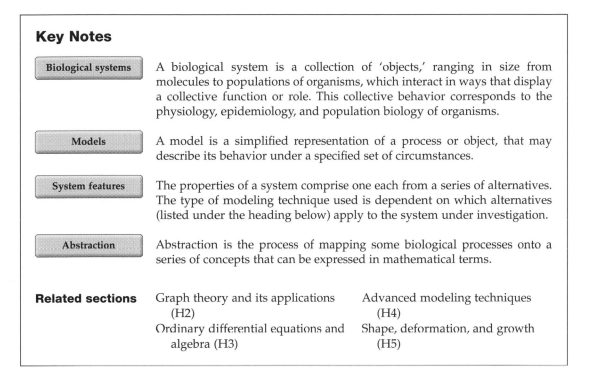

Key Notes

Biological systems

A biological system is a collection of 'objects,' ranging in size from molecules to populations of organisms, which interact in ways that display a collective function or role. This collective behavior corresponds to the physiology, epidemiology, and population biology of organisms.

Models

A model is a simplified representation of a process or object, that may describe its behavior under a specified set of circumstances.

System features

The properties of a system comprise one each from a series of alternatives. The type of modeling technique used is dependent on which alternatives (listed under the heading below) apply to the system under investigation.

Abstraction

Abstraction is the process of mapping some biological processes onto a series of concepts that can be expressed in mathematical terms.

Related sections

Graph theory and its applications (H2)

Ordinary differential equations and algebra (H3)

Advanced modeling techniques (H4)

Shape, deformation, and growth (H5)

The previous chapter dealt with statistical techniques that can be used to derive meaning from a population of biological objects. In the following five sections, we shall examine the broad selection of mathematical techniques used to represent (i.e., to model) the processes that take place in biological systems. However, before looking at these in any detail, it is important to grasp the general concepts relating to **systems.**

Biological systems

Organisms share various characteristics of life: digestion, excretion, respiration, sensitivity, reproduction, growth, and movement, which, at the cellular level, involve a wide variety of interactions and chemistry between molecules and multi-molecular complexes. Also, the physiology of higher organisms is mediated by various interactions, including the impact that the environment has upon organisms and populations of organisms at a range of levels. An individual physiological process can be thought of as a **biological system**. At its simplest level, the system might simply involve a single enzyme or environmental receptor, but more often biological systems are more **complex** both in the general and specific mathematical meaning of the term. To confound matters, processes at one physical scale (e.g., an animal or plant) can influence and be influenced by processes at a different scale (e.g., an enzyme activity). Contrary to the perception of some biologists, no specific physical scale is always the determining factor. A similar situation arises with respect to timescales, a specific biochemical reaction might take place in microseconds but its cumulative effect over years can lead to physiological changes that eventually block the reaction from taking place.

Models

When proposing mechanisms that describe specific aspects of the characteristics of life, biologists devise some simplified representation of the organism that focuses on the mechanism under consideration. A **model** is a simplified representation of the system being described, which like all biological 'facts' is valid under certain conditions. Nowadays, biologists usually develop conceptual models that appear as figures in papers and textbooks. These are very useful for capturing the general qualitative idea of how a system behaves, but often do not provide a quantitative idea of how much of something is present, or being changed, or affecting something else. For this, quantitative models must include mathematics in some form. The features of the system or the detail in which it is being studied dictate the types of mathematics to be used.

System features

Depending on what is known and what features of the system particularly need to be modeled, these representations can take a series of alternative forms.

- *Static* and *dynamic* models show respectively either a snapshot of the process and how the system changes over time, or in response to some specific perturbation. However, a series of static models could represent a dynamic process in the same way as a series of images can be used to produce an animated cartoon.
- *Mechanistic* and *phenomenological* models respectively represent the mechanism of the process, and simply its overall behavior under one set of circumstances. If one wishes to understand how regulatory processes function, so that predictions can be made, then mechanistic models are required.
- *Discrete* and *continuous* models comprise changes in a step-wise or smooth manner respectively. For example, a metabolic pathway could be viewed as a series of discrete reaction steps or smooth decreases in the concentration of substrates leading to a smooth increase in the products.
- *Deterministic* and *stochastic* models respectively represent systems where the behavior always results in the same outcome every time a simulation is performed, and a range of different simulation outcomes (because the system and model have random components or intrinsic variability).
- *Single-* and *multi-scale* models represent systems that either take place on similar physical or time scales, and very different scales respectively. For example, biochemical reactions, gene expression, and diseases, respectively, take microseconds, minutes, and up to decades. These can be modeled at their respective scales, but studies of genetic predisposition to disease require models that span the different scales. A similar situation arises with different physical scales; for example, differential tissue growth happens at the millimeter-to-meter scale but is mediated by molecular interactions and gene expression which occur at a nanometer scale.

There are also other features of a system that can influence the type of modeling approach that can be used.

- *Equilibrium* or *non-equilibrium* conditions define whether the ratios of components remain static or not. A system at equilibrium is also described as being in a **steady state**. For example, a eukaryotic cell in G_0 phase may be in a steady state because the amounts of RNA, protein, and lipid remain the same even though they are being synthesized and broken down at fixed rates. However, if this cell is activated by some external signal, then the ratios of certain mRNAs and proteins will be altered, resulting in cell growth (G_1 phase), which is not in a state of equilibrium. A slight exception to this concerns cells growing in a steady state in certain types of fermenter; where the ratios of biological molecules remain constant but the total biomass is increasing.

- *Diffusion* is a process that may be a significant component of a system and requires particular mathematics to represent it. Furthermore, where diffusion is thought to be taking place, then there may be barriers to it (known as **percolation** effects). Eukaryotic cells are cluttered with organelles and micro-filaments, so small molecules might diffuse freely whereas protein movement is much more restricted.

- Finally, the *shape* of a system might play a significant role in its behavior. For example, the behavior of the same molecules in a lymphocyte and neuron may be very different because one is spherical while the other is extremely long and thin. Likewise, the behavior of the same type of cell in different parts of an organ (e.g., hepatocytes in a liver) can be different because its local environment is different.

Different branches of mathematics (each often called a theory) and computational tools are better suited to different system features. Some representations include a mixture of these alternatives, in which case they are known as **hybrid models**. Multi-scale models are arguably the most difficult to develop.

Abstraction

The modeling process begins by identifying the key features of the system so that it can be represented in a mathematical form. This process is known as **abstraction** and can be used both to simplify the representation and to reveal defined relationships between entities.

Take the very simple example: $y = 2x + 2$. The quantitative relationship between x and y remains the same irrespective of what they actually are (in biological terms) and how much of them is present. Once the relationship has been identified, it means that the amount of one can be calculated from a knowledge of the other. When biologists work with modelers, they often face the challenge of distilling down the biological detail into the key steps that most affect the behavior of the system. This is especially true when many of the molecular steps are known. Incorporating every interaction might make the model more fully representative, but the model will be slower to build, be more prone to error during construction, and dynamic simulations will be considerably slower. The trade-off between simplicity and degree of detail is a constant tension.

H2 GRAPH THEORY AND ITS APPLICATIONS

Key Notes

Graphs	Mathematicians describe a graph as a hypothetical structure that consists of a series of nodes connected by edges. Graphs containing cycles or alternative paths are increasingly referred to as **networks**. The nodes and edges can have a range of properties defined as **colors**, which may have quantitative values, referred to as **weights**. When describing metabolic signal transduction of gene-regulatory pathways, biologists are usually thinking in terms of graphs.
Computer representation of graphs	Within a computer, graphs are almost invariably represented as matrices of various types, including adjacency matrices, adjacency lists, and stoichiometric matrices. There are also many ways in which computers can be programmed to lay out graphs so that their structure and noteworthy features can be more readily seen.
Network topology and properties	The overall topology of networks may be random or have particular properties that make them robust and stable to perturbation. Biological networks, like the internet, have a scale-free topology, with remarkably short network radii given the number of nodes involved.
Multi-scale representation	Nodes do not have to represent biological objects on the same scale. Thus one node (for a molecule) may have an edge connecting it to a node representing a cell or tissue. The edge indicates that the molecule exerts an effect on the cell/tissue.
Petri-nets and P-systems	These are special types of graphs for which a range of specific software has been developed. They are popular among computer scientists who wish to model a range of biological processes.

Related sections	Analyzing differential gene expression (K3)	Network studies of metabolism (O1)
	Interaction databases and networks (L3)	Physiology (P)

Graphs

In mathematical terms, graphs are structures that consist of a series of places – each referred to as a **node** or **vertex** (plural vertices) – which are connected by a series of lines (referred to as **edges**, **links** or **arcs**). A **self-edge** is one that sets out from a node and returns immediately to that node. Many molecular biology textbooks depict metabolic and regulatory processes in this way; nodes can represent metabolites that are connected by enzymes that comprise the edges.

Indeed, in mathematical terms, the Boehringer metabolic pathways chart is one large graph. Signal transduction and gene regulatory pathways can also be viewed in this way. When a graph includes cycles or alternative paths between two nodes, then it is more common to refer to it as a **network**.

Both nodes and edges can have properties (which mathematicians refer to as **color**) that may relate to some aspect of their behavior. The quantity of that property is referred to as its **weight.** Thus, networks derived from the Biocyc databases can have two colors depending on whether the node is a metabolite (or enzyme that is itself a substrate for post-translational modification) or a reaction intermediate. Nodes (representing proteins) might also be colored by the extent to which their gene is being expressed (as determined say by microarray analyses – see Section K3). A time series dataset shows the changes in the network over time. A prime example of this has been a study of the inflammatory response in leukocytes.

Edges also have properties. The simplest is whether or not they have a direction associated with them. In biological terms, a directed edge can correspond to an irreversible enzyme reaction, allostery, the effect of a regulatory subunit of an enzyme on its catalytic activity, post-translational modification, or gene regulation by some transcription factor. Undirected edges can correspond to reversible reactions and multi-subunit proteins. Computationally, it is often easier to encode an undirected edge between two nodes as two directed edges going in opposite directions.

Edge properties include whether or not the target node is stimulated or inhibited, which substrates and products of a reaction should be associated with each other because they share many of the same atoms (e.g., in amino-transferase reactions), or a reaction rate. Where alternative paths exist, data on reaction rates (weights) can be used to calculate which is the fastest path. Edges might also have properties associated with the ratios of the interacting partners. The interaction might or might not be stoichiometric; i.e., in the interactions between the interacting partners taking place at a fixed ratio of one partner to the other. This leads to four types of interaction that can be edge properties. *Table 1* shows the biological equivalents of these alternatives.

Table 1.   Interaction classes

	Stoichiometric	Non-stoichiometric
Directed	Regulatory subunits (e.g., in protein kinase A)	Enzyme/substrate relationships (e.g., protein tyrosine kinases)
Undirected	Multi-subunit complexes (e.g., ribosomal sub-units)	Filaments (e.g., actin or tubulin)

Depending upon the topology of the node and edge connections, a range of features can be defined (see *Fig. 1*). These features have counterparts in biological networks. Thus, **hubs** are often essential molecules for cell survival, and correspond to molecules that engage in a very large number of reactions; for example, water, ATP, NAD; or protein phosphatase; or the TATA-box binding protein (which is required for the expression of most eukaryotic genes). Cellular machinery feature on protein-interaction networks as **cliques**, and **cycles** with one or more directed edges correspond to physiological regulation mechanisms providing feedback inhibition, stress responses, or cell growth/differentiation.

Such cliques and cycles are also sometimes known as **modules**, because they operate together to execute some physiological function. Critical nodes also may play a crucial role but this is more in terms of properly coordinating the activities of different modules.

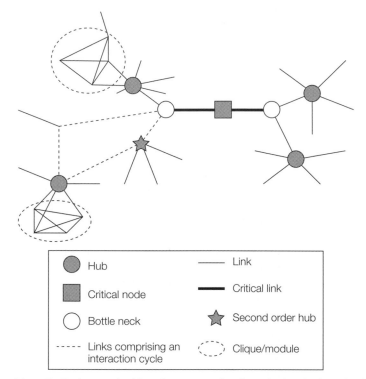

● Hub	—— Link
■ Critical node	—— Critical link
○ Bottle neck	★ Second order hub
---- Links comprising an interaction cycle	Clique/module

Fig. 1. A hypothetical network with various types of node and edge depicted by the symbols shown below.

Computer representation of graphs

The most common way of representing a graph is as a matrix in which there is one row and column for every node. An **adjacency matrix** is the simplest form of graph representation, in which a 0 or 1 denotes whether or not there is a directed edge from one node to the other. *Fig. 2* shows a hypothetical cycle involving five metabolites and three enzymes of which one is irreversible. This metabolic process (graph) can be represented by the adjacency matrix in *Table 2*.

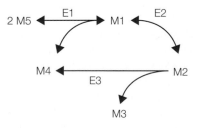

E3 reaction irreversible

Fig. 2 A small hypothetical metabolite network involving five metabolites M1–M5 and three enzymes E1, E2, and E3, of which only the latter catalyzes an irreversible reaction.

Table 2. Adjacency matrix for Figure 2

	M1	M2	M3	M4	M5
M1	0	1	0	1	1
M2	1	0	1	1	0
M3	0	0	0	0	0
M4	1	0	0	0	0
M5	1	0	0	0	0

Note that because M3 is the product of an irreversible reaction it cannot connect to anywhere so no elements on that row have a value of 1. For colored and weighted graphs, the cells of the matrix can be themselves arrays, matrices or hashes. When the number of nodes reaches a large number and the number of edges between the average number of edges per node remains low, then most of the elements in the array will be empty (i.e., have a zero). In these circumstances, it is computationally more efficient to represent the graph as an **adjacency list**. The simplest form is a two-column matrix in which the first and second columns represent the start and destination nodes and each row is a directed edge. Undirected edges are represented by two rows with the nodes reversed between one row and the other (see rows 1 and 4 in *Table 3*). Colors and weights are added to the list by adding extra columns to the matrix. Thus the adjacency list for the process in *Fig. 2* can be seen in *Table 3*.

Table 3. Adjacency list

M1	M2
M1	M4
M1	M5
M2	M1
M2	M3
M2	M4
M4	M1
M5	M1

There are several variations on this type of list.

Software libraries have been developed in various programming languages to construct, manipulate, and interrogate these structures, such as GRAPH (in perl), BOOST (C++) and NetworkX (python). Typical queries include the shortest distance between two nodes (this may be in terms of number of intervening edges or taking into account the weight of edges), which nodes are the furthest apart, how far are they apart, and what is the mean path length between all the nodes in a graph.

A final way of representing processes that captures the ratios of materials involved is a **stoichiometric matrix**. In such matrices there is a row for each reaction and a column for every node (usually a metabolite). For a given reaction, an integer is put into a cell corresponding to the number of molecules consumed (negative numbers) or produced (positive numbers) by the reaction. These matrices have particular uses in metabolic bioengineering, and are described further in Section O1. The matrix for *Fig. 2* is shown in *Table 4*.

Table 4. Stoichiometric matrix from Figure 2

	M1	M2	M3	M4	M5
E1f	1	0	0	-1	-2
E1r	-1	0	0	1	2
E2f	-1	1	0	0	0
E2r	1	-1	0	0	0
E3	0	-1	1	1	0

A final way of representing graphs or networks is in the form of a 2-D or 3-D diagram. The computer generation of such diagrams is an art form and science in its own right. There is a growing range of generic software for this – GraphViz, Pajek, Cytoscape, and Tom Sawyer are amongst the most popular.

Network topology and properties

The nodes of a graph may be connected through edges in various defined ways, for which it helps to understand some further terms from graph theory. The number of edges connecting to a node is known as its **vertex degree**. The value can be split into the *in*-degree and *out*-degree corresponding to the number of edges coming into and out from the node, respectively. A **path length** is defined as the number of edges to cross to get from one node to another. The **diameter** of a graph is the longest minimum path between any pair of nodes in the network, while the **radius** is the average minimum path length, which will usually end up being a decimal number. Biological networks are not arranged in a regular or symmetrical pattern. In fact, it is clear that the vertex degrees vary widely with many having only 1–3 edges ranging up to perhaps a few dozen that have very large vertex degrees (the hubs).

For many years, it was thought that biological networks had a **random** topology, in which the nodes are connected to each other in a random order. In random networks, the network radius increases as the number of nodes increases. However, in the late 1990s, a careful study of their topology showed that biological networks are not random, because their radii did not increase as the total number of nodes increased. On the contrary, for the metabolic networks studied, their radii had a value ~3 almost regardless of the organism studied. Furthermore, they were found to have a **scale-free** topology (see *Fig. 3*), in common with the WWW and other networks in major and reliable use. This topology ensures that there are mostly very short paths between any given pair of nodes, allowing rapid communication between otherwise distant parts of the network. The whole network can respond to any given change. Furthermore, removing any node at random will usually have no detectable effect, whereas as removing a hub would be catastrophic – what would happen to a cell if you remove all its water or ATP?

Multi-scale representation

By definition, graphs comprise a series of discrete steps. One graph on its own is a static representation, though dynamics can be implied through the use of colors or depiction of a series of related graphs. Since a node may represent anything, the developers of such models may quite legitimately have nodes corresponding to objects on different physical scales. For example, one could develop a molecular-scale graph of the biochemical pathway in one cell type (say for anabolic steroid synthesis) that results in a node representing the hormone, which has edges to nodes representing cells or entire tissues. The edge connecting the molecule to the tissue denotes the effect of the hormone on tissue mass as a whole.

(a)

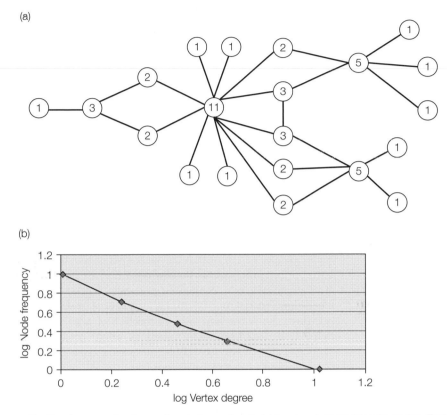

(b)

Fig. 3. An example of a small scale-free network. (a) A network comprising 21 nodes. Each node is labeled with its vertex degree. Counting the frequencies, there are 10, 5, 3, 2, and 1 nodes, respectively, having vertex degrees of 1, 2, 3, 5 and 11. When these numbers are plotted on a log-log we find the straight line shown in (b).

Petri-nets and P-systems

Petri-nets (or P-nets) are a special class of graph which has undergone considerable development by computer scientists. P-nets consist of two general classes or node: place and transition nodes. Nodes of the same type must always connect to the opposite type and never have an edge connecting each other. The mathematical definition for this is a **bipartite graph**. Place nodes may be thought of as static and have a range of properties relating, for example, to the quantity of the item which the node represents. The transition nodes are active and are said to **fire** under defined circumstances. This has the effect of altering the values or properties of the connected place nodes, again in defined ways. Petri-net models are developed to have a particular topology and initial state, then the computer systematically and repeatedly fires the transition nodes for some defined number of times, after which the states of the place nodes are investigated. A Petri-net model of yeast in a brewing vessel would show a progressive depletion of malt sugars and an increase in the concentration of ethanol.

P-nets have undergone considerable refinement since their invention in the 1960s. Place nodes might simply have a Boolean value (see Section D), which can represent some component of the system being active or inactive. However, it is more common for them to have a number of tokens, which correspond to the amount of whatever the place node represents. With regard to refinements in transition nodes, they may fire

- at different rates, so that fast and slow processes might be modeled simultaneously;
- according to some probability, so that stochastic processes might be modeled;
- solve a (possibly differential) equation based on the values in a defined set of parameters and reassign place node values accordingly.

The latter is an example of a **hybrid model**, which mixes both discrete and continuous elements, and which enables modeling of multi-scale systems.

Although Petri-nets are a well-developed modeling approach, simulation runs can be very slow when the models become moderately large. A recent development to address this issue is the approach known as P-systems. Another discrete-modeling approach that might be viewed as an extension of the Petri-net concept is **cellular automata**. A cellular automata model consists of a mesh of 'cells,' each of which is encoded with a set of functions (automaton) that change in response to the state of themselves and their neighbors. The simplest models arrange the cells in a rectangular grid (which can be programmed very easily as a two-dimensional matrix), though hexagonal grids and even user-defined grids (in which the dimensions and connection of one cell to another are specifically defined) are used also. The functions might comprise logical rules, algebraic or stochastic equations. Cellular automata have been used to model various physiological systems (see Section P).

H3 ORDINARY DIFFERENTIAL EQUATIONS AND ALGEBRA

Key Notes

Continuous changes
This section is concerned with smooth rather than step changes. It deals with quantities that may be expressed in real rather than purely integer numbers.

Mathematical representation of rates of change
This takes the form of **derivatives**, which are the stuff of differential calculus. The mathematical notation for this is introduced below.

Systems of equations
Where there is a set of interacting entities, these can be represented by a set of equations (one for each reaction), also known as a system of equations. These look quite daunting to assemble, but modern software makes it much easier to assemble these systems and can hide much of the algebra.

Solution of systems of equations
A variety of computational methods exist to solve these equations. Some determine exact solutions, others use progressive approximation techniques technically known as **numerical methods**. In models where all the interacting components are created as well as consumed, known as **balanced** models, the solution will yield steady-state concentrations. Unbalanced models have one or more components progressively generated until the **system substrates** are completely consumed.

Related sections
Use of calculus and algebra (O2)

Continuous changes

The previous section outlined the ways to represent biological systems using discrete techniques. This section is concerned with continuous change and begins the study of the quantitative behavior of biological systems. This is the difference between the qualitative knowledge that an enzyme catalyzes a specific reaction (and in which numbers of molecules can be counted; for example, through Petri-net tokens), and a description of how quickly a given number of moles of substrate are catalyzed to product. At the other end of the biological scale, one can study rates of change in population sizes in the predator-prey relationships implicit in food webs.

Mathematical representation of rates of change

Rates of change of the amount of one thing compared to another or over time are best represented using differential equations. Something whose value can change is known as a **variable** (e.g., a metabolite concentration, cell volume, rate of nerve impulse conduction, population size, etc.). In high school mathematics, x and y are usually used to represent variables as these are totally abstract, but other symbols of more meaning may be used. The relationship between variables is defined by some **function**. This consists of a set of arithmetically linked **terms** of the **independent variable** (typically x) that correspond to the value of the **dependent variable** (usually y). The terms containing the independent variable usually also have a set of constants, which are known as **parameters**. If we consider the case of a ball being thrown up into the air at a defined speed, the height of the ball (h) is dependent upon the amount of time that has elapsed (t). Thus h is a function of t, taking the general form

$$h = (a \times t^2) + (b \times t) + c$$

In this equation, there are three terms of which only two contain the independent variable, but there are three parameters (a, b, and c). If the values of these parameters have been determined, then it will be possible to specify the height at any specified time. Applied mathematicians spend a substantial amount of time requesting or seeking parameter values and, of course, it might not be possible to determine these values directly. In these situations, one must turn to methods for parameter estimation (see below).

However, we may wish to consider rates of change of a variable (also known as a **derivative**) rather than its absolute amount. In the case of throwing a ball, the derivative will show us the point at which the ball reached its highest point, as that corresponds to when the value of the derivative equals zero. At this point, introducing some mathematical shorthand makes it quicker to represent increasingly sophisticated scenarios. We see above how h is a function of t. This abbreviates to $h = f(t)$. In other situations, h may equal some other function of t, in which case we could write: $h = g(t)$ or $h(t)$ or $i(t)$, etc. This shorthand means that we do not have to specify the relationship between the variables, which anyway could be almost anything.

The rate of change of h with respect to t is represented by

$$\frac{dh}{dt}, \frac{d(h)}{dt}, h_t, \dot{h} \text{ or } \dot{h}$$

where there is mostly only one independent variable

If the derivative itself is not a constant, then we have the situation where

$$h = f(t) \text{ and } \frac{dh}{dt} = f(t) = g(t) \text{ [or } h(t) \text{ or } j(t), \text{ etc.]}$$

This second equation is also known as a **differential equation** (sometimes abbreviated to DE). There are many different types of these. The rate of change of a derivative is called a **second derivative**, which gives rise to a **second order differential equation**, and is usually represented as

$$\frac{d^2h}{dt^2} \text{ or } \frac{d^2(h)}{dt^2}$$

Of course, there may be any number of higher derivatives, leading to the general case of the nth derivative (where n replaces 2 in the above shorthand). Equations in which only one independent variable is involved are known as **ordinary differential equations** (ODEs). Equations involving two or more independent variables are known as **partial differential equations** (PDEs), of which more will be described in Section H4.

Systems of equations

Fig. 2 in H2 showed the biochemical interactions of three enzymes, which can be represented by the stoichiometric matrix in *Table 4* in H2. Each row in the matrix represents a biochemical reaction that will have its own rate constant. The cycle can be redrawn as in *Fig. 1* as a linear set of reactions, each with its own rate constant.

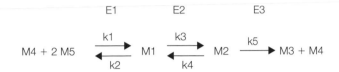

Fig. 1.   A hypothetical biochemical cycle involving five metabolites (M1–M5), three enzymes (E1–E3), and the rate constants for each reaction (k1–k5). E3 has only one rate constant because it is irreversible.

In its simplest form, an enzyme catalytic activity follows the physical Law of Mass Action, and we can convert each reaction into a rate equation. The biochemical convention for representing the concentration of something is through the use of square brackets – [A] means the concentration of A. Thus the forward reaction of E1 yields the equations

$$\frac{d[M1]}{dt} = k1 \times [M4] \times [M5]^2, \frac{d[M4]}{dt} = -K1[M4] \times [M5]^2, \text{ and } \frac{d[M5]}{dt} = -2 \times [M5]^2 \times k1 \times [M4]$$

Of course, other reactions lead to the synthesis and loss of these metabolites, so we can end up with a set of rate equations for each metabolite

$$\frac{d[M1]}{dt} = k1 \times [M4] \times [M5]^2 - k2 \times [M1] - k3 \times [M1] + k4 \times [M2]$$

$$\frac{d[M2]}{dt} = k3 \times [M1] - k4 \times [M2] - k5 \times [M2]$$

$$\frac{d[M3]}{dt} = k5 \times [M2]$$

$$\frac{d[M4]}{dt} = k5 \times [M2] - k1 \times [M4] \times [M5] + k2 \times [M1]$$

$$\frac{d[M5]}{dt} = 2 \times (k2 \times [M1] - [M5]^2 \times k1 \times [M4])$$

This is an example of a **system of equations**, more specifically a system of ODEs. It may look like hard work to write out all these equations, but software can derive these from the stoichiometric matrix plus the parameters for the rate constants and initial metabolite concentrations, which are usually input separately. Many enzyme activities do not observe simple mass-action kinetics, but instead display some form of cooperative activity that depends on substrate concentrations. Furthermore, reactions involving multiple substrates and/or multiple products involve a range of different possible mechanisms, each with their own kinetic equation, but these have been encoded so that the software can construct them and request any extra parameters associated with them. A model in this form can be used to show what the changes in the variables (in this instance the concentrations of the different components) will be over time, starting from a defined set of initial conditions or as a result of different kinetic constants.

These ODEs can be refined in various ways. For example, if there is a time delay in the activity of some part of the system, it can be represented by **delay equations**, which take the form

$$y = f(t,T), \text{ such as } y' = a \times (t - T)^2 + b \text{ where } T \text{ is the delay in t.}$$

Differential delay equations sometimes can lead to variables of the system oscillating. Expanding on this idea, they can also be used in a different way as a crude approximation of a homoeostatic mechanism. Let us suppose that a cell requires a particular concentration of some protein P to function properly. Owing to protein turnover, some of it will be broken down and some undefined mechanism senses how much needs to be produced at any given time.

This can be captured by the equation

$$\frac{d[P]}{dt} = k \times (p - [P])$$

If [P] is above the steady-state value p, then the value of $(p - [P])$ will be negative, reducing the overall [P]. Likewise, a low [P] results in a positive value for this term, leading to a higher [P]. The rate at which this mechanism restores P after any perturbation is given by the kinetic constant k.

Solution of systems of equations

One might wish to solve a system of equations, which yields various results depending on the model's details. If the model is analytically tractable, then software packages, such as Maple, will carry out the algebra and calculations for you (http://www.maplesoft.com/). If every variable can both increase and decrease in value – say all the metabolites can be synthesized as well as consumed, or the phosphorylation of proteins in a signal transduction cascade is also accompanied by dephosphorylation – then the system can reach a dynamic steady state. This is sometimes referred to as a **balanced** system, and solving the system of equations will give you the steady-state concentration of all the components. However, attempting to solve unbalanced systems, such as that shown in *Fig. 1*, may show some initial damped oscillations while the concentrations of the metabolites M1, M2, and M4 settle down, but then show a progressive loss of M5, leading to an increase in M3.

However, before a model can be solved, it must first be constructed. There are many software packages and applications for this and a detailed study is beyond the scope of this work. The most common tools are matlab, cell designer, and copasi. It is worth investigating also what can be obtained from the SBML (Systems Biology Mark-up Language) consortium web site (http://sbml.org/), which is the best software catalogue for bio-systems modeling.

H4 ADVANCED MODELING TECHNIQUES

Key Notes

Multiple dependencies

When one variable is dependent upon more than one other variable, then partial differential calculus is required. This in combination with other forms of algebra can be used to model processes such as diffusion, advection, and reaction-diffusion systems.

Stochastic interactions

Processes are known as stochastic, when they contain uncertainty or a genuine random chance of one thing happening as opposed to other things. Such processes may be represented by both discrete and continuous models. However, simulations with such models must be run multiple times to gain a statistically representative set of behaviors.

Percolation and related processes

Percolation is the flow of objects through a medium that contains obstacles at substantial density. This process may be modeled using so-called fractional differential equations, but forms of mathematics also exist for variants of this, such as combining percolation with enzyme kinetics.

Related sections

Ordinary differential equations and algebra (H3)   Use of calculus and algebra (O2)

Multiple dependencies

When one variable is dependent upon more than one independent variable, then a different type of mathematics is required: **partial differential equations** (PDEs). They may be used to model a range of biological phenomena including diffusion, fluxes of materials (metabolites, cells, etc.) through systems such as connected cells or blood vessels, and deformation or growth of cells and tissues. To illustrate the use of PDEs, let us consider a practical example, namely the diffusion of antibiotic away from a fungal colony on a Petri dish. The concentration of antibiotic in any given place is dependent on many variables including the distance away from the colony, the concentration of antibiotic immediately outside the fungus, rate of diffusion, time, and temperature. In mathematical shorthand, this becomes

[antibiotic] = f(distance, surface concentration, diffusion rate, time, temperature)
or $A = f(x, C, D, t, T)$

where $f(x,C,D,t,T)$ is an unknown function.

A model may provide a PDE that must be solved to determine f; however, often the process of solving the partial differential equation can be very complex. As with laboratory experiments, mathematical investigations become easier by keeping as many variables constant as possible; for example, with constant surface concentration, diffusion rate, and temperature, $A = f(x, t)$. Then in this example f is a solution of

$$\frac{\delta A}{\delta t} = D \cdot \frac{\delta^2 A}{\delta x^2}$$

This partial differential equation is second order (which means the derivative of a derivative), and is often called the **diffusion equation.** Note that the symbol 'd' in derivatives has now become 'δ.' When looking at equations, the appearance of 'δ' normally indicates that you are looking at partial differentials. PDEs, like ordinary differential equations (ODEs), are deterministic and always produce the same result.

There are certain system behaviors that follow on from this simple description of diffusion and partial differentials. If the medium in which the materials of interest (or even heat) is itself flowing, then we have a process known as **advection**. Where u is the velocity of the medium, this can be represented by a refinement of the above PDE to

$$\frac{\delta f}{\delta t} + u \cdot \frac{\delta f}{\delta x} = D \cdot \frac{\delta^2 f}{\delta x^2}$$

A biological example is cytoplasmic streaming in plant and other cells. In an analogous manner, when enzymes are diffusing while they interact with substrates, then the terms representing the diffusion can be brought into the system of ODEs which govern the enzyme kinetics. This results in a system of PDEs, so-called **reaction-diffusion equations.**

Solving partial differential equations analytically is often not possible, and in such cases, the equations could be solved numerically. A 3-D plot is often used to display solutions, where the x and y axes represent the independent variables; for example, time and distance, and the z axis gives the value of the function, $f(x,t)$.

Stochastic interactions

For a given individual molecule/species, the interaction with another or the result of the interaction might be a random event defined by some probability. This **stochastic** behavior may be a significant feature of a system when either the numbers of molecules in the system are low, or short timescales are being considered. The expression of genes in individual cells is a stochastic process. This was implicitly thought to be the case for many years for certain genes (e.g., lysogeny versus lysis of *E. coli* by bacteriophage γ). However, it has been substantiated in a colorful way by Elowitz *et al.* (2002), who arranged for the expression of two different fluorescent proteins at equal levels in the population of *E. coli* and then examined the varying levels in individual bacterial cells using fluorescent microscopy. It was clear that a substantial percentage of cells in the population expressed only one of the proteins, but the numbers of cells expressing each protein were very nearly the same.

Stochasticity can be built into differential equations by adding 'noise' terms that contain a random variable, which is sampled from a prescribed distribution. Such stochastic differential equations (SDEs) must be solved multiple times to discover the common as well as extreme behaviors of the system. A simple computer-encoded random-number generator can be used to obtain values for the random variables when there is an equal probability of all numbers occurring. However, sampling from alternative distributions might be more biologically realistic, such as Gaussian or inverse-square distributions. Random movement, for example Brownian motion, can also be modeled using SDEs. Stochasticity can also be introduced into discrete models, for example stochastic Petri-nets, and Wilkinson (2006) is an excellent book to find out more about stochastic modeling.

Percolation and related processes

The movement of particles through a medium is affected by the density of obstacles in the medium. Specific biological examples could be the movement of macromolecules through a cell and between compartments, or nutrients or organisms through pores in soil. When the density of particles reaches a certain threshold, then particle movement can suddenly become very restricted. This impedance of flow is known as **percolation**, and is a significant effect in eukaryotic cells where the density of micro-filaments, organelles, and other multi-protein complexes substantially impedes the movement of proteins, while affecting primary metabolites far less. This phenomenon is highly significant as it will influence cell physiological responses, such as protein-kinase signal-transduction cascades. One way to represent such processes is through the use of fractional differential equations, as championed by K. Burrage (Burrage *et al.*, 2007). Other approaches also exist, such as dimension-restricted reaction kinetics (Hiroi & Funahashi, 2007), which brings percolation and enzyme kinetics together.

H5 SHAPE, DEFORMATION, AND GROWTH

Key Notes

Shape	The established way of approximating complex shapes is through the definition of nodes and edges that cross the surface or through the body of the object. These nodes and edges break up the shape into a number of finite elements, which can then embody sets of equations to form a **finite element model**.
Deformation	These models can then be used to study the effects of imposing physical forces on the object, which result in a change in shape.
Growth	This complicates the deformation models by making the underlying algebra more complex as terms for controlling the rate of growth must be included. Furthermore, in biology, we may also wish to model cell (or population) division and differentiation into different cell types.

Related sections	Data categories (E)	Integrative biology and plant
	Advanced modeling techniques (H4)	modeling (P2)

Shape

Representation of shape was introduced in Section E. Taking this further, complex curved shapes and objects are most easily approximated graphically. That is, the object is represented by a series of nodes defined by 2-D or 3-D coordinates, which are joined by straight lines/edges. For a surface, the nodes are connected by triangles or quadrilaterals, but for a solid object the nodes are usually connected to form cuboids. This graph is sometimes referred to as a **mesh**. The sharper the curvature, the closer the nodes need to be for the shape to be satisfactorily represented, so there is always a trade-off between the number of nodes (the fewer, the faster the model analysis or simulation) and the accuracy of the model.

Much work has been done to automate the process of defining the sizes and shapes of components (elements) in the mesh, and this forms part of the activities known as **finite-element analysis** – so-called because only a finite number of elements are generated to model the shape to some defined degree of accuracy; only an infinite number of elements would fully define the shape of an object. In cases where only the surface of an object is of interest, the analogous approach is known as **boundary-element analysis**. Of course, each element in a mesh can have different properties, which results in finite-element (FE) and boundary-element (BE) models. These techniques are very widely used in mechanical engineering (Becker, 2003), but have been applied to biology in only a small number of instances (see Section P).

Deformation

When objects are subject to physical forces, they may deform in some way that is a function of the strength of the force and the **constitutive properties** of the object. In an FE model, each element may have different values and functions for the constitutive properties, which can be encoded using sets of partial differential equations (see Section H4). These constitutive properties should include data about the boundary between elastic and plastic behavior and what form the plastic behavior follows. Perhaps the best application of this has been in the models of a beating mammalian heart.

Growth

FE modeling becomes substantially more challenging when the volume is not constant; that is, the object is growing. This is because more factors/variables must be taken into consideration. First, are there any constraints to the growth? These may cause an increase in internal pressure as the increasing volume pushes against them, and this can result in growth taking place in certain directions rather than others. Where the object has a skin, then one needs to know if the skin itself is increasing in volume. If not, then the skin will become thinner, as happens in an inflating balloon. Growing systems are also amenable to FE analysis, albeit with extra equations defining the rate of volume increase and its associated internal pressures.

Even greater challenges arise when one considers a growing mass of cells, such as a bacterial colony or tumor. In this situation, cells will grow, and at some stage divide to become two autonomous cells, each with their own properties. This scenario calls for **adaptive methods**, where usually some rule defines when the cell division event will take place and what form it should take, for example

when (volume = x) {divide cell into two cells of volume a and (x-a)}

In practice, the rules will be far more complex, including definitions of the geometry of how and where the dividing septum will be placed, and in which daughter cells the components (organelles) will be placed. A biological example of this is the work on snap-dragon flower development (Section P2).

The final level of complexity probably pertains to organogenesis. In this situation, on top of cell growth and division, the daughter cells may have radically different properties from the others as they progressively differentiate into different tissues.

I1 INTRODUCTION TO ARTIFICIAL INTELLIGENCE AND MACHINE LEARNING

Key Notes

Artificial intelligence This refers to the building of 'intelligent' machines or software. The term 'intelligence' has various specific meanings, and such software has found applications in many areas of bioinformatics.

Related sections Essentials of physics (C)

Artificial intelligence

'Artificial intelligence' (AI) refers to the engineering of 'intelligent' machines or software. How intelligence itself is defined depends on the application. While some of the goals of AI are still a long way off (the aims of replicating human thought and visual capabilities are still at least decades away), aspects of the discipline have provided valuable insights and technologies. When the field started several decades ago, tackling a problem such as winning a chess game was considered 'intelligent.' Since then, computers have beaten chess masters, but it has been recognized that winning chess games does not reveal intelligence by itself. Playing chess is a particular type of problem which can be solved using brute force searching to explore all possible moves, and heuristics (i.e., rules guiding the search) to shrink the search space. Replicating more natural human processes, such as interpreting a visual scene or learning to read are actually much harder challenges. Things that we consider simple tasks are often very challenging to replicate in software. While the ultimate aim of AI might be to fully replicate our intelligence processes, on the way to that goal we can learn about how our thought processes work, and use this knowledge to inspire algorithm development that mimics a particular process or replicates a specific outcome. These algorithms, as we shall see, can be valuable for searching large amounts of data, or recognizing patterns, which is of course useful to bioinformaticians.

While the algorithms developed so far have often found practical use in many disciplines, the goal of producing a fully artificially intelligent machine is still in its infancy. Methods have been proposed to assess the intelligence of a machine. One common example uses the capability of a computer to have a conversation. The Turing test requires a user to have textual conversations on two remote terminals. Controlling one is a human operator, and controlling the other is computer software. If the user cannot tell which is the human and which is the computer after conversing with each, then a system can be thought of as having passed the Turing test.

A common theme in AI today is the accumulation and organization of data. Many areas of research, especially bioinformatics, involve enormous datasets, and ways are needed to extract information from these data. Therefore there has been substantial work on developing learning and search algorithms. In this section, we will examine some examples of these, and we will see how they are useful within bioinformatics research.

The algorithms can be broadly divided into two categories: those with a statistical basis (such as Bayesian methods and Markov models), and those which are computational (such as evolutionary algorithms and neural networks). Examples from these two approaches will be presented in Sections I2 and I3, respectively.

These artificial intelligence-led approaches provide techniques for modeling and bioinformatics that can be used for parameter optimization, searching large amounts of data, pattern recognition, data clustering, and many other important problem areas. These algorithms have been crucial to bioinformatics, having been applied to such diverse areas as the prediction of bimolecular function from sequences, exon prediction, protein fold recognition, transcription-factor binding-site prediction, transcriptome, proteome and metabolome analysis, and regulatory network inference (Sections J–M, O).

I2 STATISTICAL APPROACHES TO ARTIFICIAL INTELLIGENCE AND MACHINE LEARNING

Key Notes

Bayesian reasoning	Bayesian reasoning enables us to incorporate prior knowledge into calculations of probability. We are interested in calculating the probability that event A occurs given that we know that event B occurs. Bayes' rule allows us to calculate the probability of event A given event B when the information we have is the unconditional probabilities of events A and B, and the conditional probability of B, given A.
Bayesian networks	These provide a way of graphically displaying probabilistic problems, using graph structures, which are useful for representing complex conditional relationships. Examples of use within bioinformatics include identifying transcription factor binding sites, and inferring the structure of gene regulatory networks.
Markov models	This family of models model changes in state, where the move to a new state is determined solely by the state the system is currently in. The state changes are defined using a transition matrix. Using Markov models it is possible to generate a sequence of states to mimic a process. It is also possible to use them as a model for how a given process was generated.
Hidden Markov models	These are similar to the Markov models, except that the state the model is in is not known for sure; instead probabilities represent the possible states (hence 'hidden'). As well as a transition matrix, these models have a confusion matrix, or matrix of observational probabilities, for each state. For example, we might not be able to directly see what a system is doing, but we can use sensor readings to infer the state of the hidden system.
Monte Carlo methods	These random sampling methods simplify complex problems by randomly sampling from a particular space of solutions rather than using the complete space. Markov chain Monte Carlo methods construct a Markov chain such that repeated sampling from it approximates samples drawn from a required distribution.
Related sections	Probability and statistics (G) Domain families and databases (J5)

Bayesian reasoning	Bayesian statistics were introduced in Section G. Bayesian inference enables us to incorporate prior knowledge into calculations of probability. Our knowledge of the probability of a particular event may be influenced by knowledge of what has already taken place.

In a simple formulation of Bayesian inference, we are interested in calculating the probability that event A occurs given that we know that event B occurs. This quantity is called the conditional probability of A given B, and we write $P(A \mid B)$. For example, we may wish to know the probability of it being sunny tomorrow given that it is sunny today. This is likely to be different from the probability of it being sunny tomorrow given that it is snowing today. This is different again from the probability of it being sunny tomorrow if we do not know today's weather.

In order to explain the meaning of the term $P(A \mid B)$, consider the following Venn diagram

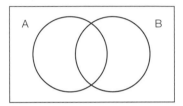

Fig. 1.   Venn diagram representing the world of all possible events (the rectangle) and specific events, event A and event B, represented by circles.

The area of each circle in *Fig. 1* gives the **marginal** (meaning unconditional) probability of a particular event, and the rectangle represents everything that can possibly happen, so its area is 1.

If we wish to calculate the conditional probability of A given B, we know that event B has definitely taken place. Therefore the circle labeled B becomes our whole world, and we wish to express the area of the circle B in which A also occurs as a fraction of the whole circle B.

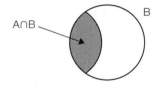

*Fig. 2.   If we **know** event B took place, the effect on Fig. 1 is that our world of possible events is reduced to those where event B definitely took place, i.e., to circle B in the Venn diagram.*

The shaded part of the circle in *Fig. 2* above corresponds to events A and B both taking place. We write $A \cap B$ to mean the **intersection** of A and B, which represents this outcome. The conditional probability of A given B is simply the ratio of the area of $A \cap B$ to the area of circle B; that is, P(B). Thus we have

$$P(A \mid B) = \frac{P(A \cap B)}{P(B)}$$

We can see from *Fig. 2* that $A \cap B$ is identical to $B \cap A$, and that $P(B \cap A) = P(B \mid A) \times P(A)$, so by substitution we arrive at Bayes' rule

$$P(A \mid B) = \frac{P(B \mid A) \times P(A)}{P(B)}$$

P(A | B) is known as the **posterior probability** that takes into account the updated knowledge currently available; that is, that *B* has occurred. *P(A)* is the prior probability of event A, and $P(B)$ is the prior probability of event B. $P(B \mid A)$ is the conditional probability of B *given* A. Bayes' rule allows us to calculate $P(A \mid B)$, which may not be possible directly, from $P(B \mid A)$.

Please refer to Section G2 for more information and a worked example using Bayesian statistics.

Bayes' rule is renowned for being hard to understand and can produce counter-intuitive results. It has even caused controversy in court cases involving the combination of multiple pieces of evidence such as DNA evidence and eye witness accounts, where it has been ruled that the jury would not understand it, despite it being theoretically correct. The confusion comes in describing the chance of DNA evidence being wrong as one in many millions – even though other evidence may refute it.

However, Bayesian reasoning has wide application, and is worth understanding – there are many good and diverse tutorials available on the WWW and in textbooks.

Bayesian networks

Bayesian networks are a way of organizing complex conditional probability information into a graph-like structure that can simplify certain calculations. For a simple example, let us again consider the probabilities of having sun two days in a row (day 1 and day 2). There are four possible outcomes and four associated probabilities, as illustrated in *Fig. 3* below.

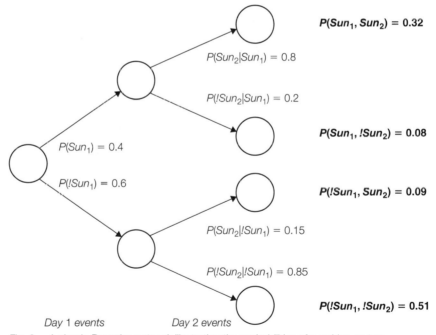

Fig. 3.   A simple Bayesian network illustrating the probabilities of sunshine on two consecutive days (The '!' character is often read as 'not' in a probability context).

The final probabilities representing each of the four possible outcomes are calculated by multiplying together the probabilities along the branches of the tree structure that lead to each particular outcome. It can be easily seen from *Fig. 3* that, unfortunately, the most likely outcome is no sun on either day! While this is obviously a simple example, this representation aids the evaluation of outcomes when there are many parameters and many conditional situations. Clearly for a particular outcome it is not necessary to evaluate **all** the probabilities in the structure; only the ones that lead to the outcome. Branches that represent probabilities independent from those in which we are interested need not be evaluated; if we know day 1 is sunny, we do not need to evaluate the bottom half of the graph structure at all.

The network layout itself is often initially determined by using causal relationships as a guideline. For quantitative parameters, the probabilities are based on expert knowledge and statistical information from the domain in question. For example, in a gene regulatory network model we can calculate the conditional probabilities of target genes in relation to their potential regulator genes from microarray data.

Bayesian networks have been used in a variety of areas within bioinformatics, including splice site prediction identification of transcription factor binding sites, and inferring the structure of gene regulatory networks.

Markov models

Markov models were introduced in Section G2 as a way of modeling patterns of letter pairs in DNA sequences. Here, we will consider a simple example followed by a look at the hidden version of Markov models.

A Markov model is a model of changes from one state to another, with a probability of transition to the new state at the next time-step determined solely from the current state. While sounding complex on first impression, this principle is easily clarified with a simple example. Suppose there are three 'states': A, B, and C. A state is simply a label for a situation. The three states could represent three weather conditions, the amino acid found at three positions in a protein sequence, three gene expression profiles or, in fact, any 'state' relevant to a particular scenario. For this example, the three states are

A = running
B = walking
C = standing still

A Markovian model describes the probabilities of changing from each state to the next in one time-step. So, we might have the following transition matrix to describe the state changes

Table 1.   Transition matrix representing the transition probabilities from each state to the next

t+1 \ t	A	B	C
A	0.5	0.3	0
B	0.5	0.5	0.75
C	0	0.2	0.25

The probabilities for transitions can be inferred from empirical data. For example, we can assume that by observing a sample of people performing tasks of interest, we could calculate the probabilities shown in *Table 1* from looking at how frequently they switch from one particular behavior to another. From *Table 1*, it can be seen that if a person is walking (B), in this world there is a 20% chance that at the next time-step he/she will be standing still (C). Likewise it can be said that if someone is standing still(C) at time *t*, it is impossible for that person to be running at time *t+1* (he/she must go through a walking phase first). This information can also be visualized as a network, with the nodes and edges representing the states and possible state transitions, respectively (see *Fig. 4*)

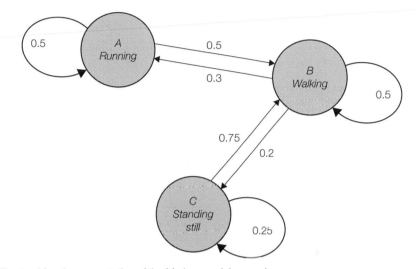

Fig. 4.   Visual representation of the Markov model example.

It should be even clearer from *Fig. 4* that it is impossible to change the state directly from running to standing still, and vice versa, as there are no edges directly connecting them. Using Markov models it is possible to generate a sequence of states to mimic a process. It is also possible to use them as a model for how a given process was generated.

Hidden Markov models

Hidden Markov models (HMMs) are similar to Markov models with an added layer of complexity. Let us suppose in the example above that instead of being able to directly see whether someone is walking, running or standing still, the only thing we have access to is a remote feed of a heart rate sensor. In other words, the actual state is **hidden** from us. All we have are the observations of the heart rate, which for this example can be slow, medium or fast. While it is likely that a fast heart rate indicates a person is running, this is not a certainty. In fact, these observations are linked to the actual states by probabilities (see *Table 2*)

Table 2.   Table of observation probabilities, known as the confusion matrix

State Heart rate	A Running	B Walking	C Standing still
Slow	0.05	0.2	0.8
Medium	0.15	0.7	0.2
Fast	0.80	0.1	0

Thus, if we are in state C (standing still), there is an 80% chance we will observe a slow heart rate and a 20% chance we will observe a medium one. To fully define a HMM, we need to define the number of states, the number of possible observations for each state, the probabilities for transitions between states and the probabilities of observing the hidden state. This is often summarized as

$$\lambda = (A, B, \pi)$$

where

A = state transition matrix
B = observation probability distribution for each state (confusion matrix)
π = initial state probabilities

As a further example, consider a hidden Markov model for generating protein sequences (*Fig. 5*). The model consists of delete states (d), match states (m), and insert states (i), and arrows which show the possible transitions between the states. Each arrow has a probability associated with it representing the transition probability of moving to the indicated state. The 'hidden' aspect is represented by the fact that match and insert states generate one of a number of amino acids according to a probability distribution. So, the **hidden** state we are in might be a match state, but the thing we *observe* is one of 20 amino acids. The final protein sequence is derived by moving through the model, following the arrows, and generating amino acids.

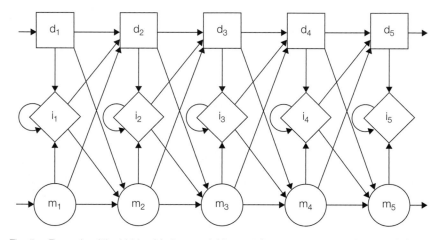

Fig. 5. *Example of the hidden Markov model for protein sequence generation, consisting of delete states (d), match states (m), and insert states (i).*

There are three main classes of problem that we wish to address using HMMs. First, we may wish to determine the probability of a given sequence of observations using a particular model. This can be used to select one model from a set that is most likely to generate that sequence. The **forward algorithm** is used to calculate the probability of a sequence given a model. This is often used to select which model best represents the input data; for example, fitting models of words to speech data in speech recognition.

The second kind of problem involves determining the most likely sequence of hidden states given a sequence of observations. The **Viterbi algorithm** is used to determine the sequence of hidden states given a set of observations and a particular model.

The final problem is to actually generate a HMM from a sequence of observations of hidden states. That is, how do we determine the model parameters (A, B, π) to maximize the probability of a sequence of observations? This is by far the most challenging of the three problems.

A common use of HMMs is speech recognition where each state can represent, for example, one phonetic unit. Applications specific to bioinformatics include gene finding with a hidden Markov model of gene structure and evolution, splice site recognition and protein function/structure prediction.

Monte Carlo, and Markov chain Monte Carlo (MCMC) methods

At the heart of **Monte Carlo** (MC) methods is the principle of replacing a very complex problem with a statistical alternative. The name Monte Carlo refers to the famous gambling location of the same name; the idea of drawing individual samples is akin to the concepts of randomness and odds used in casinos. The approach has been applied to a wide range of problems, such as optimization and the simulation of physical systems, which often take place in a high-dimensional space and thus have a large number of feasible solutions.

The basic idea behind MC methods is that from a typically large dimensional space (e.g., the set of all possible states of the system, or all possible solutions), a finite set of individual samples are taken. These samples can be used to approximate the original space, making complex problems computationally tractable. The quality of this approximation increases as the number of samples increases (*Fig. 6(a)*). A number of extensions to this simple idea exists. For example, defining a probability distribution across the original sample space is called **importance sampling**, and allows us to draw more samples from the more important or significant areas of the space, making the representation more accurate (*Fig. 6(b)*).

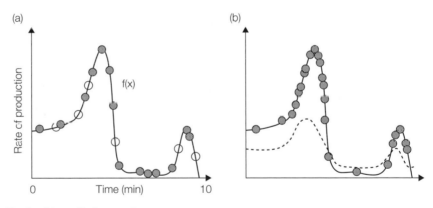

Fig. 6.   Monte Carlo sampling.

(a) In this figure there is a curve whose shape is some unknown function (f(x)) which corresponds to the rate of production of some biological molecule over time. At the outset, the curve is unknown and the hypothetical laboratory technique is very expensive, so it would be unfeasible to measure the rates at many time points. In the MC approach to sampling, the effect of measuring values at an initial 6 and a further 14 random time points is shown (as denoted by empty and filled circles, respectively). This illustrates how more samples produce a better representation of the function; the initial set would underestimate the peaks and overestimate the trough, whereas all 20 samples represent the space far better.
(b) In this figure, the dashed line denotes a probability distribution, derived from some prior information, so that importance sampling may be carried out. More samples are taken from positions that have a higher probability, leading to more accurate representation in those areas.

A common application of the MC technique is to find the area under a curve or volume under a surface; that is, to evaluate the **integral** of a function. For the hypothetical example in *Fig. 6*, the area under the curve corresponds to how much of the biological molecule has been synthesized during this time period. We can use Monte Carlo integration to estimate this area. One way to do this is by measuring values at randomly selected points and multiplying the average of these values by the time interval over which the samples were taken. It should be clear from *Fig. 6(a)* that the greater the number of samples, the better the approximation.

Furthermore, using importance sampling, the sample times can be selected using a probability distribution function which concentrates them in the most informative areas of the space (*Fig. 6(b)*). The places where the samples have larger values contribute more to the area under the curve, and so it is useful to sample these areas more accurately. In this example, it is known that there are peaks at about 3 minutes and 8 minutes after the start of the experiment, so the probability density function concentrates random samples around those time points (*Fig. 6(b)*). However, to prevent this technique skewing the final approximation to an abnormally high value, every observed value (and hence the final average) must be adjusted by dividing the sample value by the value of the probability density function at that time point.

The **Markov chain Monte Carlo** (MCMC) methods implement a technique for drawing samples and exploring the state space using a Markov chain mechanism. The technique constructs the chain in such a way that the samples produced mimic samples drawn directly from a desired distribution. This is useful when we cannot obtain samples from the original distribution directly. Over many iterations, the probability distribution for the chain will converge to an invariant solution, no matter what the initial distribution was. This means that a distribution that will not change over time will be reached. One of the key design issues with MCMC techniques is the time the chain takes to converge on the invariant distribution.

MCMC methods have been applied to a wide range of problems such as the visual tracking of a large number of targets, and estimating genome rearrangement. Monte Carlo techniques have also found application in calculating protein structure.

I3 COMPUTATION APPROACHES TO ARTIFICIAL INTELLIGENCE AND MACHINE LEARNING

Key Notes

Inductive logic programming	Languages such as Progol use inductive logic to infer hypotheses from sets of examples and background knowledge.
Neural networks	These are biologically inspired methods of pattern classification analogous to networks of collections of brain neurons.
Evolutionary algorithms	This general class of algorithms is inspired by evolution and nature, including genetic algorithms (GAs) as the most prominent example. GAs simulate recombination and mutation to breed populations of individuals that are subject to selection on the basis of a fitness function. Progressive breeding and selection produce successively 'fitter' generations.
Kernel methods	These are a general method which maps data into high-dimensional spaces to classify and elucidate relationships within the data.
Overfitting	This is an ongoing hazard associated with many learning techniques. When the reference input data are not representative of the entire class, the learned relationships may be too specific to the data presented, and will not generalize to include unseen data in the same class.
Optimization	Precise solutions to a computational problem might be intractable because the search space is too large or complicated. In these circumstances, a range of approximation methods may be used, some of which are described above. There are others, however, such as tabu searching, which relies on exploring the locality of a start point and excluding particular steps for various reasons.
Related sections	Essentials of physics (C) Introduction to artificial Data and databases (D) intelligence and machine learning (I1)

Inductive logic programming

Inductive learning refers to the practice of inferring general concepts from specific examples. A simple example is to say that as we have seen the sun rise every morning of our lives, we can use induction to infer that the sun will continue to rise **every** morning.

Here is an everyday example of an inductive learning problem. Adapted from an example in Muggleton and DeRaedt (1994), let us consider trying to learn family relationships. Your beliefs might consist of the following

1. If X is the father of Z, and Z is the parent of Y, then X is the grandparent of Y.
2. Henry is the father of Jane.
3. Jane is the mother of John.
4. Jane is the mother of Alice.

You are also given the following positive examples of grandfathers

1. Henry is the grandfather of John.
2. Henry is the grandfather of Alice.

And you have these **negative** examples

1. John is not the grandfather of Henry.
2. Alice is not the grandfather of John.

Note that we do not know yet what a parent is. Using this information, it may be possible to guess the following hypothesis

If X is the mother of Y, then X is the parent of Y.

But how is it possible to arrive at this hypothesis? Inductive learning techniques address such challenges as these, where we have some background knowledge, some examples (in this case, positive and negative), and we wish to produce a new hypothesis. Inductive logic programming (ILP) is able to generalize from such individual instances and background knowledge, and produce hypotheses from this information. A practical complication is that the positive and negative examples might be noisy – that is, we might not be able to guarantee that they are definitely **true**. ILP problems can be described using programming systems such as Progol (www.doc.ic.ac.uk/~shm/progol.html).

Broadly speaking ILP systems can be categorized into those that learn one hypothesis or more than one, those where interaction is required from an expert user, and those where either all examples are required from the start or where they can be presented incrementally. Current work is integrating probabilistic techniques with ILP to produce probabilistic ILP methods. This can be used, for example, to represent probabilistic clauses such as: If person X is a carrier for a certain disease, the chance that they suffer from it may be 70%.

Within bioinformatics, example applications of ILP include being used to predict gene function and drug design, among many others. It has been integrated into other life sciences software frameworks as well.

Neural networks   Artificial neural networks are inspired by the structures and networks of the nervous system. One of the earliest examples was the Perceptron, which was a hypothetical model for the nervous system. Some of the assumptions on which Perceptron theory was based include: at birth, the connection of such networks is largely random (subject to genetic constraints); when exposed to similar stimuli, the response will form similar pathways through the network; the presentation of positive and negative examples may facilitate or hinder the current formation of connections in progress.

Artificial neural networks model the dendrite and axon connections between neurons, and the electrical impulses that result from and cause the firing of these neurons. The combination of neurons and connections forms a network (*Fig. 1*). Input is fed into the network by activating certain input connections, and an output signal is produced at the output connections. The unique aspect of neural networks is that they do not use a formulaic approach to problem solving –

there is no high-level algorithm imposed on them; rather, there is a simple set of low-level rules determining how neurons fire signals and how the network is connected.

When a neuron receives sufficiently higher excitation signal than inhibition signal, it fires, producing a signal that is conducted to more neurons or that forms an output signal. It is the interconnections between neurons that allow these signals to be carried around the network. Neurons themselves are defined both by their connections and a formula that dictates how they fire. This can be as simple as saying that the neuron should fire if the total input signal from all incoming connections is greater than a threshold, t (see *Fig. 1(a)*). In its simplest form, the output of a firing neuron is 1, and a non-firing neuron is 0. However, more complex output functions which are more smoothly varying can allow networks to learn more complex patterns.

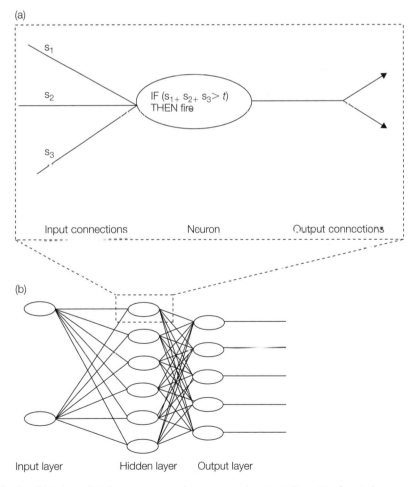

Fig. 1. (a) Schematic of one neuron and some example connections. The inputs (s_1, s_2, s_3) are summed and if their total exceeds a threshold t, the neuron fires, producing an output signal. (b) Example network, consisting of several artificial neurons and connections. Networks may, in fact, consist of many thousands of neurons and connections.

Learning occurs in a neural network by learning the firing rules at each neuron. This usually involves learning the signal input threshold needed for activation, and the weights of each incoming connection. Typically, input patterns are presented to the network and the rules must be adjusted such that the correct output classification is produced. The power of neural networks stems from their ability to classify correctly patterns that are close to, but not the same as, the learned patterns. This makes them powerful at solving pattern-recognition and classification problems.

There are several variations of networks which can aid classification and learning. The earliest networks were single-layer, feed-forward networks. Feed-forward refers to the lack of cycles in the networks; that is, there is no feed-**back**. Single-layer alludes to the lack of hidden layers (for example, see *Fig. 1(b)*). Single-layer, feed-forward Perceptron networks can only learn linearly separable problems, and cannot correctly resolve certain logical input configurations, such as the XOR ('Exclusive OR') function (see *Table 1*).

Table 1. Inputs and outputs of a logical XOR function

Inputs	Desired output
True, True	False
True, False	True
False, True	True
False, False	False

Adding multiple hidden layers increased the range of problems the networks could classify; for example, solving the XOR problem. Adding feed-back to the network to assist learning (a term referred to as 'back propagation') improved the efficiency of learning with large networks with many inputs.

Evolutionary algorithms

This term covers a broad range of machine learning techniques, the most common of which are **genetic algorithms** (GAs). These attempt to find results by simulating the process of evolution. Initially, a random population of individuals is created. New individuals (the next generation) are then created from the existing population using processes that are equivalent to mutation and recombination, and sometimes replication. The population represents the set of solutions (typically one or more hundreds), and many GA variants keep this number constant. An individual in this context represents one version of a system or solution. Individuals are evaluated according to a **fitness function**, whose value for the individual determines its fate. Over time, the aim of the process is to increase the fitness of the population, and thus arrive at a solution to the search.

It is common to represent the underlying data as a string of bits (see Section D). Mutation occurs by randomly reversing certain bits. Recombination (sometimes called crossover) is achieved by first selecting a position within the string. Two new strings are created then, the first contains the first part of the first string before the selected position, and the second half of the second string after the selected position (see *Fig. 2*). The second new string represents the remaining combination: the first half contains the first part of the second string before the selected position, and the second half of the first string after the selected position.

```
String 1   11001010
String 2   11100000
```

Example of recombination:
Random position selected: 4, therefore the strings are split at the asterisk:
```
String 1   110*01010
String 2   111*00000
```

Two new strings:
```
11000000
11101010
```

Fig. 2.   Illustration of the principle of recombination.

The quality of an individual is evaluated using a **fitness function**. Individuals are selected for 'breeding' based on their fitness. As successive generations are produced, the overall fitness of the population should tend to increase ('survival of the fittest'). Exactly how the fitness function is defined depends on the particular problem domain and on the desired result. An example might include looking for the minimum value, in which case smaller solutions should be given a higher fitness rating than larger numbers. It may be desirable to select a small number of less fit individuals, as this serves to keep the population diverse and can help alleviate the problem of solutions converging to local minima (cf, simulate annealing in Section C).

The algorithm can continue until either a desired fitness level is reached (suggesting a sufficient quality of result has been found), the rate of increase in fitness is down to some minimum value, or a fixed number of generations has been created. In the latter case, there is no guarantee that a suitable solution has been found, but it does guarantee the procedure to have finished within a particular time frame. Example uses of evolutionary algorithms in bioinformatics are diverse, but include the prediction of transcription factor binding sites. Other evolutionary algorithms include particle-swarm optimization (derived from animal flocking habits), and ant route-finding behavior.

Kernel methods   Kernel-based learning methods cluster data by first mapping data points into higher-dimensional spaces. The analysis is then carried out in this new high-dimensional space. There are various advantages to this approach, which allow the data to be processed efficiently. One such trick is to work with the similarity measures between points rather than absolute coordinates; these are often cheaper to compute. By using the correct kernel function, these approaches can discriminate non-linear separations in the data using linear discrimination algorithms; in other words, they make the mathematics substantially easier. The approach is appealing in that in can be applied to general types of data, and classical algorithms can be used to cluster the newly remapped data.

Fig. 3 illustrates the basic concept behind kernel methods. In practice, the remapping can happen in any number of dimensions (e.g., the number of different genes on an array chip might lead to many thousands of dimensions); consequently the dividing dashed line in *Fig. 3* would be replaced with a dividing **hyperplane** (i.e., a plane defined in multiple dimensions – a high-dimensional equivalent of a straight line in 2-D or a plane in 3-D, which divides the space into two halves).

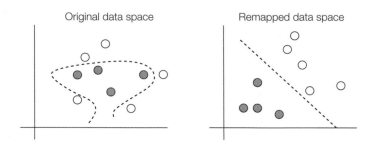

Fig. 3.   Remapping the data allows in this case a linear discrimination to be made to separate the two classes.

The most common kernel method used in bioinformatics is called 'Support Vector Machines.' They are particularly useful when the data have a high number of dimensions; for example, multiple sequence alignments, biomolecular profiles, and word frequencies in sets of journal abstracts.

Overfitting

There are two constant issues with machine learning techniques. The first is the extent to which the training data are representative of the class. The second and sometimes related issue is that the learned rules fit the training data so well that they are unfit for discovering new members of the same class. This is known as **overfitting**.

Fig. 4 shows a hypothetical example of this. The four protein sub-sequences are known to bind nucleoside triphosphates (usually ATP) and have generated the rule shown. However, other such binding including the two shown would not be identified by this rule. The classification is clearly over-specific to the training data. Overfitting is a greater issue where either the training data set is too small or the learning continues for too long, leading it to regard some specific quirks of the training set as general rules. The problem illustrates the need to have a representative training set. There are also extensions to learning algorithms to help avoid this problem, such as cross-validation and regularization.

> VDGVAGCGKTTNIK
> VDGVAGCGKTTAIK
> VDGVPGCGKTKEIL
> VTGTAGAGKTSSIQ

> Rule   VxGxxGxGKTxxIx

> False -ve   VQGPPGSGK**SHFA**I
> False -ve   VLGAPGVGK**S**TSIK

Fig. 4.   An example of overfitting. The top four protein sequences are part of nucleoside-triphosphate binding sites from structurally related proteins. These could lead to the rule shown where V, G, K, T or I are the amino acids that must be at the specified positions, otherwise an 'x' denotes any amino acid is accepted. Beneath the rule are two more binding sites, which fail this rule owing to the presence of the underlined amino acids shown.

Optimization

There are many problems for which an exact solution is unfeasible in a realistic time frame using a computer. In the simplest case, if a comparison were made of every position of a DNA sequence (of length n) with every other position, then the amount of time needed to compute the result increases with the sequence length multiplied by itself. This is known as an Order n^2 problem, or $O(n^2)$ for short. Many other problems may be $O(n^3)$ or even higher, such as calculating the shortest distance between a set of points on a road network (also known as the **travelling-salesman** problem). **Optimization** techniques, also referred to as **meta-heuristics**, aim to provide adequate solutions to otherwise intractable problems in short timescales. A given technique might not give a satisfactory result and might not necessarily run in an acceptable timescale. Different techniques are better for different types of problem, and computer science has a research field in trying to work out which ones to use when (known as **hyper-heuristics**).

Many of the algorithms described in Sections I1 and I2 come into this category, as does simulated annealing (Section C). However, there are many others. Perhaps the simplest to understand is the **steepest descent** algorithm: from a given starting point, find out if the next step in a given direction is better and, if so, then take it, otherwise try a different direction. This is very quick to execute but will always result in finding the local minimum.

A more refined variant is **tabu** (a variant spelling of **taboo**) searching. In this approach one examines the locality of a given point before deciding the next step. Certain things are not permitted (**taboo**); for example, one may not step to a place that the search has already visited, or one might not cross a path that one has already taken. These involve keeping a record of where one has been. **Aspiration criteria** help to keep the algorithm heading in the right direction, sometimes by overriding places that have been defined as tabu. This is a more directed search technique than steepest descent that often finds the local minimum sooner, and is often used for tackling challenges related to the travelling-salesman problem.

This Section completes the part of the book devoted to 'building blocks,' and the subsequent Sections examine how these have been used in different 'application areas' in life science research.

J1 DATABASES AND DATA SOURCES

Key Notes

Principles of DNA sequencing

Most DNA sequencing is performed using an automated version of the chain termination reaction, in which limiting amounts of dideoxyribonucleotides generate nested sets of DNA fragments with specific terminal bases. Four reactions are set up, one for each of the four bases in DNA, each incorporating a different fluorescent label. The DNA fragments are separated by polyacrylamide gel electrophoresis and the sequence is read by a scanner as each fragment moves to the bottom of the gel.

Types of DNA sequences

DNA sequences come in three major forms. Genomic DNA comes directly from the genome and includes extragenic material as well as genes. In eukaryotes, genomic DNA contains introns. Complementary DNA (cDNA) is reverse transcribed from mRNA, and corresponds only to the expressed parts of the genome. It does not contain introns. Finally, recombinant DNA comes from the laboratory and comprises artificial DNA molecules such as cloning vectors.

Genome sequencing strategies

Only short DNA molecules (<800 bp) can be sequenced in one read so large DNA molecules, such as genomes, must first be broken into fragments. Genome sequencing can be approached in two ways. Shotgun sequencing involves the generation of random DNA fragments, which are sequenced in large numbers to provide genome-wide coverage. Conversely, clone contig sequencing involves the systematic production and sequencing of subclones.

New generation sequencing technologies

Speed, accuracy, and cost of DNA sequencing have been improved recently with the introduction of parallel sequencing technologies. Solexa and 454 sequencing methods allow many DNA fragments to be sequenced at the same time in one machine.

Sequence quality control

High quality sequence data are generated by performing multiple reads on both DNA strands. The preliminary trace data then are base called and assessed for quality using a program such as Phred. Vector sequences and repeated DNA elements are masked off and then the sequence is assembled into contigs using a program such as Phrap. Remaining inconsistencies must be addressed by human curators.

Single-pass sequencing

Sequence data of lower quality can be generated by single reads (single-pass sequencing). Although somewhat inaccurate, single-pass sequences such as expressed sequence tags (ESTs) and genome survey sequences (GSSs) can be generated in large amounts very quickly and inexpensively.

RNA sequencing

Most RNA sequences are deduced from the corresponding DNA sequences but special methods are required for the identification of modified nucleotides. These include biochemical assays, nuclear magnetic resonance spectroscopy, and mass spectrometry.

Protein sequencing

Most protein sequencing now is carried out by mass spectrometry (MS), a technique in which accurate molecular masses are calculated from the mass/charge ratio of ions in a vacuum. Soft ionization methods allow MS analysis of large macromolecules such as proteins. Sequences can be deduced by comparing the masses of tryptic peptide fragments to those predicted from virtual digests of proteins in databases. Also, *de novo* sequencing can be carried out by generating nested sets of peptide fragments in a collision cell and calculating the difference in mass between fragments differing in length by a single amino acid residue.

Primary sequence databases

The three primary sequence databases are GenBank (NCBI), the nucleotide sequence database of EMBL, and the DNA Databank of Japan (DDBJ). These are repositories for raw sequence data, but each entry is extensively annotated and has a features table to highlight the important properties of each sequence. The three databases exchange data on a daily basis.

Subsidiary sequence databases

Particular types of sequence data are stored in subsidiaries of the main sequence databases. For example, ESTs are stored in dbEST, a division of GenBank. There are also subsidiary databases for genome survey sequences and unfinished genomic sequence data.

Submission of sequences

Sequences may be submitted to any of the three primary databases using the tools provided by the database curators. Such tools include WebIn and BankIt, which can be used over the internet, and Sequin, a stand-alone application.

SWISS-PROT, TrEMBL, and UniProt

SWISS-PROT is a collection of confirmed protein sequences with annotations relating to structure, function, and protein family assignments. The related database TrEMBL is a translation of all coding sequences in the primary nucleic acid databases. The entries in TrEMBL are annotated less extensively than those in SWISS-PROT, but are moved to SWISS-PROT when reliable annotations become available. Both TrEMBL and Swiss-Prot have been incorporated into the UniProt (Universal Protein Resource), which also incorporates the PIR database.

Database interrogation

All the databases discussed in this section can be searched by sequence. However, detailed text-based searches of the annotations are possible using tools such as Entrez. The simplest way to cross-reference between the primary nucleotide sequence databases and SWISS-PROT is to search by accession number, as this provides an unambiguous identifier of genes and their products.

Organism-specific resources

As well as general databases that serve the entire biology community, there are many organism-specific databases that provide information and resources for those researchers working on particular species. The number of organism-specific databases is growing as more genome projects are initiated, and many can be accessed from general genomics gateway sites such as the Genomes OnLine Database (GOLD).

Database formats

There is no universally agreed format for genome databases, and several viewers and browsers have been developed with graphical displays for genomic sequence analysis and annotation. EnsEMBL is highly regarded. One of the more versatile formats is ACeDB (originally designed for the nematode *Caenorhabditis elegans*), which has an object-orientated database architecture and is now used in many applications outside the field of genomic bioinformatics.

Finding organism-specific databases

Organism-specific data are widely distributed on the internet. In order to find and interrogate databases on specific organisms, it is necessary to use a gateway site to access relevant databases and information resources. Worked examples are provided, using GOLD as the gateway and illustrated with Ebola virus, the bacterium *Escherichia coli*, the fruit fly *Drosophila melanogaster*, and the human genome.

Database resources

There are many types of database available to researchers in the field of biology. These include primary sequence databases for the storage of raw experimental data, secondary databases that contain information on sequence patterns and motifs, and organism-specific databases tailored for researchers working on a particular species. Other miscellaneous databases are discussed in this section.

Specialized sequence databases

A number of databases have been developed for the storage and analysis of particular types of sequence, e.g., rRNA and tRNA sequences, introns, promoters, and other regulatory elements.

OMIM

OMIM is the Online Mendelian Inheritance in Man database, a powerful resource for the study of human genetics and human molecular biology. Each OMIM entry has a full text summary of information known about a particular gene or trait, with links to primary sequence databases and other human genetics resources.

UniGene

UniGene is an experimental facility for the clustering of GenBank sequences and is related to EST data. Currently six vertebrate and five plant species are covered by UniGene.

Structural databases

The primary resource for protein structural data is the Protein DataBank (PDB), which contains data derived from X-ray crystallography and NMR studies. Another structural database, the Molecular Modeling Database (MMDB) can be accessed at the NCBI web site using Entrez.

Proteins and higher-order functions

Many databases have been set up to store information on particular types of proteins, such as receptors, signal transduction components, and enzymes. The compilation of data on different types of proteins, their functions, and interactions makes it possible to deduce higher-order functional networks in the cell, such as biochemical pathways, signal transduction systems, and regulatory hierarchies. An example of such a combined database is the Kyoto Encyclopedia of Genes and Genomes (KEGG).

Literature databases

Literature databases store scientific articles and allow various fields (title, authors, key words, abstract) to be searched using text strings. Among the most widely used literature resources on the internet are MEDLINE and

PubMED, which cover the scientific literature from the 1960s to the present day.

Access to distributed data

Biological data are widely distributed over the WWW. As an alternative to standard search engines, dedicated data retrieval tools such as Entrez, DBGET, and SRS can be used to search multiple biological databases and retrieve relevant information.

Entrez

Entrez is a WWW-based data retrieval tool developed by the National Center for Biotechnology Information (NCBI), which can be used to search for information in 11 integrated NCBI databases, including GenBank and its subsidiaries, OnLine Mendelian Inheritance in Man and the literature database MEDLINE, through PubMED.

Getting started with NCBI and Entrez

Entrez is accessed via the NCBI homepage and is a simple, user-friendly system. Text search terms (words or Boolean phrases) can be used to search individual databases, and sequences can also be used as queries with utilities such as BLAST. Hits are listed in order of relevance or similarity, with hits on the target database known as neighbors and hits on other databases known as links.

DBGET/LinkDB

DBGET is a data retrieval tool maintained by Kyoto University and the University of Tokyo. It covers more than 20 databases and is closely associated with the Kyoto Encyclopedia of Genes and Genomes. A related system LinkDB finds relationships between entries in the various databases covered by DBGET and others. DBGET has a simpler and more limited search format than Entrez.

SRS and Entrez/DBGET

SRS (sequence retrieval system) is a data retrieval tool that, like Entrez and DBGET, can be used over the WWW. However, unlike these other systems, SRS is open source software and can be installed and run on a local computer network.

Using SRS

SRS databases are grouped but use different principles to those used by Entrez and DBGET. For example, all sequences (nucleic acid and protein) are grouped together, while these are separated by Entrez. The use of SRS involves selecting one or more of these groupings and, within each selected group, selecting one or more of the available databases. Queries can be submitted using two styles of query form, standard or extended.

Installing SRS

The advantage of SRS is that it can be installed locally. This allows SRS to be tuned to local databases that use novel data formats. The tuning of SRS to deal with local databases involves programming in SRS's own scripting language, Icarus.

Related sections

Data and databases (D)	Transcript profiling (K1)
Sequence analysis (J3)	Proteomics techniques (L1)
Domain families and databases (J5)	

Prinoiplcs of DNA sequencing

The order of nucleotides in DNA can be determined by **chain termination sequencing** (also called **dideoxy sequencing** or the **Sanger method** after its inventor). This is the method by which most genome sequences have been obtained. The basic sequencing reaction, which is summarized in *Fig. 1*, consists of a **single-stranded** DNA template, a **primer** to initiate the nascent chain, four deoxyribonucleoside triphosphates (dATP, dCTP, dGTP, and dTTP), and the enzyme **DNA polymerase**, which inserts the complementary nucleotides in the nascent DNA strand using the template as a guide. Four DNA polymerase reactions are set up, each containing a small amount of one of four dideoxyribonucleoside triphosphates (ddATP, ddCTP, ddGTP or ddTTP). These

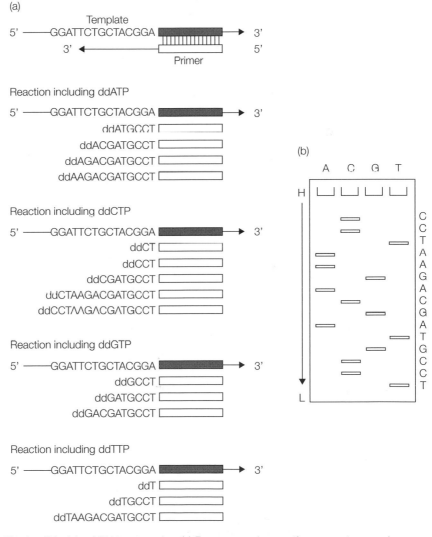

Fig. 1.   *Principle of DNA sequencing. (a) Four sequencing reactions are set up, each containing a limiting amount of one of the four dideoxynucleotides. Each reaction generates a nested set of fragments terminating with a specific base as shown. (b) A polyacrylamide gel is shown with each reaction running in a separate lane for clarity. In a typical automated reaction, all reactions would be pooled prior to electrophoresis and the terminal nucleotide determined by scanning for a specific fluorescent tag.*

act as chain-terminating competitive inhibitors of the reaction, but are present in limiting amounts so there is only a small chance the growing chain will terminate at any given position. Therefore, each of the four reaction mixtures generates a **nested set** of DNA fragments, each terminating at a specific base.

Now, most DNA sequencing reactions are **automated**. Each reaction mixture is labeled with a different fluorescent tag (on either the primer or on one of the nucleotide substrates), which allows the terminal base of each fragment to be identified by a scanner. All four reaction mixtures then are pooled and the DNA fragments separated by polyacrylamide gel electrophoresis (PAGE). Smaller DNA fragments travel through the gel faster than larger ones so the nested DNA fragments are separated according to size. The resolution of PAGE allows polynucleotides differing in length by only one residue to be separated. Near the bottom of the gel, the scanner 'reads' the fluorescent tag as each DNA fragment moves past, and this is converted into **trace data**, displayed as a graph comprising colored peaks corresponding to each base (*Fig. 2*).

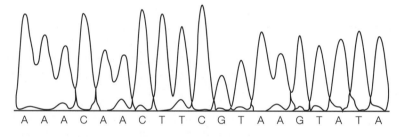

Fig. 2.   A sample of high-quality sequence trace data.

Types of DNA sequences

DNA sequences are stored in databases such as GenBank (see below) and are generally of three types: genomic DNA, cDNA, and recombinant DNA. **Genomic DNA** is taken direct from the genome and contains genes in their natural state, which in eukaryotes includes introns, regulatory elements, and large amounts of surrounding intergenic DNA. In contrast, **cDNA (complementary DNA)** is prepared by reverse transcribing mRNA. It is useful often to focus on cDNA because this provides direct access to genes, which may be difficult to find in genomic DNA. In the human genome for example, genes represent only 3% of the sequence. **Recombinant DNA** includes the sequences of vectors such as plasmids, modified viruses, and other genetic elements that are used in the laboratory. It is essential for these sequences to be stored in databases because linked vector sequences must be removed from new genomic and cDNA sequences prior to alignment to avoid spurious matches.

Genome sequencing strategies

Using current technology, only about 800 nucleotides of DNA sequence can be read in a single reaction. If a larger DNA molecule, such as an entire genome, is to be sequenced, it must be broken up first into smaller pieces, each of which has to be sequenced individually. There are two strategies for assembling genome sequences from short reads. In the **shotgun sequencing** strategy, the large DNA molecule is broken up randomly and many sequencing reactions are performed to cover the entire molecule. The entire sequence is reassembled by using a computer to search for overlaps. This method generates a large amount of sequence data very quickly, but difficulty may be experienced in closing the final gaps, a process termed **finishing**. In the alternative **clone contig** approach, DNA fragments are subcloned in a rational manner and systematically sequenced until the entire sequence is completed. Using this method, sequence data accumulate more slowly at the beginning of a project, but there are fewer sequence gaps at the end so finishing is easier.

New generation sequencing technologies

Solexa is a massively parallel high-throughput-sequencing platform capable of generating one billion bases in a single run at a fraction of the cost of conventional sequencing. It is also highly accurate. Essentially the method involves randomly fragmenting the genomic DNA and attaching the fragments in a flow cell. Amplification of the DNA fragments results in dense clusters of double stranded fragments in each channel of the flow cell. Laser excitation and measurement of the emitted fluorescence identify the sequence of each cluster base-by-base.

A related parallel sequencing method has been introduced recently. **454 sequencing** uses a technology known as **pyrosequencing**. Here clonally amplified DNA fragments are immobilized on beads. This enables the DNA on each bead to be sequenced in parallel.

Sequence quality control

Genome sequencing is subject to strict **quality control** measures. For each clone, the sequencing reaction is performed several times on each strand. Then these primary trace data are subject to **pre-assembly processing**, which includes base calling and quality assessment, the removal of vector sequences, and the masking of repeats. Finally, the sequence reads are assembled into **contigs** (contiguous sequence reads). **Base calling** means deciding to which base a particular trace peak corresponds. Ideally, each peak is clear and unambiguous (*Fig. 2*), but the quality of trace data can vary considerably. Several computer programs are available that automatically make base-calls from sequence traces, and assess the quality (i.e., confidence) of each call. One of the most widely used utilities is **Phred**, which uses dynamic programming (Section J3) to match observed and predicted peak locations and evaluate the quality of the trace data for each base at each position.

The next processing step is to remove vector sequences. Sequence reads usually start with part of the sequencing vector and it is important to remove this sequence because it can generate false overlaps. Programs such as **vector_clip** and **CrossMatch** have been developed, which align new sequence reads to a database of known vector sequences, and then mask off the vector sequence from further analysis. An essential step in genome sequence assembly is the identification of repetitive DNA elements. Genomic DNA is rich in repetitive elements, which can be arranged in tandem or dispersed throughout the genome. Such repeats are extremely troublesome when it comes to reassemble individual reads into contigs because they generate false overlaps. The program **RepeatMasker** is widely used to identify and tag such elements.

Finally, the assembly process itself is carried out using programs such as **Phrap**. This is a computationally difficult task owing to the large number of possible overlaps. For example, if there are 100 reads, there are nearly 40 000 different arrangements. The Phrap assembly method is iterative and based on calculating the log likelihood ratio for each possible match taking into account both the sequence and the trace quality data. Despite the high level of automation, human curation to resolve disputed base calls and other discrepancies is required. Editing and contig assembly can be carried out using programs such as **Staden Gap4**, which allow trace data to be evaluated and the results of multiple sequence reads to be displayed.

Single-pass sequencing

As discussed above, multiple sequencing reactions on both strands of a cloned DNA molecule are usually required to ensure accuracy. However, where large amounts of sequence data are generated (e.g., in genome projects) **single-pass sequencing** is quicker, cheaper, and provides more data at the expense of some accuracy. GenBank has subsidiary databases for such sequences. For example, **dbEST** is a database for **expressed sequence tags (ESTs)**, which are generated by rapid, single-pass sequencing of random clones from cDNA libraries. Although short (2–300 bp) and inaccurate, very large numbers of sequences have been collected rapidly and inexpensively, and can be used to identify genes in genomic DNA, and to prepare large clone sets for DNA microarrays (Section K1). The genomic equivalent of dbEST is **dbGSS**, a database for **genome survey sequences (GSSs)**, which are random, single-pass sequences of genomic DNA. Large amounts of preliminary genomic sequence data (**unfinished sequence**) are stored in the **HTG (high-throughput genomic)** division of GenBank.

Note that the vast majority of ESTs are eukaryotic sequences. This is because cDNA synthesis generally is carried out using a primer that anneals to the **polyadenylate tail** of the eukaryotic mRNA. Most bacterial mRNAs lack a polyadenylate tail and cannot be copied using the same strategy. Furthermore, many bacterial transcripts are never synthesized as full-length molecules owing to their operon organization and the coupling of transcription and translation. However, bacteria also tend to have compact genomes and genes that lack introns, therefore genomic DNA fragments already are very similar to cDNAs.

RNA sequencing

RNA sequencing is less straightforward than DNA sequencing since there are large numbers of **minor nucleotides** (chemically modified nucleotides) (e.g., in tRNA and rRNA). An RNA sequence lacking data on minor nucleotides can be deduced from either the DNA sequence of the corresponding gene or from a cDNA sequence (the sequence of the reverse transcript). In many cases, cDNA sequences are more informative than gene sequences because the RNA molecule may be extensively processed during synthesis. For example, **introns** may be spliced out of a primary transcript to generate a mature mRNA, and extra nucleotides may be inserted by **RNA editing**. Direct RNA sequencing involves the chemical characterization of modified nucleotides. One type of modification, 2'-O-methylation, is easy to pick up, as the phosphodiester bond on the 3'-side of such a nucleotide is not alkali-labile. Other types of modification can be identified by mass spectrometry (see below).

Protein sequencing

Protein sequences can be deduced from DNA sequences, but as with RNA sequencing this does not provide information on modified residues or other types of post-translational protein modification (such as cleavage or the formation of disulfide bonds). In the past, direct protein sequencing was carried out using a process called **Edman degradation**, in which the terminal residue of a protein was labeled, removed, and then identified using a series of chemical tests. Current methods for protein sequencing rely on **mass spectrometry (MS)**, a technique in which the **mass/charge ratio** (m/e or m/z) of ions in a vacuum is accurately determined, allowing molecular masses to be calculated at a resolution in the order of 10^{-5}. Only extremely small quantities of material are required for analysis and the resolution of the technique is such that, for small molecules, the accurate m/e value directly reveals the chemical formula. Hence the technique can be used to characterize minor nucleotides in nucleic acids (see above).

Mass spectrometry of macromolecules such as proteins relies on so-called **soft ionization** methods, which then allow ionization of large molecules without degradation. Two methods are commonly used: **ESI (electrospray ionization)** and **MALDI (matrix assisted laser desorption/ionization)** and these are discussed in more detail in Section L1. The simplest application of MS is to catalogue the m/e values for the proteolytic (usually tryptic) digests of a protein. Then these m/e values are compared with either a database of the predicted tryptic digests of proteins or, given a suitable database of protein sequences, with 'virtual digests' performed in real time on a computer. However, this requires that the protein in question has already been characterized. For the complete sequencing of an unknown peptide by mass spectrometry, the protein can be fragmented randomly in a **collision cell**, within the mass spectrometer, generating a nested set of fragments. Given a knowledge of the molecular masses of all amino acids, calculating the m/e ratios of all these fragments (and more specifically, the differences between them) allows the sequence to be deduced. In reality the problem is complicated by the fact there are two series of fragments generated at the same time: these are known as the **B series** (N-terminal fragments) and the **Y series** (C-terminal fragments). The principle is summarized in *Fig. 3*. The problem can be addressed by labeling the protein with, for example, an N-terminal tag, which would add recognized mass to all fragments in the B series. Note that the amino acids leucine and isoleucine cannot be distinguished by MS because they have the same molecular mass.

Y series

```
                          <--------------- Y2 --------------->
                                         <-------Y1 ------->
      H₂NCH(R¹H)C(=O)---NHCH(R²H)C(=O)--- NHCH(R³H)C(=O)OH

      <---- B1 ---->
      <------------ B2 ------------>
```

B series

Fig. 3.   Illustration of the origin of B and Y series ions in the fragmentation of a hypothetical tripeptide. Peptide bonds are shown as lines. B1 and Y1 are the two smallest fragments (single amino acids) and B2 and Y2 are the dipeptides. In general, a peptide of N residues contains (N-1) peptide bonds and, assuming the collision cell only breaks one peptide bond in each molecule, the Y and B series will each contain (N-1) peptides.

Primary sequence databases

The **primary sequence databases** are repositories for raw sequence data, and can be accessed freely over the World Wide Web. There are three such databases, comprising the **International Nucleotide Sequence Database Collaboration**. These are **GenBank**, maintained by the **National Center for Biotechnology Information (NCBI)**, the **Nucleotide Sequence Database** maintained by the **European Molecular Biology Laboratory (EMBL)**, and the **DNA Databank of Japan (DDBJ)**. New sequences can be deposited in any of the databases since they exchange data on a daily basis.

The databases contain not only sequences but also extensive annotations. As an example, *Fig. 4* shows part of a GenBank file, in this case for the human gene *BTEB*. Much of the introductory part of the file is self-explanatory, containing information such as the locus name, the accession number, the source species, literature references, and the date of submission. An important section of the file is the **features table**, which describes interesting features of the sequence. Since GenBank is a nucleic acid repository, the fact that there is a protein-coding region is an interesting feature. Note that *BTEB* is a very simple gene that has no introns. If there were introns, the CDS (coding sequence) feature would be more complicated: the entry would be extended to indicate the base positions of the exons, delimited by commas. For example, if there were a second exon encoding a further 20 amino acids residues, the CDS feature would read as follows: 1265..19992100..2159.

```
LOCUS       HUMBTEB      4859 bp    mRNA              PRI       07-FEB-1999
DEFINITION  Human mRNA for GC box binding protein, complete cds.
ACCESSION   D31716
VERSION     D31716.1  GI:505081
KEYWORDS    GC box binding protein; zinc finger.
SOURCE      Homo sapiens germline cDNA to mRNA, clone_lib:placenta.
  ORGANISM  Homo sapiens
            Eukaryota; Metazoa; Chordata; Craniata; Vertebrata; Euteleostomi;
            Mammalia; Eutheria; Primates; Catarrhini; Hominidae; Homo.
REFERENCE   1
            {.........}
REFERENCE   2   (bases 1 to 4859)
  AUTHORS   Ohe,N., Yamasaki,Y., Sogawa,K., Inazawa,J., Ariyama,T., Oshimura,M.
            and Fujii-Kuriyama,Y.
  TITLE     Chromosomal localization and cDNA sequence of human BTEB, a GC box
            binding protein
  JOURNAL   Somat. Cell Mol. Genet. 19 (5), 499-503 (1993)
  MEDLINE   94120483
COMMENT     Submitted (31-May-1994) to DDBJ by:
            Yoshiaki Fujii-Kuriyama
            {.........}
FEATURES             Location/Qualifiers
     source          1..4859
                     /organism="Homo sapiens"
                     /db_xref="taxon:9606"
                     /clone_lib="placenta"
                     /germline
     gene            1265..1999
                     /gene="BTEB"
     CDS             1265..1999
                     /gene="BTEB"
                     /note="three-times repeated zinc finger motif"
                     /codon_start=1
                     /product="GC box binding protein"
                     /protein_id="BAA06524.1"
                     /db_xref="GI:1060891"
                      translation="MSAAAYMDFVAAQCLVSISNRAAVPEHGVAPDAERLRLPEREVT
                     KEHGDPGDTWKDYCTLVTIAKSLLDLNKYRPIQTPSVCSDSLESPDEDMGSDSDVTTE
                     SGSSPSHSPEERQDPGSAPSPLSLLHPGVAAKGKHASEKRHKCPYSGCGKVYGKSSHL
                     KAHYRVHTGERPFPCTWPDCLKKFSRSDELTRHYRTHTGEKQFRCPLCEKRFMRSDHL
                     TKHARRHTEFHPSMIKRSKKALANAL"
BASE COUNT     1285 a    1111 c    1193 g    1270 t
ORIGIN      Chromosome 9, q13.
        1 cacgttgggt gacataatgg ggtttttta attatagatt cacactgcat ttattcatca
        {...........}
     4801 ttcaccattg tggaatgatg ccctggcttt aaggtttagc tccacatcat gcttctctt
//
```

Fig. 4.   GenBank entry for the human gene BTEB. Nonessential information that has been deleted from the file for the sake of brevity is indicated thus '{........}.'

Subsidiary sequence databases

The main sequence databases have a number of subsidiaries for the storage of particular types of sequence data. For example, **dbEST** is a division of GenBank, which is used to store **expressed sequence tags (ESTs)**, and an example entry in dbEST is shown in *Fig. 5*. Other divisions of GenBank include **dbGSS**, which is used to store single-pass genomic sequences (**genome survey sequences**), **dbSTS**, which is used to store **sequence tagged sites** (unique genomic sequences that can be used as physical markers), and the **HTG (high-throughput genomic) division**, which is used to store unfinished genomic sequence data.

```
LOCUS       T48601           355 bp     mRNA            EST        06-FEB-1995
DEFINITION  yb01a01.s1 Stratagene placenta (#937225) Homo sapiens cDNA clone
            IMAGE:69864 3' similar to similar to gb:S71043_rna1 IG ALPHA-2
            CHAIN C REGION (HUMAN), mRNA sequence.
ACCESSION   T48601
VERSION     T48601.1  GI:650461
KEYWORDS    EST.
SOURCE      human.
  ORGANISM  Homo sapiens
            Eukaryota; Metazoa; Chordata; Craniata; Vertebrata; Euteleostomi;
            Mammalia; Eutheria; Primates; Catarrhini; Hominidae; Homo.
REFERENCE   1  (bases 1 to 355)
  AUTHORS   {....}
TITLE       Generation and analysis of 280,000 human expressed sequence tags
JOURNAL     Genome Res. 6 (9), 807-828 (1996)
MEDLINE     97044478
COMMENT     Other_ESTs: yb01a01.r1
            Contact: Wilson RK
            Washington University School of Medicine
            4444 Forest Park Parkway, Box 8501, St. Louis, MO 63108
            Tel: 314 286 1800
            Fax: 314 286 1810
            Email: est@watson.wustl.edu
            High qality sequence stops: 277
            Source: IMAGE Consortium, LLNL
            This clone is available royalty-free through LLNL ; contact the
            IMAGE Consortium (info@image.llnl.gov) for further information.
            Seq primer: -21m13
            High quality sequence stop: 277.
FEATURES            Location/Qualifiers
     source         1..355
                    /organism="Homo sapiens"
                    /db_xref="GDB:490761"
                    /db_xref="taxon:9606"
                    /clone="IMAGE:69864"
                    /sex="male"
                    /clone_lib="Stratagene placenta (#937225)"
                    /lab_host="SOLR cells (kanamycin resistant)"
                    /note="Organ: placenta; Vector: pBluescript SK-; Site_1:
                    EcoRI; Site_2: XhoI; Cloned unidirectionally.  Primer:
                    Oligo dT. Caucasian. Average insert size: 1.2 kb; Uni-ZAP
                    XR Vector; ~5' adaptor sequence: 5' GAATTCGGCACGAG 3' ~3'
                    adaptor sequence: 5' CTCGAGTTTTTTTTTTTTTTTTTT 3'"
BASE COUNT        62 a      117 c       98 g       69 t        9 others
ORIGIN
        1 ggcggctcag tagcaggtgc cgtccacctc cgccatgaca acagacacat tgacatgggt
       61 gggtttacca ccaagcgtcc gatggtcttc tgtgtgaagg ccagccaggc gcctccatgg
      121 caccatgcag gagaaggnct cccccttctt ccagtcctcg gctgccacgc gcagtatgct
      181 ggtcacacga aggtcgtggt gccctggctg gntcctncan ggatgcccaa gtcaggtact
      241 tntcgcgggg cagctcctgt gacccctgca gccagcgaac cagcacgtcc ttggggcttn
      301 aagcngcgct accaggcact tcaaccgttc nccagcttcg ttcagggcca ncttc
//
```

Fig. 5. GenBank (dbEST) entry for a human EST clone. Nonessential information that has been deleted from the file for the sake of brevity is indicated thus '{........}.'

Submission of sequences

The robustness of data submitted to the primary sequence databases is important in the context of bioinformatics software. Clearly the integrity of the scientists who submit the data is not checked readily by computers but errors must be avoided in database consistency. It is essential that the data are submitted in a supported format and that the submission is carried out by means of software provided by the database curators. Examples are **WebIn** provided by EMBL (www.ebi.ac.uk/embl/Submission) and **BankIt** provided by the NCBI (http://www.ncbi.nlm.nih.gov/BankIt/), each of which can be used to submit sequences to the databases over the WWW. A powerful stand-alone software tool **Sequin** is provided by the NCBI, and can be used on UNIX, PC/Windows, and MAC systems for sequence submission for those with no WWW access (http://www.ncbi.nlm.nih.gov/Sequin/index.html).

SWISS-PROT, TrEMBL, and UniProt

SWISS-PROT and the related database **TrEMBL** are repositories for annotated protein sequences. *Fig. 6* shows the SWISS-PROT entry for the BTEB protein, corresponding to the GenBank entry in *Fig. 4*. The entry contains large numbers of annotations including a **features table** before the sequence. Each line begins with two letters, many of which are self-explanatory; for example, ID (identity), AC (accession number), DT (date), DE (description), GN (gene name), CC (comment). Continuation lines are indicated by the symbols '-!-' at the start of a section and indents thereafter (this is shown in the CC field in *Fig. 6*). Characteristic features of SWISS-PROT entries include the DR (reference), KW (key words), and FT (features) fields. It is the presence of these careful and extensive annotations that makes SWISS-PROT so popular with biochemists. For example, in *Fig. 6*, there is a fairly comprehensive description of the protein and its function, but also (in the DT field) cross-references to the relevant entries in the secondary databases PROSITE, PRINTS, and Pfam (Section J5).

SWISS-PROT provides the most up-to-date and extensively annotated information on protein sequences and its quality reflects its active management by human curators. TrEMBL (translated EMBL) is another database in the same format. The entries in TrEMBL are derived from translation of all coding sequences in the EMBL Nucleotide Sequence Database that are not already in SWISS-PROT. As further data ensure the reliability of annotations, TrEMBL entries are moved to SWISS-PROT.

SWISS-PROT and TREMBL have been incorporated now into the UniProt (Universal Protein Resource) along with the PIR database. UniProt has three components. The first (UniRef) improves search speeds by combining similar sequences together in a single record. The second (UniParc) keeps a record of the history of the sequences. The third (UniMes) contains a record of metagenomic and environmental data. The resulting UniProt database provides a resource for curated protein information, including functions, classifications, and cross-references.

Database interrogation

Detailed queries of the text annotation in the databases discussed above can be carried out using tools like SRS and Entrez. However, a comparison of Figs 4 and 6 shows how a user can cross-reference between these databases. Let us assume, for example, that searching GenBank with a new sequence obtained in the laboratory identifies the gene in *Fig. 4* as particularly interesting for a research project. How do we find the corresponding entry in SWISS-PROT? Note that the IDs are different and, although SWISS-PROT has alternative GN entries (gene names), they do not correspond to the name on the GenBank file. The way to

```
ID   BTE1_HUMAN      STANDARD;      PRT;    244 AA.
AC   Q13886; Q16196;
DT   15-DEC-1998 (Rel. 37, Created)
DT   15-DEC-1998 (Rel. 37, Last sequence update)
DT   20-AUG-2001 (Rel. 40, Last annotation update)
DE   TRANSCRIPTION FACTOR BTEB1 (BASIC TRANSCRIPTION ELEMENT BINDING
DE   PROTEIN 1) (GC BOX BINDING PROTEIN 1) (KRUEPPEL-LIKE FACTOR 9).
GN   BTEB1 OR BTEB OR KLF9.
OS   Homo sapiens (Human).
OC   Eukaryota; Metazoa; Chordata; Craniata; Vertebrata; Euteleostomi;
OC   Mammalia; Eutheria; Primates; Catarrhini; Hominidae; Homo.
OX   NCBI_TaxID=9606;
RN   [1]
RP   SEQUENCE FROM N.A.
RX   MEDLINE=94120483; PubMed=8291025;
RA   Ohe N., Yamasaki Y., Sogawa K., Inazawa J., Ariyama T., Oshimura M.,
RA   Fujii-Kuriyama Y.;
RT   "Chromosomal localization and cDNA sequence of human BTEB, a GC box
RT   binding protein.";
RL   Somat. Cell Mol. Genet. 19:499-503(1993).
RN   [2]
RP   SEQUENCE OF 1-31 FROM N.A.
RX   MEDLINE=94327649; PubMed=8051167;
RA   Imataka H., Nakayama K., Yasumoto K., Mizuno A., Fujii-Kuriyama Y.,
RA   Hayami M.;
RT   "Cell-specific translational control of transcription factor BTEB
RT   expression. The role of an upstream AUG in the 5'-untranslated
RT   region.";
RL   J. Biol. Chem. 269:20668-20673(1994).
CC   -!- FUNCTION: TRANSCRIPTION FACTOR THAT BINDS TO GC BOX PROMOTER
CC       ELEMENTS. SELECTIVELY ACTIVATES MRNA SYNTHESIS FROM GENES
CC       CONTAINING TANDEM REPEATS OF GC BOXES BUT REPRESSES GENES WITH
CC       A SINGLE GC BOX.
CC   -!- SUBCELLULAR LOCATION: NUCLEAR.
CC   --------------------------------------------------------------------------
CC   This SWISS-PROT entry is copyright. It is produced through a collaboration
CC   between  the Swiss Institute of Bioinformatics  and the  EMBL outstation -
CC   the European Bioinformatics Institute.  There are no  restrictions on  its
CC   use  by  non-profit  institutions as long  as its content  is  in  no  way
CC   modified and this statement is not removed.  Usage  by  and for commercial
CC   entities requires a license agreement (See http://www.isb-sib.ch/announce/
CC   or send an email to license@isb-sib.ch).
CC   --------------------------------------------------------------------------
DR   EMBL; D31716; BAA06524.1; -.
DR   EMBL; S72504; AAD14110.1; -.
DR   MIM; 602902; -.
DR   InterPro; IPR000822; Znf-C2H2.
DR   Pfam; PF00096; zf-C2H2; 3.
DR   PRINTS; PR00048; ZINCFINGER.
DR   SMART; SM00355; ZnF_C2H2; 3.
DR   PROSITE; PS00028; ZINC_FINGER_C2H2_1; 3.
DR   PROSITE; PS50157; ZINC_FINGER_C2H2_2; 3.
KW   Transcription regulation; DNA-binding; Nuclear protein; Repeat;
KW   Zinc-finger; Metal-binding.
FT   DOMAIN        84     116       ASP/GLU-RICH (ACIDIC).
FT   DOMAIN       143     225       ZINC FINGERS.
FT   ZN_FING      143     167       C2H2-TYPE.
FT   ZN_FING      173     197       C2H2-TYPE.
FT   ZN_FING      203     225       C2H2-TYPE.
SQ   SEQUENCE   244 AA;  27234 MW;  2D1B5A5BB9D42221 CRC64;
     MSAAAYMDFV AAQCLVSISN RAAVPEHGVA PDAERLRLPE REVTKEHGDP GDTWKDYCTL
     VTIAKSLLDL NKYRPIQTPS VCSDSLESPD EDMGSDSDVT TESGSSPSHS PEERQDPGSA
     PSPLSLLHPG VAAKGKHASE KRHKCPYSGC GKVYGKSSHL KAHYRVHTGE RPFPCTWPDC
     LKKFSRSDEL TRHYRTHTGE KQFRCPLCEK RFMRSDHLTK HARRHTEFHP SMIKRSKKAL
     ANAL
//
```

Fig. 6. SWISS-PROT entry for the human protein BTEB, equivalent to the GenBank entry shown in Fig. 4.

find the correct SWISS-PROT file is to search the SWISS-PROT database for the accession number D31716. Although SWISS-PROT has its own accession number, D31716 can be found as a DR field entry. The SWISS-PROT site or one of its mirrors will locate the entry successfully with D31716 as the search string.

Organism-specific resources

The annotated sequence databases discussed in Sections J1 and J5 are general to all organisms, and contain data relevant to viruses, bacteria, microbial eukaryotes, animals and plants, as well as recombinant molecules produced in the laboratory. However, there are also many databases devoted to particular organisms, and their numbers are increasing as further genome projects are initiated. Typically such databases contain not only sequence data but also information on gene expression, mutant phenotypes, genome maps, genome sequencing projects, and relevant scientific literature, and provide links to resources for obtaining clones and mutants, and for contacting researchers. A selected list of organism-specific databases is provided in *Table 1*, but this represents only a small fraction of the resources available. The user interested in an organism that is not listed in *Table 1* could try using a search engine to find a useful resource, but there are also a number of excellent gateways available on the WWW that provide information on multiple organism-specific resources and links to the relevant sites. A number of these gateways are listed in *Table 2*.

Table 1. A small selection of organism-specific genomic databases available on the WWW. These databases are curated actively by members of the research community working on the particular organism of interest, and generally include links to organism-specific resources such as clone sets and mutant strains

Organism	Database/Resource	URL
Escherichia coli	EcoGene	http://ecogene.org/
	EcoCyc (Encyclopedia of *E. coli* genes and metabolism)	http://www.ecocyc.org/
	Colibri	http://genolist.pasteur.fr/Colibri/
Bacillus subtilis	SubtiList	http://genolist.pasteur.fr/SubtiList/
Saccharomyces cerevisiae	*Saccharomyces* Genome Database (SGD)	http://genome-www.stanford.edu/Saccharomyces/
Plasmodium falciparum	PlasmoDB	http://PlasmoDB.org
Arabidopsis thaliana	MIPS *Arabidopsis* thaliana Database (MAtDB)	http://mips.gsf.de/proj/thal/db
	The *Arabidopsis* Information Resource (TAIR)	http://www.arabidopsis.org/
Drosophila melanogaster	FlyBase	http://flybase.bio.indiana.edu/
Caenorhabditis elegans	A *C. elegans* DataBase (ACeDB)	http://www.acedb.org/
Mouse	Mouse Genome Database (MGD)	http://www.informatics.jax.org/
Human	OnLine Mendelian Inheritance in Man (OMIM)	http://www.ncbi.nlm.nih.gov/omim

Database formats

Genomics databases need to facilitate the storage and analysis of large amounts of data, but must also have a user-friendly front-end graphical display to allow relevant data to be displayed and analyzed. A number of viewers and browsers have been developed for genomic sequence analysis and annotation (*Table 3*). These include Artemis, Apollo, EnsEMBL, TAIR, and GoldenPath. EnsEMBL is a widely used resource hosting data for a broad number of genomes and has a

Distributed Annotation System (DAS), so that data maintained and updated by groups around the world (on their local servers) can be accessed from the one genome browser. TAIR is a highly regarded resource for *Arabidopsis* and other plant science researchers, and GoldenPath is used particularly by human and medical scientists.

Table 2. Useful gateway sites providing information and links to multiple, organism-specific, and genomic resources

Gateway site	URL
NCBI Genomic Biology	http://www.ncbi.nlm.nih.gov/Genomes/index.html
GOLD (Genomes OnLine Database)	http://www.genomesonline.org/
TIGR Microbial Database	http://cmr.jcvi.org/tigr-scripts/CMR/CmrHomePage.cgi
Bacterial genomes	http://genolist.pasteur.fr/
Yeast databases	http://genome-www.stanford.edu/Saccharomyces/yeast_info.html
EnsEMBL Genome Database Project	http://www.ensembl.org/
MIPS (Munich Information Center for Protein Sequences)	http://mips.gsf.de

One of the more versatile database formats is **ACeDB**. This was originally designed for research on the nematode worm *Caenorhabditis elegans* (**A *C. elegans* DataBase**). AceDB has an object-orientated database architecture (Section D) rather than a simple collection of data, and it has been used to handle data on other organisms including the yeast *Schizosaccharomyces pombe* and humans. AceDB has a graphical user interface with displays and tools designed for genomic data. Other features of AceDB include **AQL (AceDB query language)**, interfaces with Perl and Java, WWW interfaces (of which AceBrowser is the current and supported version), CITA (a CORBA interface to the database), and Acembly (a sequence assembly system). The URL is http://www.acedb.org/

Table 3. Database tools for displaying and annotating genomic sequence data

Viewer format	URL for further information and tutorials
Artemis	http://www.sanger.ac.uk/Software/Artemis
ACeDB	http://www.acedb.org/Tutorial/brief-tutorial.shtml
Apollo	http://apollo.berkeleybop.org/current/install.html
EnsEMBL	http://www.ensembl.org
NCBI map viewer	http://www.ncbi.nlm.nih.gov/mapview/
GoldenPath	http://genome.ucsc.edu/

Finding organism-specific databases

This section provides a number of worked examples of how to find organism-specific databases, resources, and information. A good starting point for this type of search is the Genomes OnLine Database (GOLD). Once the GOLD (http://www.genomesonline.org/) top page has been accessed, the site can be searched for information on any organism by clicking on the 'GOLD tables' link; for example, the bacterium *Escherichia coli*, the fruit fly *Drosophila melanogaster*, and humans.

Ebola virus
Information on Ebola virus can be found by clicking on the c*EBI Viruses* hyperlink under 'Links' on the GOLD top page. This accesses the European Bioinformatics

Institute page of completed viral genomes, which currently lists over 1500 viruses whose genomes have been sequenced, together with the corresponding EMBL Nucleotide Sequence Database file, the sequence in FASTA format. The URL is http://www.ebi.ac.uk/genomes/virus.html

Escherichia coli

Entering *Escherichia coli* as a search term after accessing the SEARCH GOLD query form under the 'GOLD tables' link finds records on 64 different bacterial strains. Resources are listed in a table with the following headings: *Organism, Type, Information, Size, DATA, Institution, Genome Database,* and *Project Status*. Of these, the links provided under the headings *Information* and *Genome Database* are the most useful. There are more than 20 resources listed including some general ones (e.g., NCBI and SWISS-PROT) and some specific ones (e.g., EcoCyc, EcoGene, Colibri, which are also shown in *Table 1*). The *E. coli* resources can be accessed directly via hyperlinks, and many of the sites have lists of further resources. For example, the EcoCyc page, the Encyclopedia of *E. coli* genes and metabolism, contains a link to '*E. coli web links*' which is a page containing further links to *E. coli* resources on the WWW. Similarly, the Colibri site has a link 'other sites related to *E. coli*' which lists 18 additional resources.

Drosophila melanogaster

When *Drosophila melanogaster* was used as a search term, the *Genome Database* links provided by GOLD were as follows: NCBI, BDGP (Berkeley *Drosophila* Genome Project), IMG (Integrated Microbial Genomics), FlyBase-USA, J. Craig Venter Institute, and IBM-Annotation. FlyBase-USA links to http://flybase.bio. indiana.edu/ and provides a large number of *Drosophila* resources, including annotated sequences, raw sequence data in FASTA format, sequence sets, and access to sequence analysis tools.

Human

When *Homo sapiens* was used as a search term, links provided by GOLD were as follows: NCBI, ORNL, EnsEMBL, Sanger Centre, IMG, HOWDY, J. Craig Venter Institute, and IBM-Annotation. Clicking on NCBI links to the NCBI human genome resources page, which contains images of human chromosomes, the user can select from a drop-down menu of clones, genes, physical maps, genetic maps, and variation, then click on any of the chromosomes to see the information available for the chromosome chosen under that heading. The NCBI server also provides access to UniGene, OnLine Mendelian Inheritance in Man (OMIM), and other miscellaneous databases.

Database resources

Databases are essentially large storage devices for scientific and other data. They can be searched and cross-referenced either over the Internet or using downloaded versions on local computers or computer networks. Specific types of database are discussed in different sections throughout this book. For example, the three primary nucleic acid databases (GenBank, the EMBL Nucleotide Sequence Database, and the DNA Databank of Japan) are called **primary databases** because they store raw sequence data. Similarly, SWISS-PROT and TrEMBL are the major primary databases for the storage of protein sequences. There are also secondary databases of protein families and sequence patterns, such as PROSITE, PRINTS and BLOCKS (Section J5). These are called **secondary databases** because the sequences they contain are not raw data, but have been derived from the data in the primary databases. There are also many organism-specific databases containing information, links, and resources dedicated to particular species. In this section, we discuss some of the remaining database resources available, which can be grouped under the description **miscellaneous databases**. Note that the journal *Nucleic Acids Research* devotes its first issue every year to articles describing new databases and updates to existing ones. These articles can be accessed online at the following URL: http://www.nar.oupjournals.org/

Specialized sequence databases

The primary sequence databases are unbiased as to the type of sequence data they contain. However, a number of more specialized databases have been developed with particular types of nucleic acid or protein sequence in mind. For example, there are databases specifically for rRNA and tRNA sequences (e.g., the database of 5S ribosomal RNA sequences) (http://biobases.ibch.poznan.pl/5SData/), and the database of small subunit rRNA sequences (http://bioinformatics. psb.ugent.be/webtools/rRNA/ssu/). Further examples include databases for promoter sequences and other transcriptional regulatory elements, databases for regulatory elements in the non-coding region of mRNAs, a database of scaffold/ matrix attached regions (S/MARt D), and InBase, a database of inteins, which are small peptides that are spliced out of some microbial proteins (http://www.neb. com/neb/inteins.html).

OMIM

OMIM (Online Mendelian Inheritance in Man) is a comprehensive database of human genes and genetic disorders maintained by the NCBI, and can be accessed at the following URL: http://www.ncbi.nlm.nih.gov/omim or through Entrez (the NCBI query system). Each OMIM entry has a full text summary of a gene or genetically determined phenotype and has numerous links to other databases, such as the primary sequence databases SWISS-PROT, PubMed references, general and locus-specific mutation databases, gene nomenclature databases, and mapviewer. OMIM is an excellent starting point to find information on human genetics. An example of an OMIM file is shown in *Fig. 7*.

UniGene

UniGene is another resource for genome research. In the words of its developers: 'UniGene is an experimental system for automatically partitioning GenBank sequences into a non-redundant set of gene-oriented clusters. Each UniGene cluster contains sequences that represent a unique gene, as well as related information such as the tissue types in which the gene has been expressed and its map location.' UniGene incorporates about 10^5 ESTs and is used by experimenters to design probes and reagents for gene mapping and expression analysis. The organisms included in UniGene were chosen on the basis of the availability of large amounts of EST data, and to give a reasonable coverage of the vertebrate and plant kingdoms with examples of closely and distantly related species. These include human, mouse, cow, rat, zebrafish, *Xenopus*, wheat, rice, barley, maize, and *Arabidopsis*.

Structural databases

Structural databases store data on protein (and nucleic acid) structure. The primary resource for protein structure data is the **Protein Data Bank (PDB)** available at the following URL: http://www.pdb.org/ This is the single worldwide archive of structural data and is maintained by the **Research Collaboratory for Structural Bioinformatics (RCSB)**, at Rutgers University. The associated **Nucleic Acid Data Bank (NDB)** is also maintained there. Data from both X-ray crystallography and NMR spectroscopy studies can be deposited in the PDB, using a web-based

```
1: *602902
BASIC TRANSCRIPTION ELEMENT-BINDING PROTEIN 1; BTEB1

Alternative titles; symbols
BTEB

Gene map locus 9q13

TEXT
The GC box is a common regulatory DNA element of eukaryotic genes. The
promoter region of rat CYP1A1 (108330) contains a single GC box within a
basic transcriptional element (BTE) required for constitutive expression of
the gene. By screening a liver library for the ability to bind BTE, Imataka
et al. (1992) isolated rat cDNAs encoding Sp1 (189906) and a protein that
they designated BTEB(BTE binding protein). Sequence analysis revealed that,
like Sp1, BTEB contains 3 consecutive zinc finger motifs. In transient
transfection experiments both BTEB and Sp1 stimulated promoters with
repeated GC boxes. However, the CYP1A1 promoter with only 1 GC box was
activated by Sp1 and repressed by BTEB. Ohe et al. (1993) used a rat BTEB
cDNA to screen a human placenta library and isolated cDNAs encoding BTEB1.
The sequences of the predicted 244-amino acid rat and human proteins are 98%
identical. Imataka et al. (1992) and Ohe et al. (1993) found that the mRNAs
encoding BTEB and BTEB1 contain a GC-rich leader sequence in the 5-prime
untranslated region that has the potential to form stem-loop structures and
that may control translation.

By analysis of a somatic cell hybrid panel and by fluorescence in situ
hybridization, Ohe et al. (1993) mapped the BTEB1 gene to 9q13.


REFERENCES
1. Imataka, H.; Sogawa, K.; Yasumoto, K.; Kikuchi, Y.; Sasano, K.;
Kobayashi, A.; Hayami, M.; Fujii-Kuriyama, Y. : Two regulatory proteins that
bind to the basic transcription element (BTE), a GC box sequence in the
promoter region of the rat P-4501A1 gene. EMBO J. 11: 3663-3671, 1992.
PubMed ID : 1356762

2. Ohe, N.; Yamasaki, Y.; Sogawa, K.; Inazawa, J.; Ariyama, T.; Oshimura,
M.; Fujii-Kuriyama, Y. : Chromosomal localization and cDNA sequence of human
BTEB, a GC box binding protein. Somat. Cell Molec. Genet. 19: 499-503, 1993.
PubMed ID : 8291025


CREATION DATE
Rebekah S. Rasooly : 7/29/1998

EDIT HISTORY
alopez : 7/29/1998


Copyright (c) 2000 Johns Hopkins University
```

Fig. 7. The OMIM file for the human gene BTEB.

interface called the **AutoDep Input Tool (ADIT)**. The data are extensively checked and verified by human curators before acceptance. An equivalent European database is the Macromolecular Structure Database (MSD) maintained by the European Bioinformatics Institute. The RSCB and MSD databases contain the same data.

There are also other structural databases such as Entrez's **Molecular Modeling Database (MMDB),** which aims to provide information on sequence and structure neighbors, links between the scientific literature and 3-D structures, and sequence and structure visualization.

Proteins and higher-order functions

One important aim of bioinformatics is to use biological data to understand the higher-level functions of the cell; that is, biochemical pathways, regulatory networks, signal transduction pathways, and how these influence cell and organism behavior. A number of databases have been established with this goal in mind. Several databases have been designed, for example, to provide information on the functional annotation of proteins. Such databases include **PIR (Protein Information Resource),** which can be accessed at http://pir.georgetown.edu/ Another valuable resource is the **Kyoto Encyclopedia of Genes and Genomes (KEGG)**, which is the primary resource of the Japanese GenomeNet service. KEGG is available at the following URL: http://www.genome.ad.jp/kegg/ The main database integrates a number of subsidiaries including **PATHWAY** (which stores data on molecular pathways and complexes), **GENES** (which stores functional information about genes and their products), and **LIGAND** (which stores data about chemical compounds and reactions occurring in the cell). Together, these data can be used for functional annotation and the grouping of genes and proteins into common pathways, networks, and hierarchies.

There are also many databases that provide information on specific aspects of protein function. For example, **DIP (the Database of Interacting Proteins)** and **BIND (Biomolecular Interaction Network Database)** provide functional annotations of proteins on the basis of their interactions with each other and with other ligands. There are also databases for particular types of protein; for example, **ReBase**, a database of restriction endonucleases and their target sites; **TRANSFAC**, a database of transcription factors; **Sentra**, a database of signal transduction proteins; and **NUREBASE**, a database of nuclear receptors.

Literature databases

A literature database contains the abstracts and, in some cases, the full text and figures of published scientific articles. Such databases can be searched using text strings to find words in the title, abstract, keywords, or main text, or by author or author's institution. One of the earliest comprehensive online library resources was **Medline**, which has been incorporated into a large resource called **PubMed** maintained by the NCBI. They are integrated into the NCBI's Entrez suite of databases. **Web of Science** is another useful resource but requires an institutional subscription.

Access to distributed data

A large amount of biological information is available over the WWW, but the data are widely distributed and it is therefore necessary for scientists to have efficient mechanisms for **data retrieval**. One approach is to use standard **search engines** to find relevant web pages. However, it is sometimes difficult to find the desired information using this method, especially if the chosen search term has other connotations and pulls out many irrelevant sites. Alternatively, there are a number of dedicated **data retrieval tools** that can be used to access information for molecular biologists. The most widely used of these are **Entrez**, **DBGET**, and **SRS (Sequence Retrieval System)**. Each of these tools allows text-based searching of a number of linked databases as well as sequence searching with utilities such as BLAST (Section J3). They differ in the databases they cover and how the retrieved information is accessed and presented.

Entrez

Entrez is a WWW-based data retrieval system, developed by the National Center for Biotechnology Information (NCBI), which integrates information held in all NCBI databases. These databases include nucleotide sequences (from GenBank and its subsidiaries), protein sequences, macromolecular structures, and whole genomes. Other resources linked to the NCBI can also be searched using Entrez. These include OnLine Mendelian Inheritance in Man (OMIM), and the literature database MEDLINE, through PubMed. Entrez can be accessed via the NCBI web site at the following URL: http:///www.ncbi.nlm.nih.gov/Entrez/ In total, Entrez links to 11 databases, which are listed in *Table 4*.

Table 4.   *The databases covered by Entrez, listed by category*

Category	Database
Nucleic acid sequences	Entrez nucleotides: sequences obtained from Genbank, RefSeq, and PDB. Also UniGene, PopSet, Probe, Trace Archive, PA, UniST, dbEST, dbGSS, dbSNP, dbST, HomoloGene, and MGC
Protein sequences	Entrez protein: sequences obtained from SWISS-PROT, PIR, PRF, PDB, and translations from annotated coding regions in GenBank and RefSeq. Also 3D domains, Protein Clusters, and PROW
3D structures	Entrez Molecular Modeling Database (MMDB). Also 3D domains
Genomes	Complete genome assemblies from many sources
OMIM	OnLine Mendelian Inheritance in Man
Taxonomy	NCBI Taxonomy Database
Books	Bookshelf
Expression databases	Gene Expression Omnibus (GEO), SAGE
Literature	PubMed

Getting started with NCBI and Entrez

Entrez is the common front-end to all the databases maintained by the NCBI and is an extremely easy system to use. The Entrez main page, as with all NCBI pages, is undemanding in its browser requirements and downloads quickly. Part of the front page is illustrated in *Fig. 8*. The databases available for searching can be accessed by hyperlinks, or by using the search box as shown. The search term

Fig. 8.   *The Entrez main page, showing the drop-down menu of available databases and the search field.*

may be a single word or a Boolean phrase. Clicking on 'GO' initiates the search. The number of matching records in the Entrez databases are shown. Clicking the number takes you to the database records.

For the newcomer, the following URL provides an overview of Entrez and a useful tutorial: http://www.ncbi.nlm.nih.gov/Database/index.html This page is shown in *Fig. 9*, and includes a diagram illustrating the connectivity between eight of Entrez's databases.

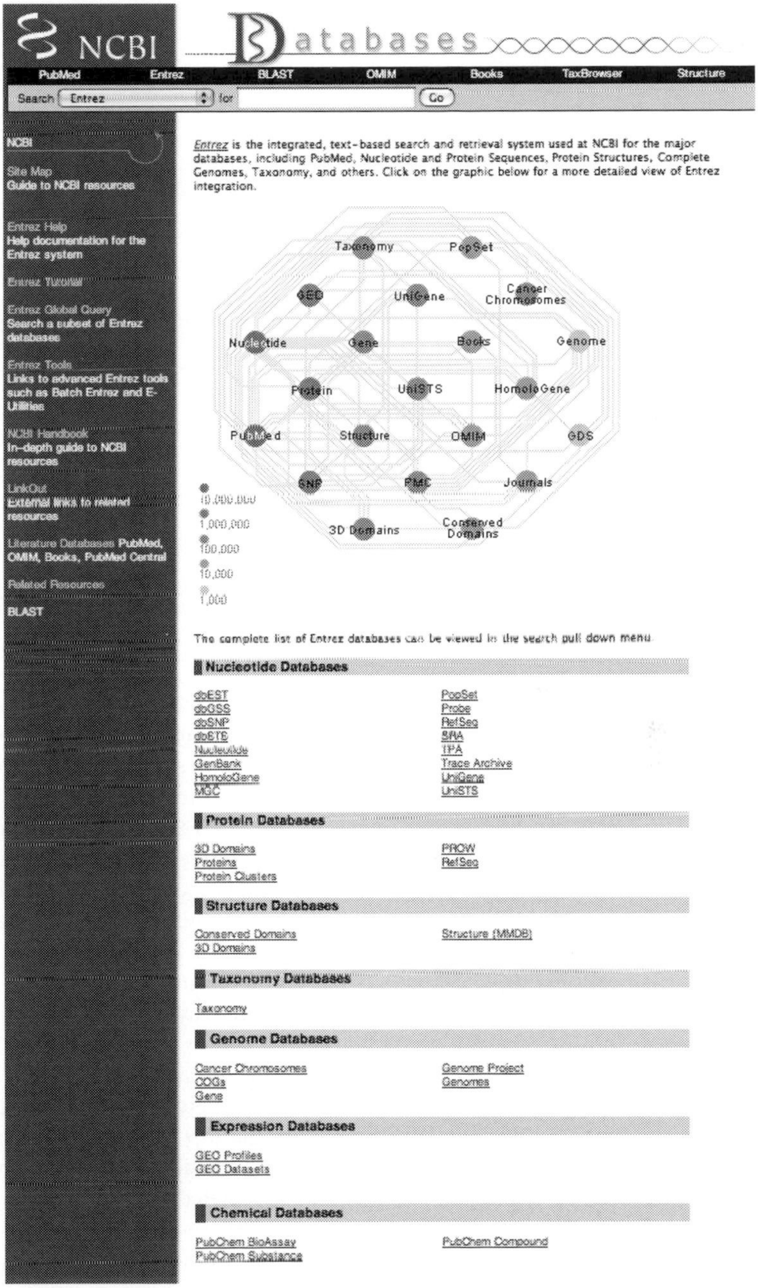

Fig. 9.   The Entrez overview page, showing the tutorial link and the relationship between eight Entrez databases.

DBGET/LinkDB   DBGET is an integrated data retrieval system developed and jointly maintained by the Institute for Chemical Research (Kyoto University) and the Human Genome Center (University of Tokyo). It is integrated with more than 30 databases (*Table 5*), which can be searched one at a time or in combination. Hits are presented as a list of results together with any available associated information. **LinkDB** is an associated database of links (**binary relationships**) between entries in the different databases available to DBGET and also further organism-specific databases, such as AceDB, Flybase, and SGD. DBGET is associated closely with KEGG, the Kyoto Encyclopedia of Genes and Genomes, which is maintained by the same group.

Table 5.   *The databases covered by DBGET/LinkDB, listed by category*

Category	Database
Nucleic acid sequences	GenBank, EMBL, RefSeq
Protein sequences	SWISS-PROT, PIR, PRF, PDBSTR, UniProt
3D structures	PDB
Sequence motifs	PROSITE, EPD, TRANSFAC, BLOCKS, PRODOM, PRINTS, PFAM
Enzyme reactions	LIGAND
Metabolic pathways	KEGG
Amino acid mutations	PMD
Amino acid indices	AAindex
Genetic diseases	OMIM
Literature	LITDB Medline
Organism-specific gene catalogs	*E. coli, H.influenzae, M.genitalium, M.pneumoniae, M.jannaschii, Synechocystis, S.cerevisiae*

SRS and Entrez/ DBGET   SRS (Sequence Retrieval System) is a retrieval tool developed by the European Bioinformatics Institute (EBI) that integrates over 80 molecular biology databases. These are listed at http://srs6.ebi.ac.uk/srs6bin/cgi-bin/wgetz?-page+databanks+-newId and are summarized in *Table 6*. Like Entrez and DBGET, SRS can be used over the WWW. However, the difference between SRS and Entrez/DBGET is that SRS is **open source software** that can be downloaded and installed locally. The result is that the databases and utilities that are available to the end user are not restricted by the activities of curators (at the NCBI in the case of Entrez, or GenomeNet in the case of DBGET). Several large sites, including SWISS-PROT, use SRS as standard. The main ExPASy site (www.expasy.ch), one of the most useful bioinformatics gateways, is an example of SRS in action.

Using SRS   To start using SRS, access the appropriate site on your local network (if SRS has been installed locally) or alternatively try the SRS homepage at http://srs6. ebi.ac.uk On the WWW version, the top page lists 14 **classifications** under the 'library page' link. By clicking on the adjacent '+' symbol, each classification can be expanded to reveal the associated databases, and by clicking on the '–' symbol, the classifications can be collapsed. It is possible to expand or collapse all classifications. Most of the classifications are self-explanatory, but new users may find some unfamiliar. For example, *Protein function, structure and interaction databases* (*see Fig. 10*) refers to the secondary databases of protein motifs.

Table 6.   The databases covered by the SRS at http://srs6.ebi.ac.uk, listed by category

SRS description	Examples
Literature	MEDLINE TAXONOMY, OMIM
Nucleotide sequence databases	EMBL, RefSeq,
Nucleotide sequence related	TFSITE, TFFACTOR, REBASE
Protein sequence databases	UNIPROT, REFSEQ, IPI
Protein function, structure and interaction databases	INTERPRO, PRODOM, PRINTS, BLOCKS, PFAMHMMFS, PROSITE, PDB, FSSP, EXPERIMENT, INTERACTION, INTERACTOR
TransFac	TFSITE, TFFACTOR, TFCELL, TFCLASS, TFMATRIX ,TFGENE
Enzymes reactions and metabolic pathways	ENZYME, UPATHWAYM UENZYME, UREACTION, UPATHWAY
Mutation and SNP databases	HGVBASE
User-owned databanks	USERDNA, USERPROTEIN
Application results	FASTA, FASTX, FASTY, NFASTA, BLASTP, BLASTN, CLUSTALW, NCLUSTALW, PPSEARCH, RESTRICTIONMAP, PSIBLAST, HMMPFAM
EMBOSS result databases	Including: ANTIGENIC, BACKTRANSEQ, BIOSEDN, RESTRICT, MERGER, ETC.

To search with SRS, expand the relevant classifications and check boxes corresponding to the required databases as shown in *Fig. 10*. Clicking on the 'Query Form' tab then brings up the query form. Search terms can be entered in a number of fields by selecting from the drop-down menu (e.g., accession number,

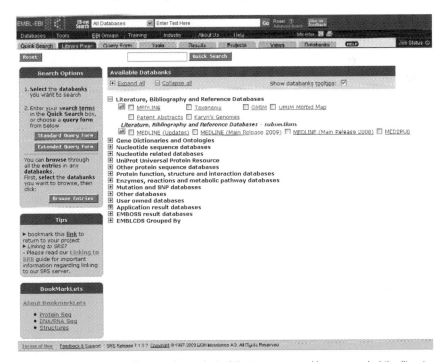

Fig. 10. The SRS top page at http://srs6.ebi.ac.uk The user has selected the top page and has expanded the literature classification. One or more of the databases (MEDLINE, MEDLINENEW, and GO) can be selected and searched now.

description, keywords, organism) or alternatively it is possible to search all fields simultaneously. Up to four different search terms can be used, linked by Boolean operators across multiple fields, so quite specific searches can be carried out. After clicking the 'Search' button, the query results are displayed as a table, with hyperlinks to files in the appropriate databases. A helpful tutorial for SRS users is available at http://srs6.ebi.ac.uk/srs6bin/cgi-bin/wgetz?-page+docoPage+-id+4uG9h1IIeZt+-e+[srsbooks-id:usr1_1].

Installing SRS

SRS is available to be installed locally. Essentially it provides both the graphical interface and files to allow databases (such as those in *Fig. 10*) to be accessed. SRS works by constructing indexing files, and should be updated regularly. The installation set comes with several configuration files and module files for the most popular bioinformatic databases ('databanks' in SRS). SRS configuration files establish which databases are available to SRS. Some of these files are used to add databases and others specify a grouping mechanism for databases or create new built-in views for data results. Module files correspond to the databases to be indexed.

Thus an SRS system can be installed and operated with the minimum of effort. However, there are several points to consider when planning (or asking for help with) an SRS installation. Although the databases may all be remote, the indexing files are very large, and on a PC or other small system, the indexing process can take a lot of processor time (days). Nevertheless a modest SRS system using example configuration module files is easily maintained and updated. The strength of SRS lies, in part, with providing the installer the capacity to use databases of his or her own, even if they use a novel format. In order to do this, SRS has an associated scripting language called **Icarus**.

J2 GENOME ANNOTATION

Key Notes

Annotation	Annotation means obtaining useful information; that is, the structure and function of genes and other genetic elements, from raw sequence data. Owing to differences in gene structure and genome organization, the annotation of prokaryote and eukaryote genomes involves different problems.
Finding genes by computer	Computers can be used to predict where the genes are in genomic DNA. This is achieved by a combination of signal sensing (looking for conserved motifs), content sensing (looking for regions with gene-like sequence context), and homology searching (looking for regions that match the sequences of previously discovered genes). However, no gene-finding method is 100% reliable.
Detecting open reading frames (ORFs)	In bacterial genomes, genes are rarely interrupted by introns. Therefore, a useful way to detect genes is to carry out a six-frame translation of the genome sequence and identify long open reading frames. This method detects most genes, but more sophisticated approaches are required to detect short genes, genes that use variations of the genetic code, and overlapping genes.
Detecting exons and introns	In higher eukaryote genomes, genes are widely dispersed and interrupted by large and numerous introns. Exons are too short to identify by ORF searching alone, so a combined approach involving exon detection by content sensing, the identification of splice signals, and the incorporation of auxiliary information such as cDNA sequences, is required to build a whole-gene model. Even so, many genes remain undetected or incorrectly delimited, and nonfunctional full or partial gene copies (pseudogenes) may be incorrectly identified as genuine genes.
RNA genes	Special methods are required to identify RNA genes since these do not contain a coding region. Such methods are based on homology searches and the prediction of secondary structure.
Annotating the human genome	The number of genes in the human genome has been the subject of much study and debate. Since the publication of the genome sequence it seems likely the number is substantially smaller than some of the estimates, which were 120 000 or more. A more likely figure is around 20 000–25 000.
Gene-prediction software	Programs for gene prediction use *ab initio* methods and/or homology searches to identify genes in genomic DNA. Simple programs such as NCBI ORF finder identify open reading frames by carrying out six-frame translations. For complex eukaryotic genomes, more sophisticated statistical analysis methods are required. These may incorporate simple rule-based algorithms (e.g., GeneFinder), neural nets (e.g., GRAIL, FGENESH), or Markov models (e.g., Genie, GenScan). Neural nets and Markov model algorithms must be trained on data from specific organisms.

Measuring predictive accuracy	No gene-finding program is 100% accurate and it is better to use several programs to annotate the same genomic sequence. Predictive accuracy is measured in terms of sensitivity (ability to correctly predict genuine genes, or exons), and specificity (ability to correctly eliminate false genes or exons). Common annotation errors include the detection of false coding regions, the merging of exons, failure to detect true exons (especially 5′ and 3′ exons, or very short exons), and the misplacing of intron/exon boundaries. Not all annotation errors are a result of limitations of gene-prediction software.
Annotation pipelines	The only way to handle the large amount of data from genome projects is to annotate 'on the fly' using a continuous pipeline. Raw sequence data are fed in at one end. They are assembled, verified, and annotated using one or more gene-prediction programs, and then the results are fed into a database that allows sequence and annotations to be viewed using a graphical interface. Suitable formats include Artemis, Apollo, EnsEMBL, GoldenPath, or ACeDB.
Related sections	Databases and data sources (J1) Sequence analysis (J3) Sequence families, alignment, and phylogeny (J4) Domain families and databases (J5)

Annotation

The term annotation means obtaining useful biological information from raw sequence data. Essentially this means finding genes and other functional elements in genomic DNA (structural annotation) and then assigning functions to these sequences (functional annotation). Since it became possible to sequence whole genomes, deriving biological meaning from the long sequences of nucleotides has been a crucial biological research problem. The methods discussed in Sections J3–J5 are applied to all genome sequences for functional annotation. In this and the following section, we consider the impact of bioinformatics on structural annotation.

A point that must be emphasized early on is that annotation involves different problems in prokaryote and eukaryote genomes. In prokaryotes, the gene density is high (i.e., there is little intergenic DNA), and the vast majority of genes have no introns. In eukaryotes, gene density falls and gene complexity increases as the organism itself becomes more complex. In the yeast *Saccharomyces cerevisiae*, for example, about 70% of the genome is made up of genes, and most of these have no introns; about 5% of *S. cerevisiae* genes have a single, small intron. In *Drosophila*, only 25% of the genome is made up of genes. About 20% of genes have no introns, and most genes have between one and four introns. In mammals and higher plants, gene density drops to 1–3%. Only 6% of human genes have no introns, and there are approximately equivalent numbers of genes with between one and 12 introns. A few human genes have more than 50 introns! In the human genome, exons are typically about 150 bp long but introns are more likely to be in the range 1–5 kb. For this reason, finding genes in higher eukaryotic genomes can be very difficult.

Finding genes by computer

There are many hybridization-based techniques and other experimental methods that can be used to detect genes in genomic DNA. However, these can only be applied on a clone-by-clone basis and cannot possibly keep pace with the rapid accumulation of genomic sequence data. Bioinformatics provides a way to fill this information gap by rapidly mining sequence data to identify potential genes. Note the use of the qualifier 'potential.' No computer-based gene-finding method is 100% accurate, and each annotation should be tested in the laboratory.

Many computer algorithms have been developed to annotate genomic sequence data and some of these are discussed in the following section. But what features of genes do these programs identify that lets them discriminate between genes and the surrounding DNA? Essentially, three types of feature are recognized: signals, contents, and homologies.

Signals are discrete, local sequence motifs such as promoters, splice donor and acceptor sites, start and stop codons, and polyadenylation sites. Such motifs tend to have consensus sequences, and algorithms can be developed to search for these either singly or in the context of other surrounding signals. Such algorithms are known as signal sensors. The most sensitive of these employ weight matrices that assign a cost to each of the four nucleotides when they occur at any given position in the motif. The overall score generated by a weight matrix sensor reflects the sum of all costs at all positions, and there will be a threshold value that determines whether the candidate motif is a real signal.

Contents are extended sequences of variable length. The most important content of a gene is usually the coding region, although other sequences may be relevant, such as CpG islands in vertebrate genes. Contents do not have consensus sequences, but they do have conserved features that distinguish them from surrounding DNA. For example, nucleotide frequencies and nucleotide dependencies (the likelihood of two particular nucleotides occurring at two particular positions at the same time) differ between exons and introns or intergenic DNA. This can be tested by content sensor algorithms using statistical models.

Homologies are matches to known genes. More detail about sequence homology can be found in Section J3. With the large amount of sequence data available, especially ESTs that represent genes (Section J1), many programs now incorporate database homology searches (Section J3) as part of the gene-finding algorithm.

The most effective programs incorporate elements of signal and content sensing as well as homology searching to produce an integrated gene-finding package.

Detecting open reading frames (ORFs)

Protein-encoding genes have an open reading frame (ORF), a long series of sense codons bracketed by a start or initiation codon (usually but not exclusively ATG) and a stop codon (or termination codon, or nonsense codon), which may be TGA, TAG or TAA. In bacterial genomes, ORFs are easy to detect generally because they are uninterrupted by introns. The normal procedure is to carry out a so-called six-frame translation (i.e., a translation of the genomic sequence in all six possible reading frames, three forwards and three backwards), and identify the longest ORF in the six possible protein sequences. Long ORFs tend not to occur by chance, so the existence of 300 bp or more of uninterrupted coding sequence in any one reading frame is good evidence for a gene.

While this method is generally sound, some genuine genes may be overlooked completely or the boundaries incorrectly specified. This applies to genes that are shorter than 300 bp (or any other length criterion applied) and genes that use rare variations of the genetic code. The standard genetic code[1] is shown in *Fig. 1*. In most bacterial genes, the initiation codon is ATG, but in some instances the codons GUG or UUG may be used. The frequency of alternative initiation codon usage varies in different bacterial species, so the results for any new bacterial genome sequence cannot be anticipated. ATG is used also as an internal codon, so the misidentification of an internal ATG as a start codon, especially where GUG or UUG is the genuine start codon, may lead to gene truncation from the 5′ end. In such situations, the genuine start site may be indicated by the presence of other signals, such as the Kozak sequence (ribosome binding site). Another example involves the stop codon TGA. In the vast majority of genes, TGA means stop, but in a very few it leads to the incorporation of the rare amino acid selenocysteine. This is dictated by the context of the sequence surrounding the codon, and if not recognized by a gene-finding algorithm, could lead to gene truncation from the 3′ end.

Other quirks of nature that may be missed by gene-finding algorithms are intentional frameshifts, which allow the production of two or more proteins from the same gene by reading through a termination codon, and the existence of shadow genes (overlapping open reading frames on opposite DNA strands). Content sensors, which detect differences in nucleotide frequencies and dependencies, are essential for finding these genes. It is important to note that the genetic code is degenerate (i.e., most amino acids are specified by more than one codon), and that different species show different codon bias (i.e., a preference for particular codons). Content-sensors must therefore be trained using the codon frequency tables appropriate for each species.

Detecting exons and introns

In higher eukaryotes, the detection of protein coding sequences is complicated by the presence of introns, which may interrupt the open reading frame one or more times. Introns are best described as unhelpful to genome annotation. They may separate exons neatly; that is, leaving individual codons intact, but they may also split a codon between the first and second or second and third positions. Exons are generally small (average size 150 bp), which means that they cannot be identified simply by the detection of a long uninterrupted series of sense codons. The annotation of eukaryotic genomes therefore involves a more complex process of predicting exons using content sensors; predicting the positions of splice donor, acceptor, and branch sites to identify introns, and building these into a whole gene model. The use of auxiliary information; for example, cDNA sequences and the presence of promoter elements, is much more important for the correct identification and delimitation of eukaryotic genes than it is for prokaryote genes. The situation in eukaryotes is further complicated by the fact that many genes undergo alternative splicing (so that gene predictions based on cDNA sequences can miss out exons), and that eukaryotic genomes are littered with pseudogenes and gene fragments, which resemble genuine genes but are not expressed. Further data, for example from expression or mutational analysis, is required often before a gene can be confirmed as genuine.

RNA genes

The methods discussed above are suitable for the detection of genes that are translated into protein. However, a number of genes function only at the RNA

---

[1] Note that some bacteria and fungi use modifications of the standard genetic code, and the code also varies in mitochondria. These modifications must be taken into account when annotating; for example, mitochondrial genomes.

Group	Amino acids	Codons	Number	Ambiguities
1	L	CUN, UUR	6	UUG initiation (M)
	R	CGN, AGR	6	None
	A	AAN	4	None
	G	GGN	4	None
	P	CCN	4	None
	S	AGY	4	None
	T	ACN	4	None
	V	GUN	4	GUG initiation (M)
2	I	AUY, AUA	3	None
	*	UAY, UGA	3	UGA internal (Z)
	C	UGY	2	None
	D	GAY	2	None
	E	GAR	2	None
	F	UUY	2	None
	H	CAY	2	None
	K	AAR	2	None
	N	AAY	2	None
	Q	CAR	2	None
	Y	UAY	2	None
3	M	AUG	1	Initiation
	W	UGG	1	None
	Z	UGA	1	*

Fig. 1.   The standard genetic code, tabulated to emphasize degeneracies and ambiguities. N is any nucleotide, Y is any pyrimidine, and R is any purine. Z is the amino acid selenocysteine and '*' represents a stop codon. Group 1 amino acids show complete degeneracy at the third position of the codon. Group 2 amino acids show partial degeneracy (purine or pyrimidine degeneracy) at the third position of the codon. Group 3 amino acids have specific codons.

level. These include ribosomal RNA (rRNA) and transfer RNA (tRNA) genes, and specialized genes such as *XIST*, which is involved in mammalian X-chromosome inactivation, and *H19*, which is involved in genomic imprinting. How are these genes identified?

Generally, the detection of RNA genes relies on homology searching. Such genes tend to be highly conserved because the RNAs have a defined tertiary structure, which requires extensive intramolecular base-pairing. There are also specialized algorithms for detecting RNA genes, such as **tRNAScan-SE**, which search for sequences with the potential to form secondary structures such as stem-loops and hairpins.

Annotating the human genome

The number of protein-coding genes in the human genome was always a source of intense speculation and discussion right up to the publication of the first draft complete genome sequence early in 2001. Clearly the question of how many genes are needed to make an organism as complex as a human being is very interesting. Many estimates were made, some *ad hoc*, others based on statistical consideration of EST sequences and well annotated parts of the genome that became available prior to the publication of the sequence. These estimates ranged from 30 000 to 120 000 or more, with the more popular estimates between these two limits.

On the publication of the genome sequence it became clear that although some computational methods (see the next section) predicted more than 120 000 genes, the number of genes present and confirmed by EST or existing protein sequence data was quite small actually, only around 30 000. It is possible then that the human being has relatively few genes. This number is to be compared with 4 000 protein coding genes in the bacterium *Escherichia coli* and around 20 000 in the nematode *Caenorhabditis elegans*. It seems likely that biological complexity is achieved by means other than gene number; for instance, complex gene interactions, alternative splicing, and post-translational modification.

At the time of writing, the figures on the EnsEMBL WWW site (http://www.ensembl.org) are around 21 500 confidently predicted genes (backed by some form of experimental evidence that is at least a matching EST sequence), and 69 000 predicted by computational methods but with little experimental backing. Other evidence in the literature based on the sensitive method of RT-PCR (reverse transcriptase polymerase chain reaction) suggests that at least some of these predicted but unconfirmed genes will be real, and that the final number of human genes might be 20 000–25 000.

Recent detailed studies (for instance the ENCODE project) have shown that transcription from the human genome is much more extensive than first thought. In particular, it is the norm that several transcripts are produced by each protein coding gene, and transcription from the anti-sense strand is also common. Further transcripts are produced from much of the non-coding DNA in the genome, and their function is at present unclear. It seems likely that our understanding of transcriptional dynamics is at present very limited and that this will be a major research area in the future

Gene-prediction software

There are many different programs available for finding genes (*Table 1*). These range from the simple to the extremely complex, and vary according to the gene-finding method used, the basis of the gene-finding algorithm, the range of features identified, and the organisms to which they are applicable. Generally, gene-finding programs can be classed as those using *ab initio* prediction methods (identifying genes by looking for general features), and those that also incorporate homology searches.

Table 1. Gene-prediction software available over the internet

Program	URL
FGENESH	http://linux1.softberry.com/berry.phtml
GeneBuilder	http://www.itba.mi.cnr.it/webgene
GENEID	http://www1.imim.es/geneid.html
GeneMarkHMM	http://opal.biology.gatech.edu/GeneMark
GENIE	http://www.fruitfly.org/seq_tools/genie.html
GENSCAN	http://genes.mit.edu/GENSCANinfo.html
GlimmerM	http://cbcb.umd.edu/software/glimmer/
GRAIL	http://compbio.ornl.gov
HMMGene	http://www.cbs.dtu.dk/services/HMMgene
ORF finder (NCBI)	http://www.ncbi.nlm.nih.gov/gorf/gorf.html
Wise2	http://www.ebi.ac.uk/Wise2

An example of a very simple gene-finding program is ORF Finder, which is a component of the NCBI suite of programs and can be used over the internet. This program carries out a six-frame translation of any sequence, and the user is allowed to define the minimum size of the polypeptide as well as the genetic code to be used. The results can be used directly in a BLAST search to identify related sequences in GenBank (Section J1) or transferred to the Sequin database deposition tool.

Programs like ORF Finder are suitable for the detection of most bacterial genes, which tend to be densely organized and intronless. However, the genes of higher eukaryotes are much more complex, so more sophisticated algorithms are required to detect them. A number of programs are available that detect single features of eukaryotic genes; for example, MZEF and HEXON are exon-prediction algorithms that use hexamer frequency (the relative frequency with which specific groups of six nucleotides appear in coding vs. non-coding DNA) to discriminate between exons and introns/intergenic sequence. However, most eukaryotic gene-prediction programs can be used to detect and delimit whole genes.

The basis of the algorithm plays an important part in the reliability and scope of gene prediction. Early gene-prediction programs, such as GeneFinder and GeneParser, involved rule-based algorithms in which an explicit set of rules was used to determine whether a particular nucleotide was included or excluded from a predicted gene. More recently, neural nets have been used to combine content and signal sensing. Such programs are capable of learning, and must be trained using annotated sequence data in order to build a set of rules. Programs that use neural nets include **GRAIL** (Gene Recognition and Assembly Internet Link) and **FGENEH**. The difficulties associated with exon detection in higher eukaryotes stimulated the development of a new range of algorithms based on the statistical analysis of nucleotide frequencies and dependencies, and their integration with signal sensing. These algorithms use dynamic programming (Section J3), and are based on statistical models in which the probability of a given nucleotide appearing in a sequence reflects the preceding nucleotides (this being known as a Markov model). Where hidden variables are used in such algorithms, they are known as hidden Markov models (HMMs). Such algorithms are the basis of a number of highly successful gene-prediction programs such as GeneMarkHMM, HMMgene, GeneParser, GlimmerM, Genie, GenScan, and Wise2. Neural net and Markov model-based algorithms can detect genes reliably only in those organisms upon which they have been trained, since they build rules according to experience. For example, GRAIL is suitable for the bacterium *Escherichia coli*, the fruit fly *Drosophila*, the model plant *Arabidopsis thaliana*, mouse, and human DNA. GeneScan is suitable for humans (and other vertebrates), *Drosophila*, *Arabidopsis* and maize. Genie is suitable only for *Drosophila* and human sequences, while MZEF can be applied only to human DNA. Recent improvements in such programs allow automatic training on data from new organisms.

Measuring predictive accuracy

With such a large number of gene-finding programs available, how does one know which is the best to use? As discussed above, this may be dictated in part by the source of the genomic sequence, since many of the programs are trained on DNA from particular organisms. However, there is also a useful way to measure the overall accuracy of gene predictions. This is based on the concepts of sensitivity and specificity. Each nucleotide in a given test sequence will either be predicted to be part of a gene (predicted positive, PP) or predicted to be outside the boundaries of a gene (predicted negative, PN). When a gene is annotated manually, these assignments can be checked against the actual positive (AP) and actual negative (AN) nucleotides in the sequence. Such comparisons allow four values to be calculated: the number of true positives (TP), false positives (FP), true negatives (TN), and false negatives (FN) assigned by the program. Sensitivity is a measure of the ability of a gene-finding program to detect true positives, and is defined as the ratio TP/AP. Conversely, specificity is a measure of the program's ability to eliminate false positives, and is defined as the ratio TP/PP. These values are combined into an overall reliability known as the approximate correlation (AC)[2].

Gene prediction software can be scored also on its ability to correctly predict exons. Many annotation errors involve exons; for example, failure to detect genuine exons (especially very short exons, or 5'- and 3'-exons since these often contain a non-coding sequence), the prediction of false exons, the fusion of exons, and the incorrect specification of intron/exon boundaries. A predicted exon (PE) may be either a true exon (TE) or a false exon (FE), and this can be checked against actual annotated exons (AE). Sensitivity is defined as the ratio TE/AE and is a measure of the program's ability to predict true exons correctly. Specificity is defined as the ratio TE/PE and is a measure of the program's ability to eliminate false exons. Two other values are often used: the missing exons (ME) value is defined as the proportion of annotated exons that is not predicted by the software, and the wrong exons (WE) value is defined as the proportion of predicted exons that are false.

Gene-prediction programs are continually tested for their ability to identify genes correctly as part of an ongoing international project called GASP (Gene Annotation aSsessment Project). At present, even the most advanced programs can achieve only about 60% sensitivity and 50% specificity in eukaryote genomes. It is therefore best to use several programs for the analysis of the same sequence. High confidence can be placed in those predictions that correlate across several programs (especially if the algorithms use different statistical methods). Finally, it should be noted that all annotation errors are not the fault of gene-prediction software. Just as many mistakes arise upstream owing to sequencing and sequence assembly errors or because of misleading information already present in the databases!

Annotation pipelines

The large amount of sequence data generated by genome projects means that annotation must be carried out continuously. The data are fed in at one end and the annotations flow from the other. This is the nature of an annotation pipeline. The pipeline begins with the sequencing work, assembly, and quality control. Then the finished data are mined for information using the prediction tools discussed in this Section and integrated into an appropriate database. Database formats suitable for the analysis and display of annotated genome data are discussed in Section J1. These include Artemis, Apollo, EnsEMBL, GoldenPath, and ACeDB.

---

[2] For the interested reader, approximate correlation is calculated as follows:
$$AC = [(TP/(TP+FN)) + (TP/(TP+FP)) + (TN/(TN+FP)) + (TN/(TN+FN))]/2 - 1$$

J3 SEQUENCE ANALYSIS

Key Notes

Sequence similarity searches	Sequence similarity searches of databases enable us to extract sequences that are similar to a query sequence. Information about these extracted sequences can be used to predict the structure or function of the query sequence. Prediction using similarity is a powerful and ubiquitous idea in bioinformatics. The underlying reason for this is molecular evolution.
Sequence alignment	Any pair of DNA sequences will show some degree of similarity. Sequence alignments are the first step in quantifying this in order to distinguish between chance similarity and real biological relationships. Alignments show the differences between sequences as changes (mutations), insertions or deletions (indels or gaps), and can be interpreted in evolutionary terms.
Alignment algorithms	Dynamic programming algorithms can calculate the best alignment of two sequences. Well-known variants are the Smith-Waterman algorithm (local alignments) and the Needleman-Wunsch algorithm (global alignments). Local alignments are useful when sequences are not related over their full lengths, for example proteins sharing only certain domains, or DNA sequences related only in exons.
Alignment scores and gap penalties	A simple alignment score measures the number or proportion of identically matching residues. Gap penalties are subtracted from such scores to ensure that alignment algorithms produce biologically sensible alignments without too many gaps. Gap penalties may be constant (independent of the length of the gap), proportional (proportional to the length of the gap), or affine (containing gap opening and gap extension contributions). Gap penalties can be varied according to the desired application.
Measuring sequence similarity	Sequence similarity can be quantified using the score from the alignment algorithm, percentage sequence identities, or more complex measures. The most useful statistical measures are outlined below.
Similarity and homology	Similarity may exist between any sequences. Sequences are homologous only if they have evolved from a common ancestor. Homologous sequences often have similar biological functions (orthologues), but the mechanism of gene duplication allows homologous sequences to evolve different functions (paralogues).
Maximizing amino acid identities	Protein sequences can be aligned to maximize amino acid identities, but this will not reveal distant evolutionary relationships.
Evolution	Protein coding sequences evolve slowly compared with most other parts of the genome, because of the need to maintain protein structure and function. An exception to this is the fast evolution that might occur in the redundant copy of a recently duplicated gene.

Allowed changes

Changes in protein sequences during evolution tend to involve substitutions between amino acids with similar properties because these tend to maintain the structural stability of the protein.

Substitution score matrices

These matrices give scores for all possible amino acid substitutions during evolution. Higher scores indicate more likely substitutions. Example matrices are BLOSUM62 and PAM250. PAM stands for accepted point mutations, and in this case the evolutionary distance of the matrix is 250 amino acid changes per 100 residues. Dynamic programming algorithms for sequence alignment can operate using scores from these matrices.

Significance

Substitution score matrices allow detection of distant evolutionary relationships between protein sequences. It is possible to detect much more distant relationships by comparing protein sequences than by comparing nucleic acid sequences.

Visualization

Dot plots are a very good way to visualize sequence similarity and find repeats.

Database searches

Sequences similar to a query can be found in a database by aligning it to each database sequence in turn and returning the highest scoring (most similar) sequences. This can be achieved by dynamic programming algorithms but in practice faster, less exact methods are often used.

Algorithms and software

BLAST and FASTA provide very fast searches of sequence databases. Unlike dynamic programming they do not guarantee to find the best possible alignment to each database sequence, but in practice the effect on performance is usually minimal. Each operates by first locating short stretches of identically or near identically matching letters (words) that are eventually extended into longer alignments.

Statistical scores

The p value of a similarity score is the probability of obtaining a score at least as high in a chance similarity between two unrelated sequences of similar composition. Low p values indicate significant matches that are likely to have real biological significance. The related E value is the expected frequency of chance occurrences scoring at least as high as the identified similarity. A low p value for a similarity between two sequences can translate into a high E value for a search of a large database.

Sensitivity and specificity

These measures quantify the success of a database search strategy. Sensitivity measures the proportion of real biological sequence relationships in the database that were detected as hits in the search. Specificity is the proportion of the hits corresponding to real biological relationships. Changing E and p value thresholds results in a trade-off between these complementary measures of success.

Database types

Databases and query sequences can be protein or nucleic acid sequences, and different query strategies are required for different types and combinations. In general searches are more sensitive using strategies where protein coding nucleic acid database and/or query sequences are first translated to protein sequences.

Worked example and program availability	In our opinion the best BLAST server runs at the NCBI and can be used to search many general purpose sequence databases. A similar FASTA implementation is available at the EBI. BLAST and FASTA are also used on many WWW sites to search organism-specific databases; for example, at the Sanger Center. We give a worked example using NCBI BLAST and FASTA at the EBI, which shows that a list of hits ranked by E value is useful, but that it is necessary always to inspect the associated sequence alignments to discover the region or domain of similarity between the sequences.
Non-specific sequence similarity	Certain types of sequence similarity are less likely to be indicative of an evolutionary relationship than others. Examples of this are similarity between regions of low compositional complexity, short period repeats, and protein sequences coding for generic structures like coiled coils.
Similarity searches	Regions of the types mentioned above can degrade the results of similarity searches and often are filtered out of query sequences prior to searching. The programs SEG and DUST can be used to detect and filter low complexity sequences, XNU can filter short period repeats, and COILS can detect the presence of potential coiled coil structures.
Detection of evolutionary relationships	Divergent evolution can change protein sequences beyond recognition while preserving structure and function. Methods like BLAST and FASTA sometimes only detect a small proportion of the evolutionary relationships in a database, and much bioinformatics research has focused on the detection of distant evolutionary relationships between sequences.
Iterative database searches	PSI-BLAST is an iterative search method that improves on the detection rate of BLAST and FASTA. Each iteration discovers intermediate sequences that are used in a sequence profile to discover more distant relatives of the query sequence in subsequent iterations. Potential problems with PSI-BLAST are associated with the potential for unrelated sequences to pollute the iterative search, and difficulties associated with the domain structure of proteins. PSI-BLAST often detects up to twice as many evolutionary relationships as BLAST.
Related sections	Databases and data sources (J1) Domain families and databases (J5) Sequence families, alignment, and phylogeny (J4)

Sequence similarity searches

The sequences of biological macromolecules in modern databases are the products of molecular evolution. When sequences share a common ancestral sequence they tend to exhibit similarity in their sequences, structures, and biological functions. This is probably the most powerful idea in bioinformatics, because it enables us to make predictions. Often little is known about the function of a new sequence from a genome sequencing program, but if similar sequences can be found in the databases for which functional or structural information is available, then this can be used as the basis of a prediction of function or structure for the new sequence. It is very useful therefore to have search tools that take the new sequence as 'query,' and search the database for similar sequences. Examples of such tools are the BLAST, FASTA, and Smith-Waterman algorithms, which we will discuss later, but first we need to discuss how sequence similarity might be discovered, visualized, and quantified.

Sequence alignment

Any pair of nucleic acid sequences will share a degree of similarity. For instance, DNA sequences are constructed from an alphabet of only four letters, {A,T,G,C}, so any sequence that consists of a mixture of these letters will show some similarity to any other similarly constructed sequence. We need to distinguish between this type of 'chance' similarity, and the similarity that is the result of a real evolutionary and/or functional relationship. We need a way of quantifying similarity.

Quantifying similarity begins with an alignment, such as that shown for the very short DNA sequences in *Fig. 1*. In this figure, the two sequences are written one above the other and the letters (defining DNA bases) that are vertically directly above and below each other are **aligned** or **equivalenced**. Reading the alignment from left to right, the first two As in each sequence are aligned, and these are followed by a T aligned with a C, and then another pair of aligned Ts, and so on. It is important that not all letters in a particular sequence have equivalents in the other sequence. In *Fig. 1*, letters 7–9 of sequence 1 (TTG) are not aligned with any letters from sequence 2. When this happens we say that a **gap** has been introduced. The point of introducing a gap here is that it enables a better alignment of the two sequences, in which more of the aligned pairs are identical letters. In this case it enables the shared CGCAT directly after the gap to align between the two sequences.

Alignments can be interpreted in evolutionary terms. When identical letters are aligned the simplest interpretation is that these letters were part of the ancestral sequence and have remained unchanged. When non-identical letters are aligned the simplest interpretation is that a mutation has occurred in one of the sequences. Of course, in the absence of information about the ancestral sequence, it is not possible to know in which sequence the mutation actually occurred. Gaps in alignments can be interpreted in terms of the **insertion** or **deletion** of letters in one of the sequences with respect to the ancestral sequence. In *Fig. 1*, the gap could have resulted either from an insertion of three letters in sequence 1, or a deletion of three in sequence 2. Again, without the ancestral sequence these possibilities cannot be distinguished, so the gap is referred to sometimes as an **indel**.

SEQ1: AATTGATTGCGCATTTAAAGGG
SEQ2: AACTGA---CGCATCTTAAGGG

Fig. 1. An alignment of two short DNA sequences. The '-' character denotes a gap equal to one base in the sequence, so this alignment contains a gap of three bases.

Alignment algorithms

Finding the best alignment between the two sequences in *Fig. 1* was fairly straightforward, because the sequences are short and very closely related. It is much more difficult if the sequences are longer or if they are less closely related. In these cases computational methods are required to find the best alignment of the sequences. Fortunately there are known computational methods for this task, called **dynamic programming algorithms**. These algorithms take two input sequences and produce as output the best alignment between them.

Sequence similarity analyses commonly use two dynamic programming algorithms, the **Needleman-Wunsch algorithm** and the **Smith-Waterman**

algorithm. These are closely related, but the main difference is that the Needleman-Wunsch algorithm finds **global** similarity between sequences while the Smith-Waterman algorithm finds **local** similarity. An alignment from the Smith-Waterman algorithm might cover only a small (local) part of each sequence, while a Needleman-Wunsch alignment will try to cover as much of the sequences as possible, starting at the left-most end of one of the sequences and finishing at the right-most end of one of the sequences. The Smith-Waterman algorithm is the most used because real biological sequences often are not similar over their entire lengths, but only in local portions. Examples of biological reasons for this are genes from different organisms with similar exons (local similarity) but different intron structures, or proteins that share only certain domains. Even though the Smith-Waterman algorithm is able to find these local similarities, it should discover also global sequence similarity if it exists.

Alignment scores and gap penalties

In their simplest incarnation, dynamic programming algorithms find alignments containing the largest possible number of identically matching letters, as in the alignment of *Fig. 1*. The problem however is that if it is permitted to insert gaps freely into the alignment then the algorithms produce alignments containing very large proportions of matching letters and large numbers of gaps. We need to control this, because insertion and deletion of monomers is a relatively slow evolutionary process, and alignments with large numbers of gaps do not make biological sense.

Dynamic programming algorithms get around this problem by using **gap penalties**. Essentially the algorithms form a score reflecting the quality of the alignment (a high positive score indicates a good alignment). A simple score contains a positive additive contribution of one for every matching pair of letters in the alignment, and a gap penalty is subtracted for each gap that has been introduced. In *Fig. 1*, if the gap penalty were equal to 1, then because there are 16 identically matching letters in the alignment and one gap, the score would be $16 - 1 = 15$. Dynamic programming algorithms find the alignment between the two sequences that maximizes the alignment score. This alignment depends on the choice of gap penalties: high gap penalties result in shorter, lower scoring alignments with fewer gaps, and lower penalties give higher scoring, longer alignments with more gaps.

Several forms of gap penalty are used commonly. We described the simplest form above, where each gap attracted a **constant** penalty of 1, independent of the length of the gap. In general a constant gap penalty can be written as A ($A = 1$ above) where the size of A controls how strongly gaps are penalized. Sometimes a **proportional** penalty is used, where the penalty is proportional to the length of the gap. This penalty has the form Bl, where B is a constant (again controlling how strongly gaps are penalized), and l is the length of the gap. With this form longer gaps attract larger penalties than shorter ones, which makes good biological sense. Finally the most complex form of gap penalty is known as **affine**; this has both constant and proportional contributions, and it takes the form $A + Bl$. In this case the constant A is called the **gap opening** penalty, because it is applied to a gap of any length. The constant B is called the **gap extension** penalty, because it is the penalty attached to extending the length of an existing gap by one unit. The motivation for the affine gap penalty is that opening a gap (penalty = A) should be strongly penalized, but once a gap is open it should cost less (B) to extend it.

It is not always clear what values should be used for the gap penalty constants *A* and *B*, and it does depend to some extent on the intended application. Most software for sequence alignment or similarity search comes with good default values for general sequence alignment or searching. In general if you are interested in detecting only close sequence relationships then it would be a good idea to try increasing gap penalties above the default values to reduce the number of gaps. On the other hand, more distant relationships might be discovered by decreasing gap penalty values. Often it is worth experimenting with different gap penalties around the default values. In some applications gap penalties that differ significantly from default values should be used. One example would be the removal of vector sequence during genome sequencing (Section J1): here the only concern is to find exact matches to the known vector sequence and very high gap penalties are appropriate.

Measuring sequence similarity

We have explained how dynamic programming algorithms attribute an alignment score that reflects the degree of relatedness of the aligned sequences. Some other measures are used, and perhaps the most common is the percentage of identically aligned residues. In *Fig. 1* the number of identically aligned residues is 16, and the length of the alignment is 22, so the percentage sequence identity is $(16/20) \times 100 = 73\%$. This has the advantage that it is independent of the length of the alignment, so it can provide comparability between alignments of different length. However, it must always be remembered that high percentage identities are much more likely to reflect real biological or evolutionary relationships if they extend over long alignments. An alignment with 50% identity means little if its length is only 10 nucleotides, because it is likely to have occurred by chance. An alignment of 50% identity over 100 nucleotides is much more likely to be significant. Perhaps the most powerful methods of assessing sequence similarity are the statistical measures described later in this section.

Aligning sequences to maximize identical matches is reasonable for DNA sequences. However, with protein sequences it is useful to take more account of the properties of the monomers that make up the sequence. Some amino acids substitute for each other better than others, and it is useful to take account of this in protein sequence alignment algorithms.

Similarity and homology

These terms are often confused. Any set of sequences can exhibit similarity and this may be quantified as discussed above. Sequences are homologous only if they evolved by divergence from a common ancestor. To say that two sequences share 50% homology is nonsense; to say that they share 50% similarity and that this indicates possible homology is the correct usage of the terms.

Sequence similarity searches are used very commonly to predict gene or protein function. The underlying theory is that similar sequences are likely to be homologous and therefore to have similar functions. If a biological function is essential for an organism to survive and reproduce, then loss of that function would be selected against by evolution, and this is the reason why homologous genes might be expected to retain the same function in different organisms. However, the evolutionary mechanism of gene duplication allows organisms to acquire redundant copies of genes. These redundant copies are free to evolve new functions, and become homologous genes with different functions. This is the principal objection to prediction of gene function through sequence similarity, and is probably the reason for a significant amount of incorrect bioinformatics-derived annotation in the sequence databases. Examples of homologous genes with different functions are lysozyme (an enzyme) and alpha-lactalbumin (a mammalian regulatory protein). These proteins have very similar sequences, are almost certainly homologous, and yet have very different functions.

When two homologous genes in different species have the same function they are known as **orthologues**, when two genes in the same or different species have different functions they are known as **paralogues.**

Maximizing amino acid identities

Protein sequences are constructed from an alphabet of 20 naturally occurring amino acids. Like nucleic acid sequences (see previous section), protein sequences can be aligned to maximize the number of identically matching pairs within the alignment. This is a good way of aligning closely related sequences. In this case, the contribution of every aligned pair of identical amino acids to the alignment score is one, and the contribution of every aligned pair of non-identical amino acids is zero. An example alignment is shown in *Fig. 2*.

	C	S	T	P	A	G	N	D	E	Q	H	R	K	M	I	L	V	F	Y	W
C	1																			
S	0	1																		
T	0	0	1																	
P	0	0	0	1																
A	0	0	0	0	1															
G	0	0	0	0	0	1														
N	0	0	0	0	0	0	1													
D	0	0	0	0	0	0	0	1												
E	0	0	0	0	0	0	0	0	1											
Q	0	0	0	0	0	0	0	0	0	1										
H	0	0	0	0	0	0	0	0	0	0	1									
R	0	0	0	0	0	0	0	0	0	0	0	1								
K	0	0	0	0	0	0	0	0	0	0	0	0	1							
M	0	0	0	0	0	0	0	0	0	0	0	0	0	1						
I	0	0	0	0	0	0	0	0	0	0	0	0	0	0	1					
L	0	0	0	0	0	0	0	0	0	0	0	0	0	0	0	1				
V	0	0	0	0	0	0	0	0	0	0	0	0	0	0	0	0	1			
F	0	0	0	0	0	0	0	0	0	0	0	0	0	0	0	0	0	1		
Y	0	0	0	0	0	0	0	0	0	0	0	0	0	0	0	0	0	0	1	
W	0	0	0	0	0	0	0	0	0	0	0	0	0	0	0	0	0	0	0	1

```
Sequence 1       :    MILVKP VVLKGDFG
Sequence 2       :    MILLKP AIIIRAEY
Position score:       111011 00000000

Total alignment score = sum of
position scores = 5.
```

Fig. 2.   Alignment to maximize amino acid identities, and the associated substitution score matrix. The matrix is able to align the first half of sequence 1 with the first half of sequence 2 because they are closely related with many identities. The second halves are poorly aligned because the relationship between the sequences is much weaker. There are no identities and the score matrix does not show how chemically reasonable matches could be made.

Evolution

Evolution changes nucleic acid sequences, but protein coding sequences evolve under strong functional constraints. Changes to these sequences generally survive only if they do not have a deleterious effect on the structure and function of the protein, because loss of the function of a protein is usually a disadvantage to the organism. An obvious exception to this occurs after a gene duplication event, when one copy of the gene can evolve very quickly, perhaps to become a pseudo-gene, or perhaps to gain a new and useful function. Usually, however, protein-coding genes evolve much more slowly than most other parts of any genome, because of the need to maintain protein structure and function.

Allowed changes

When evolutionary changes do occur in protein sequences, they tend to involve substitutions between amino acids with similar properties, because such changes are less likely to affect the structure and function of the protein. For instance, hydrophobic amino acids of similar size tend to substitute for each other quite well, because more often than not these occupy positions within the hydrophobic core of the protein, where tight packing and hydrophobicity strongly affect the stability of the protein structure. Examples of such mutations would be changes between LEU and ILE or VAL. Protein sequences from within the same evolutionary family usually show substitutions between amino acids with similar physico-chemical properties. *Fig. 3* shows physico-chemical relationships between amino acids.

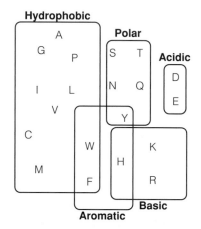

Fig. 3.   *Relationships between the physico-chemical properties of amino acids. The single-letter code for amino acids has been used.*

Substitution score matrices

Substitution score matrices are used to show scores for amino acid substitutions. For example, *Fig. 2* shows a very simple score matrix in which identical substitutions score one and non-identical ones score zero. More general matrices show the likelihood of substitution occurring between each pair of amino acids. They enable the detection of similarity between protein sequences that would be missed by the simple identity matrix of *Fig. 2*.

An example is the PAM250 substitution matrix (*Fig. 4*): its elements are scores for alignment of (i.e., substitution between) each possible pair of amino acids. For instance, the score for aligning the similar amino acids L and I is high (2), reflecting the high likelihood of this substitution as an evolutionary process, while the score for aligning the dissimilar amino acids G and W is low (-7). Notice that the scores for alignments of identical amino acids vary. These reflect the frequency of occurrence of the amino acids in natural protein sequences. The alignment of two identical uncommon amino acids (e.g., aligning W with W) is more likely to reflect an evolutionarily significant alignment than the alignment of two identical common ones (e.g., S with S), which would be likely to happen by chance in two unrelated protein sequences. Alignments of uncommon identical amino acids are therefore given higher scores.

	C	S	T	P	A	G	N	D	E	Q	H	R	K	M	I	L	V	F	Y	W
C	12																			
S	0	2																		
T	-2	1	3																	
P	-1	1	0	6																
A	-2	1	1	1	2															
G	-3	1	0	-1	1	5														
N	-4	1	0	-1	0	0	2													
D	-5	0	0	-1	0	1	2	4												
E	-5	0	0	-1	0	0	1	3	4											
Q	-5	-1	-1	0	0	-1	1	2	2	4										
H	-3	-1	-1	0	-1	-2	2	1	1	3	6									
R	-4	0	-1	0	-2	-3	0	-1	-1	1	2	6								
K	-5	0	0	-1	-1	-2	1	0	0	1	0	3	5							
M	-5	-2	-1	-2	-1	-3	-2	-3	-2	-1	-2	0	0	6						
I	-2	-1	0	-2	-1	-3	-2	-2	-2	-2	-2	-2	-2	2	5					
L	-6	-3	-2	-3	-2	-4	-3	-4	-3	-2	-2	-3	-3	4	2	6				
V	-2	-1	0	-1	0	-1	-2	-2	-2	-2	-2	-2	-2	2	4	2	4			
F	-4	-3	-3	-5	-4	-5	-4	-6	-5	-5	-2	-4	-5	0	1	2	-1	9		
Y	0	-3	-3	-5	-3	-5	-2	-4	-4	-4	0	-4	-4	-2	-1	-1	-2	7	10	
W	-8	-2	-5	-6	-6	-7	-4	-7	-7	-5	-3	2	-3	-4	-5	-2	-6	0	0	17

```
Sequence 1      :    MILVKP -VVLKGDFG
Sequence 2      :    MILLKP AIIIRAEY-
Position score:      656256 044231370

Total alignment score = (sum of
position scores) - (gap penalty) =
54 - 1 = 53.
```

Fig. 4. The PAM250 matrix and the alignment of the sequences from Fig. 2 produced with this matrix. Total alignment scores for the two matrices should not be compared, but note that the PAM matrix is able to detect a much better alignment in the second halves of these sequences than the identity matrix of Fig. 2. With the introduction of a single gap we see sensible alignments of hydrophobic amino acids, and alignments of K with R (both basic), D with E (both acidic), and F with Y (both aromatic).

The dynamic programming sequence alignment algorithms described later are not limited to aligning identical letters. They can operate with any system of scoring alignments between monomers that can be expressed as a substitution matrix. The only difference is that instead of scoring one for every match and zero for every mis-match, substitution scores are taken from the matrix. The total alignment score that the algorithms maximize is the sum of the substitution scores in the alignment, with the usual gap penalty contribution.

PAM stands for 'accepted point mutations' (the P and A are swapped to make it easier to say). PAM250 refers to an evolutionary distance of 250 accepted point mutations (PAMs) per 100 amino acid residues. PAM matrices are derived by counting observed evolutionary changes in closely related protein sequences, and then extrapolating the observed transition probabilities to longer evolutionary distances. It is possible to derive PAM matrices for any evolutionary distance, but in practice the most commonly used matrices are PAM120 and PAM250. PAM matrices with smaller numbers represent shorter evolutionary distances. Choosing the matrix for the most appropriate evolutionary distance might result in the best possible alignment of two sequences, but in practice it is rarely possible to know what the evolutionary distance is, and experience shows that the PAM250 matrix usually produces reasonable alignments.

The BLOSUM series of matrices, and in particular BLOSUM50 and BLOSUM62, have recently become popular. These matrices have some theoretical advantages (their derivation does not involve the process of extrapolation to large evolutionary distances), and there is some evidence to suggest that they do out-perform PAM matrices in practice. For this reason they are now more commonly used. The numbers attached to BLOSUM matrices do not have the same interpretation as those for PAM matrices (accepted mutations per 100 residues). In fact, BLOSUM matrices with smaller numbers represent longer evolutionary distances (BLOSUM50 represents a longer distance than BLOSUM62). There are several other matrices in fairly common use, and some specialized matrices (e.g., D. Jones and co-workers have derived PAM matrices specific to integral trans-membrane proteins).

Significance

A great deal of bioinformatics is concerned with the detection of evolutionary relationships between sequences. The use of matrices like BLOSUM62 extends our ability to detect distant relationships far beyond what could be found using the identity matrix in *Fig. 2*. This ability to detect very distant relationships, sometimes between sequences whose percentage of identical residues has diverged to below 30%, is the reason why sequence comparison should be carried out always at the protein rather than the nucleic acid sequence level, when this is possible. The ability to encode permissible changes in protein structures means that protein sequence alignment can reveal much more distant evolutionary relationships than naïve comparison of nucleic acid sequences.

Visualization

It is often useful to be able to visualize the similarity between two sequences. This can be done using dot plots, as illustrated in *Fig. 5*. In the case of *Fig. 5* the dot plot has been used to discover internal similarity between N and C terminal portions of the same sequence, but it is equally possible to use dot plots to study similarity between two different sequences.

Database searches

Sequence similarity searches aim to extract sequences from a database that are similar to a query sequence, perhaps with the aim of gathering family members, or predicting structure and function. We have discussed ways of quantifying the similarity between macromolecular sequences using sequence alignments and substitution score matrices. Similarity searches work by comparing (aligning) the query sequence to each database sequence in turn, and then ranking the database sequences with the highest scoring (most similar) at the top. This process can be carried out by the dynamic programming algorithms described below, but in practice these algorithms are often too slow for searching large databases. We will discuss some alternative methods in this section. Also discussed in this section are the statistics of sequence similarity searching, and some issues surrounding database types and search parameters.

Algorithms and software

Dynamic programming algorithms are guaranteed to find the best alignment of two sequences for given substitution matrices and gap penalties. This is

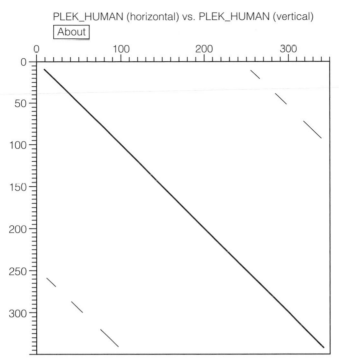

PLEK_HUMAN (horizontal) vs. PLEK_HUMAN (vertical)

Fig. 5. A dot plot of the human pleckstrin sequence against itself produced with Erik Sonnhammer's 'dotter' program. The sequence is plotted from the amino (N-) to carboxy (C-) terminus along the horizontal and vertical axes between residues 1 and approximately 350. Every residue pair with one amino acid from the horizontal sequence and one from the vertical sequence corresponds to a small square in the area of the plot. Such squares are colored according to the match score for the residue pair concerned. Darker colors correspond to better match scores and more similar residues. The dark line running diagonally from the origin to the lower right corner is made up of the squares corresponding to matches between residue 1 of the vertical sequence and residue 1 of the horizontal sequence, residue 2 and residue 2, etc., and is dark colored because the horizontal and vertical sequences are identical in this case. More interesting are the faint lines running parallel to this in the top right and lower left corners of the plot. These indicate significant similarity between approximately residues 0-100 and 250-350 of the sequence and correspond to a repeated domain. In this case, this is the pleckstrin homology (PH) domain found in both N- and C- terminal parts of the pleckstrin sequence.

impressive, but the process is often quite slow, perhaps taking hours for a search of a large database. For this reason alternative methods have been developed. Perhaps the most used of these are FASTA and BLAST; they are typically 5 (FASTA) to 50 (BLAST) times faster than dynamic programming. Unlike dynamic programming, these methods are not guaranteed to find the best alignment, and for this reason they could be less accurate. In practice however, BLAST and FASTA usually suffer a minor degradation in accuracy when compared with dynamic programming, and the increased speed is often worth this small cost.

FASTA and BLAST are both based around similar ideas, in particular they both make the reasonable assumption that high scoring alignments are likely to contain short stretches of identical or near identical letters. These short stretches are called 'words.' In the case of BLAST the first step is to look for words of a certain fixed word length (W) that score higher than a certain threshold score (T). The value of W is normally 3 for protein sequences or 11 for nucleic acid sequences. W and T are under the control of the user, but there is rarely any need to change the default values. FASTA, on the other hand, looks for identically matching words of length *ktup*. *ktup* is under the control of the user, but the default values of 2 for protein sequences and 6 for DNA sequences again are seldom changed.

Both FASTA and BLAST employ a second step where extensions to the initially identified word matches are sought. BLAST extends individual word matches until the total score of the alignment falls by a certain amount from its maximum value, producing an alignment without gaps. Early versions of BLAST always produced alignments without gaps, but later versions now produce gapped alignments. The BLAST terminology for these high-scoring local alignments is high-scoring segment pairs (HSPs).

FASTA tries to find ungapped alignments of the sequence that contain a high density of the initially identified word matches, and then attempts to join these into high-scoring gapped alignments. Finally, having identified high-scoring alignments between the sequences, FASTA produces its final alignment and score by a full dynamic programming alignment of the identified high-scoring regions of the sequences. While BLAST might identify several regions of high-scoring similarity between two sequences, FASTA only identifies a single one in the highest scoring region.

We have already commented that the principal advantage of FASTA and BLAST over dynamic programming is that of speed. This has been very important for those developing WWW access to sequence database searches. When a database is accessed regularly by users from all over the world it is not uncommon that the computers running the searches have several jobs running at the same time. When this happens even the most powerful computers can have problems, and the amount of computational resource required by each search takes on crucial importance. For this reason, WWW developers have used the BLAST and FASTA algorithms in preference to full dynamic programming with the Smith-Waterman algorithm.

Statistical scores   One of the main problems with sequence similarity searches is to know when an identified sequence similarity is significant; that is, when it represents a real biological or evolutionary relationship rather than a chance similarity between unrelated sequences. We have discussed how alignments of protein sequences could be scored using a combination of substitution matrices and gap penalties. The higher the score of an alignment by this method the more likely it is to reflect a real biological relationship. But, how high is a high score?

A major breakthrough in sequence similarity searches was the development of a statistical theory of the alignment scores. Like many theories of statistical significance, this theory centers around the calculation of p values. In the sense of a sequence similarity search, the p value of an identified similarity of score s is the probability that a score of at least s (i.e., a score of s or greater) would have been obtained in a chance match between two unrelated sequences of similar composition and length. Very low values of p therefore correspond to significant matches: in these cases it is highly improbable that the score obtained occurred by chance, and it is probable that it occurred as the result of a real biological or evolutionary relationship. All the search methods we have discussed so far report p values for identified similarities, and these can be used to judge the significance of the discovered relationships. Of course, the user still has to form a judgment about his own threshold p value for considering a similarity significant. Clearly a value of p = 0.5 does not represent a significant similarity (because it has a 50% probability of occurring by chance between unrelated sequences), but values of p less than 0.01 are much more significant.

A related quantity that is reported by most search software is the E (Expect) value. For an identified similarity of score s, the E value is the expected frequency of scores of at least s. It can be interpreted as the number of scores of at least s that would have been expected to have occurred by chance. The calculation of p and E values can be carried out for the two matched sequences in isolation (i.e., the probability of a match between the two sequences scoring at least as high by chance), or for the entire database search (i.e., the probability of a score at least as high in the entire database search). The latter are generally higher (because the database is larger than any individual sequence), and are obviously the most relevant in assessing database search results. FASTA and BLAST take slightly different approaches to whole database p and E values.

The E values reported by the FASTA software are related to the p value by the equation $E = Np$, where N is the number of sequences in the database (i.e., E refers to the database search but p to the sequences in isolation). This reflects an assumption that all the database sequences are equally likely to be related to the query sequence, which is questionable when they may differ significantly in length. The calculation of E values for BLAST does not make this assumption, but rather treats the database as one very long sequence. Nevertheless, it illustrates the following important fact: even when the p value for a match between two sequences is low, E values for a search of a large database can be quite large. For instance, a FASTA similarity score s with a p value of 0.001 might seem fairly significant, but in a database of 10 000 sequences this would correspond to an E value of $0.001 \times 10\ 000 = 10$. So we would expect the scores of at least s to occur 10 times by chance, which is clearly not a high level of significance. For this reason E values are often more informative than p values.

Sensitivity and specificity

Evaluation of the results of a database search is best viewed in terms of two complementary measures, the **sensitivity** and the **specificity**. Suppose a threshold value of E or p is chosen so that sequence similarities corresponding to p or E values lower than the threshold are considered significant. It is common to call the sequences with significant similarities **hits**. The search partitions the database into two subsets, hits (positives) and non-hits (negatives). These can be further partitioned, at least conceptually, into true and false positives and true and false negatives. A true positive is a hit with a real biological relationship to the query sequence, and a false positive is a hit without such a relationship. A true negative is a non-hit with no real biological relationship to the query, and a false negative is a non-hit with a real biological relationship to the query. The sensitivity of the search is the proportion of the real biological relationships in the database that were detected as hits, and it can be written as

$$S_n = n_{tp} / (n_{tp} + n_{fn})$$

where n_{tp} is the number of true positives and n_{fn} is the number of false negatives. The specificity of the search is the proportion of hits that correspond to real biological relationships, and it can be written as

$$S_p = n_{tp} / (n_{tp} + n_{fp})$$

where n_{fp} is the number of false positives. An ideal database search would have sensitivity and specificity both as close as possible to one, but in practice this is not attainable. Note that there is a trade-off between the two quantities. If the threshold for significance is increased, so that more hits are considered significant, then the sensitivity is likely to increase (there will be more true positives and less false negatives), but the specificity is likely to decrease (there will be more false positives).

It is of course only possible to carry out an analysis of sensitivity and specificity if the real biological relationships in the database are already known (so that they can be assigned to the categories of true and false positive and negative). Further, this information must come from a source independent of sequence similarity analyses, for instance experimental determination of protein structure and function. In general this information is not available, but it is useful to understand the trade-off between sensitivity and specificity when choosing threshold p and E values. Some investigators have used ideas like these with sets of proteins of known structure and function to compare the performance of FASTA, BLAST, and Smith-Waterman methods, and have found evidence for slightly better performance of the Smith-Waterman by comparison with the faster methods.

Database types

The database search methods we have discussed can be used with protein or nucleic acid databases and query sequences. The names of the alternative programs are given in *Table 1*. Several of these programs first translate a nucleic acid sequence in all possible reading frames to produce six possible coded protein sequences. Then sequences are compared at the protein level rather than the nucleic acid level. When these options are chosen in preference to options that compare simply at the nucleic acid level, the searches take longer to run but the results are better. This is because it is possible to detect much more distant relationships using protein sequences.

Table 1. BLAST and FASTA programs (translated means the nucleic acid sequence is translated to protein in six reading frames)

Program name	Query sequence	Database type
Blastp	Protein	Protein
Blastn	Nucleic acid	Nucleic acid
Blastx	Nucleic acid (translated)	Protein
Tblastn	Protein	Nucleic acid (translated)
Tblastx	Nucleic acid (translated)	Nucleic acid (translated)
Fasta	Protein or nucleic acid	Protein or nucleic acid
Tfastx	Protein	Nucleic acid (translated)
Fastx	Nucleic acid (translated)	Protein

Worked example and program availability

If you want to use BLAST or FASTA to search the large, general purpose sequence databases like SWISSPROT, Genbank, and EMBL, then it is probably best done using the WWW sites of the organizations that maintain these databases; for example, the EBI and the NCBI. Here we give a worked example using FASTA at the EBI site and BLAST at the NCBI site.

The results of the worked example are shown in *Figs 6–9*. The search carried out was with the sequence of human pleckstrin and similar sequences were sought in the SWISSPROT database. The searches were carried out with the BLOSUM62 amino acid substitution matrix and gap opening and extension penalties of 8 and 2. These relatively low penalties were used to achieve maximum comparability between the two search servers using the limited sets of values available at each site, and ideally the higher default penalties of 12 and 2 should be used. *Fig. 6* shows the significant similarities found by FASTA. Using 0.1 as a threshold E value below which similarity was considered significant, the algorithm found 17 significant sequence similarities. The sequence alignment associated with one of the similar sequences (human cytohesin 1, the 12th on the list of significant similarities) is shown in *Fig. 7*. Note that the alignment covers almost the entire length of the pleckstrin sequence.

Sequence	Len	opt	bits	E
SW:PLEK_HUMAN P08567 PLECKSTRIN (PLATELET P47 PRO	(350)	1854	469	7.1e-132
SW:KRAC_DICDI P54644 RAC-FAMILY SERINE/THREONINE	(444)	140	46	0.00029
SW:Y053_HUMAN P42331 HYPOTHETICAL PROTEIN KIAA005	(638)	139	46	0.00045
SW:Y041_HUMAN Q15057 HYPOTHETICAL PROTEIN KIAA004	(632)	123	42	0.007
SW:AKT3_MOUSE Q9WUA6 RAC-GAMMA SERINE/THREONINE P	(479)	115	39	0.022
SW:AKT3_HUMAN Q9Y243 RAC-GAMMA SERINE/THREONINE P	(479)	115	39	0.022
SW:CYH3_MOUSE O08967 CYTOHESIN 3 (ARF NUCLEOTIDE-	(399)	110	38	0.045
SW:SPCO_HUMAN Q01082 SPECTRIN BETA CHAIN, BRAIN (	(2364)	118	41	0.046
SW:AKT3_RAT Q63484 RAC-GAMMA SERINE/THREONINE PRO	(454)	108	38	0.07
SW:CYH1_MOUSE Q9QX11 CYTOHESIN 1 (CLM1).	(398)	107	37	0.075
SW:CYH1_RAT P97694 CYTOHESIN 1 (SEC7 HOMOLOG A) (	(398)	107	37	0.075
SW:CYH1_HUMAN Q15438 CYTOHESIN 1 (SEC7 HOMOLOG B2	(398)	107	37	0.075
SW:CYH4_HUMAN Q9UIA0 CYTOHESIN 4.	(394)	106	37	0.088
SW:SPCO_MOUSE Q62261 SPECTRIN BETA CHAIN, BRAIN (	(2363)	114	40	0.091
SW:RSG2_MOUSE P58069 RAS GTPASE-ACTIVATING PROTEI	(848)	109	38	0.096
SW:RSG2_HUMAN Q15283 RAS GTPASE-ACTIVATING PROTEI	(849)	109	38	0.096
SW:3BP2_HUMAN P78314 SH3 DOMAIN-BINDING PROTEIN 3	(561)	107	38	0.098
SW:CYH2_MOUSE P97695 CYTOHESIN 2 (ARF NUCLEOTIDE-	(400)	105	37	0.11

Fig. 6. *Results obtained by searching the SWISS-PROT protein sequence database for sequences similar to human pleckstrin with the FASTA program using the server at the European Bioinformatics Institute. This shows the highest scoring hits. Each line shows a single protein sequence. First come database accession codes and then the name followed by the length of the sequence in parentheses. Following that are two scores that are of technical interest only (opt and bits), and on the end of the line is the FASTA E value. Scores are given down to an E value cutoff of 0.1.*

Fig. 8 shows the significant similarities discovered by BLAST. Note that in this case BLAST finds more similarities below the $E = 0.1$ cutoff chosen. BLAST and FASTA use different ways of calculating E values and it is unsurprising that they produce different results. Both algorithms find the significant similarity to the human cytohesin 1 sequence featured in *Fig. 7*. The alignments from BLAST are shown in *Fig. 9*. BLAST finds two regions of similarity, but unlike FASTA does not find similarity over the entire sequence.

In the case of these sequences it is instructive to compare the results with what is already known about the sequences. It is known that the human pleckstrin sequence contains two PH (pleckstrin homology) domains (amino acids 5–100 and 245–345) separated by another unrelated domain. Cytohesin 1 contains a PH domain (amino acids 260–375) preceded by an unrelated domain. The first of the BLAST alignments corresponds to the match between the second PH domain of pleckstrin and the PH domain of cytohesin, and this similarity is also the part of the FASTA alignment where similarity is strongest. The second BLAST alignment is the similarity between the first PH domain of pleckstrin and the PH domain of cytohesin. BLAST does not discover any sequence similarity outside the PH domains. Since no similarity was expected here, it seems that this is a better result than that produced by FASTA. In this case FASTA would have performed better with higher gap penalties, and this one case does not give sufficient evidence for any real comparison of the performance of the two methods. However, it does illustrate an important general point. The list of significant similarities ranked by *E* value is useful, but it is almost always necessary to inspect the sequence alignments to see where the sequences are most similar and where they are unrelated.

```
>>SW:CYH1_HUMAN Q15438 CYTOHESIN 1 (SEC7 HOMOLOG B2-1).   (398 aa)
 initn: 111 init1:  76 opt: 107  Z-score: 155.6  bits: 37.5 E(): 0.075
Smith-Waterman score: 152;   23.864% identity (31.343% ungapped) in 352 aa overlap
(37-347:70-378)

             10        20        30        40        50        60
PLEK_H REGYLVKKGSVFNTWKPMWVVLLEDGIEFYKKKSDNSPKGMIP--LKGSTLTSPCQDFGK
                               .::  .  .::   :    .... : .  :.:...
SW:CYH EIAEVANEIENLGSTEERKNMQRNKQVAMGRKKFNMDPKKGIQFLIENDLLKNTCEDIAQ
          40        50        60        70        80        90

           70        80        90       100       110
PLEK_H RMFVFK---ITTTKQQDHFFQAAFLEERDAW-VRDINKAIKCIEGGQKFARKSTRR---S
       :..:    .. :    :      .: :::  ... ..  ..   :     .. :.    :
SW:CYH --FLYKGEGLNKTAIGD------YLGERDEFNIQVLHAFVELHEFTDLNLVQALRQFLWS
         100       110         120       130       140       150

         120       130       140       150       160       170
PLEK_H IRLP---ETIDLGALYLSMKDTEKGIKELNLEKDKKIFNHCFTGNC-VIDW--LVSNQSV
       .:::    . ::        :  ..    . .   .:  ::  : .:     .. : :.
SW:CYH FRLPGEAQKID------RMMEA---FAQRYCQCNNGVFQS--TDTCYVLSFAIIMLNTSL
         160         170       180       190       200

           180       190       200       210       220
PLEK_H RN---RQEGLMIASSLLNEGYLQPAGDMSKSAVDGTAENPFLDNPDAFYYFPDSGFFCEE
       ::   ...  .      .::.: .. .::. ...  : ..:...  .        :.
SW:CYH HNPNVKDKPTVERFIAMNRG-INDGGDLPEELLRNLYES-IKNEP---FKIP------ED
         210       220       230       240

         230       240       250       260       270       280
PLEK_H NSSDDDVILKEEFRGVIIKQGCLLKQGHRR-KNWKVRKFILREDPAYLHYYDPAGAEDPL
       ...:      :  :  .     ..: ::: :  : :.:: : ::: ..   :.:.. .  ..:
SW:CYH DGND----LTHTFFNPD-REGWLLKLGGGRVKTWKRRWFILTDN--CLYYFEYTTDKEPR
        250          260       270       280       290       300

         290       300       310       320
PLEK_H GAIHLRGCVVTSVESNSNGRKSEEEENLFEII---TADEV---------------H--YF
       : : :..  . .::.     :... : ::.   . :.:              :  :
SW:CYH GIIPLENLSIREVED------SKKPNCFELYIPDNKDQVIKACKTEADGRVVEGNHTVYR
           310         320       330       340       350

         330       340       350
PLEK_H LQAATPKERTEWIKAIQMA-SRTGK
       ..: ::.:. :::::.:. : ::
SW:CYH ISAPTPEEKEEWIKCIKAAISRDPFYEMLAARKKKVSSTKRH
           360       370       380      390
```

Fig. 7.   The FASTA alignment corresponding to one of the high-scoring similarities (actually the 12th best score) from the search in Fig. 6. Many of the details in the header are of technical interest only, but the E value is reported along with a percentage identity for the alignment.

**Non-specific
sequence
similarity**

Sequences or sequence segments may be similar to each other for reasons other than sequence conservation, and do not reflect similarity in biological function. Sequences with biased composition or containing short period repeats are perhaps the most obvious examples of this. In protein sequences it is common to find short segments whose composition is dominated by a small number of amino acids, and these segments are said to have low compositional complexity. For example, part of the sequence of the human Huntington's disease protein is shown in *Fig. 10* with low complexity regions associated with compositional bias towards glutamine and proline. It is interesting that in this case the length of the low complexity glutamine repeat, produced by unstable tri-nucleotide repeat expansion, is related to the severity of the disease. It is possible that two sequences each contain low complexity segments with similar composition (e.g., another protein containing repeated glutamine and/or proline), but this type of

```
Sequences producing significant alignments:                      (bits)  E Value
gi|4505879|ref|NP_002655.1|   (NM_002664) pleckstrin; p47 [Ho...    726   0.0
gi|7661882|ref|NP_055697.1|   (NM_014882) KIAA0053 gene produ...     50   5e-06
gi|7019505|ref|NP_037517.1|   (NM_013385) pleckstrin homology...     50   7e-06
gi|1730069|sp|P54644|KRAC_DICDI   RAC-FAMILY SERINE/THREONINE...     49   1e-05
gi|8392888|ref|NP_058789.1|   (NM_017093) murine thymoma vira...     47   5e-05
gi|7242195|ref|NP_035312.1|   (NM_011182) pleckstrin homology...     45   1e-04
gi|6680674|ref|NP_031460.1|   (NM_007434) thymoma viral proto...     45   1e-04
gi|4502023|ref|NP_001617.1|   (NM_001626) v-akt murine thymom...     45   1e-04
gi|6755186|ref|NP_035311.1|   (NM_011181) pleckstrin homology...     45   2e-04
gi|8670546|ref|NP_059431.1|   (NM_017457) cytohesin 2, isofor...     45   2e-04
gi|4758964|ref|NP_004753.1|   (NM_004762) cytohesin 1, isofor...     44   3e-04
gi|7242193|ref|NP_035310.1|   (NM_011180) pleckstrin homology...     43   7e-04
gi|13124031|sp|P97694|CYH1_RAT   CYTOHESIN 1 (SEC7 HOMOLOG A)...     43   7e-04
gi|6753032|ref|NP_035915.1|   (NM_011785) thymoma viral proto...     42   0.002
gi|4885549|ref|NP_005456.1|   (NM_005465) v-akt murine thymom...     42   0.002
gi|15100164|ref|NP_150233.1|   (NM_033230) murine thymoma vir...     41   0.002
gi|13124042|sp|O43739|CYH3_HUMAN   CYTOHESIN 3 (ARF NUCLEOTID...     41   0.003
gi|13124032|sp|P97696|CYH3_RAT   CYTOHESIN 3 (SEC7 HOMOLOG C)...     41   0.003
gi|1170702|sp|Q01314|KRAC_BOVIN   RAC-ALPHA SERINE/THREONINE ...     41   0.003
gi|400112|sp|P31748|KAKT_MUVAT   AKT KINASE TRANSFORMING PROTEIN     41   0.003
gi|400144|sp|P31750|KRAC_MOUSE   RAC-ALPHA SERINE/THREONINE K...     41   0.004
gi|4885061|ref|NP_005154.1|   (NM_005163) serine/threonine pr...     40   0.005
gi|4506927|ref|NP_003014.1|   (NM_003023) SH3-domain binding ...     40   0.007
gi|6755496|ref|NP_036023.1|   (NM_011893) SH3-domain binding ...     39   0.010
gi|13928778|ref|NP_113763.1|   (NM_031575) thymoma viral prot...     39   0.011
gi|3183205|sp|Q15057|Y041_HUMAN   HYPOTHETICAL PROTEIN KIAA00...     39   0.012
gi|14777221|ref|XP_027525.1|   (XM_027525) hypothetical prote...     38   0.021
gi|6094287|sp|P91621|SIF1_DROME   STILL LIFE PROTEIN TYPE 1 (...     38   0.030
gi|465965|sp|P34512|YMX4_CAEEL   HYPOTHETICAL 43.2 KDA PROTEI...     37   0.033
gi|6094288|sp|P91620|SIF2_DROME   STILL LIFE PROTEIN TYPE 2 (...     37   0.033
gi|6755288|ref|NP_035375.1|   (NM_011245) RAS protein-specifi...     37   0.040
gi|121515|sp|P28818|GNRP_RAT   GUANINE NUCLEOTIDE RELEASING P...     37   0.041
gi|13124259|sp|Q13972|GNRP_HUMAN   GUANINE NUCLEOTIDE RELEASI...     36   0.074
gi|121742|sp|P09851|RSG1_BOVIN   RAS  GTPASE-ACTIVATING            36   0.012
gi|121742|sp|P09851|RSG1_BOVIN   RAS  GTPASE-ACTIVATING            36   0.012
```

Fig. 8. Results obtained by searching the SWISS-PROT protein sequence database for sequences similar to human pleckstrin with the BLAST program using the server at the NCBI. This shows the highest scoring hits. Each line shows a single protein sequence. First come database accession codes (which differ from those in Fig. 6 for the same sequences), and then the name of the sequence. Following that is a score that is of technical interest only (bits), and on the end of the line is the BLAST E value. Scores are given down to an E value cutoff of 0.1.

```
>gi|2498175|sp|Q15438|CYH1_HUMAN CYTOHESIN 1 (SEC7 HOMOLOG B2-1)

 Score = 44.3 bits (103), Expect = 3e-04
 Identities = 36/121 (29%), Positives = 55/121 (44%), Gaps = 30/121 (24%)

Query: 247 KQGCLLK-QGHRRKNWKVRKFILREDPAYLHYYDPAGAEDPLGAIHLRGCVVTSVESNSN 305
            ++G LLK  G R K WK R FIL ++    L+Y++    ++P G I L    + VE
Sbjct: 263 REGWLLKLGGGRVKTWKRRWFILTDN--CLYYFEYTTDKEPRGIIPLENLSIREVED--- 317

Query: 306 GRKSEEENLFE--------------------IITADEVHYFLQAATPKERTEWIKAIQM 344
               S++ N FE                    ++  +   Y + A TP+E+ EWIK I+
Sbjct: 318 ---SKKPNCFELYIPDNKDQVIKACKTEADGRVVEGNHTVYRISAPTPEEKEEWIKCIKA 374

Query: 345 A 345
            A
Sbjct: 375 A 375


 Score = 40.4 bits (93), Expect = 0.005
 Identities = 17/50 (34%), Positives = 32/50 (64%), Gaps = 1/50 (2%)

Query: 7   REGYLVK-KGSVFNTWKPMWVVLLEDGIEFYKKKSDNSPKGMIPLKGSTL 55
           REG+L+K  G    TWK  W +L ++ + +++    +D P+G+IPL+ ++
Sbjct: 263 REGWLLKLGGGRVKTWKRRWFILTDNCLYYFEYTTDKEPRGIIPLENLSI 312
```

Fig. 9. The BLAST alignments corresponding to the twelfth highest scoring similarities from the search in Fig. 8. Query is the human pleckstrin sequence, and Sbjct is the database sequence (human cytohesin 1 in this case). This alignment is with the same database sequence as the one shown in Fig. 6. Many of the details in the header are of technical interest only, but the E value is reported along with a percentage identity (29%). Note that BLAST finds two separate regions of similarity between the sequences, the first (amino acids 247–345) scoring much more strongly than the second (amino acids 7–55).

similarity is unlikely to indicate that the proteins are homologous or that they share similar biological function. It is much more likely to reflect a common mechanism for the creation of the low complexity or repeat region, such as unstable expansion of a tri-nucleotide repeat.

Another type of non-specific sequence similarity is that associated with proteins that form coiled coil structures. These structures comprise two α helices coiled around each other and are associated with a seven-residue pseudo-repeat in the amino acid sequence, in which residues 1 and 4 are predominantly non-polar. Examples of such structures are found in keratin and myosin. It seems that this type of structure is one that is advantageous to peptide sequences in general, and has been formed by many non-homologous proteins in a process of convergent evolution to a similar structure. The presence of a shared coiled coil structure is not strong evidence for an evolutionary relationship between two proteins.

```
MATLEKLMKA FESLKSFQQQ QQQQQQQQQQ QQQQQQQQQQ PPPPPPPPPP PQLPQPPPQA
QPLLPQPQPP PPPPPPPPGP AVAEEPLHRP KKELSATKKD RVNHCLTICE NIVAQSVRNS
PEFQKLLGIA MELFLLCSDD AESDVRMVAD ECLNKVIKAL MDSNLPRLQL ELYKEIKKNG
APRSLRAALW RFAELAHLVR PQKCRPYLVN LLPCLTRTSK RPEESVQETL AAAVPKIMAS
```

Fig. 10. Part of the sequence of the human Huntington's disease protein (Huntingtin) showing low complexity regions (underlined) associated with compositional bias towards glutamine (Q) and proline (P).

Similarity searches

Similarity searches usually aim to detect homologous proteins or proteins with similar functions. Non-specific sequence similarity therefore degrades the results of such searches. It is common to remove segments of the query sequence that might be associated with non-specific sequence similarity prior to carrying out a similarity search. Several programs are available to carry this out. The programs SEG (Wootton and Federhen) and DUST (Tatusov and Lipman) will detect low complexity sequences. SEG is used mainly for protein sequences but will work with nucleic acids, and DUST is exclusively for nucleic acid sequences. The program XNU (Claverie and States) will detect short periodicity repeats. The program COILS will predict potential coiled coil structures in proteins.

Many WWW-based sequence similarity search servers give the user an option to use the standard filters (SEG, XNU or DUST) prior to carrying out the search. These typically take the query sequence and replace monomers in the identified regions with the letter X so that they are ignored by the search software.

Detection of evolutionary relationships

Divergent evolution changes the sequences of homologous proteins. In long evolutionary times these changes can be very significant, and the degree of similarity between the homologous sequences can become very small. In many cases the degree of sequence similarity falls below the level that is detectable by methods like BLAST and FASTA. This happens when the similarity between the homologous protein sequences is similar to the level of similarity that commonly occurs by chance between unrelated sequences. In the case of proteins, we can tell that this has happened by looking at the protein three-dimensional structures, because structural similarity tends to be preserved even when the sequences have diverged beyond recognition. Well-known examples of this are found within the globin family of homologous proteins. These proteins share a common function (oxygen transport) and have the same structure, and yet there are pairs of members of the family from different organisms whose sequences have less than 10% identical residues.

Using the database of three-dimensional protein structures it has been estimated that BLAST and FASTA may detect only 20% of the total number of *distant* evolutionary relationships to the query sequence within the database. Therefore, much bioinformatics research effort has been devoted to methods that are able to improve on this detection rate. One such method is PSI-BLAST.

Iterative database searches

PSI_BLAST makes use of iterated BLAST searches in order to extend the number of evolutionary relationships detected. The idea is illustrated in *Fig. 11*. An initial BLAST search is performed with the query sequence. Then each of the hits from this search (above a chosen *E* value cutoff) can be used as the query sequence in a second iteration of BLAST searching. This second iteration should detect more sequences related to the initial query sequence, and this process can be repeated (iterated) until no more significant sequence similarities are found. After a number of iterations more of the distant evolutionary relationships to the initial query sequence should have been detected. This is achieved by using information from **intermediate** sequences. The evolutionary relationship between sequences A and C may not be detectable by BLAST, but if an evolutionary relationship between A and an intermediate sequence B can be detected, along with a further relationship between B and C, then it may be inferred that A is related to C.

PSI-BLAST works in essentially this way. The only difference from the explanation above is a minor technical one. The second and subsequent iterations are not carried out as BLAST searches with each single sequence hit from the previous iteration,

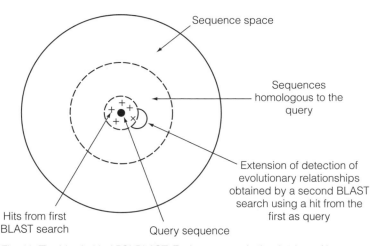

Fig. 11. *The idea behind PSI-BLAST. Each sequence in the database (the* sequence space*) is represented by a point within the outer circle, and the query sequence is represented by the large black dot. Distances between the points representing sequences are proportional to their evolutionary distance, so that points close to each other are closely related. The sequences within the large dashed circle centered on the query sequence are all homologous to the query and are targets for detection by the similarity search. Within the smaller dashed circle the crosses represent sequences detected as similar to the query by an ordinary BLAST search (the hits). The small circular arc attached to this circle represents further homologous sequences that could be detected by a second BLAST search, using one of the hits from the first BLAST search as query. In this case the query sequence used is the one marked by a diagonal (St Andrew's) cross. In this way, hits from the first and subsequent BLAST searches can be used to extend the coverage of evolutionary relationships to the original query sequence.*

but rather as a BLAST search with a **sequence profile** formed from all the hits. Sequence profiles are probabilistic modes of sets of related sequences that contain information from all the sequences. Using sequence profiles means that only one BLAST search is required for each iteration, rather than a BLAST search for each of the new hit sequences from the previous iteration. In practice, PSI-BLAST searches often detect up to twice as many evolutionarily related sequences as ordinary BLAST searches, which makes it a very useful tool indeed.

There is one potential problem with iterative database searches. This is contamination with unrelated sequences. If the hits from any one iteration of BLAST contain a false positive then this sequence will contaminate all further iterations with its own relatives that are not relatives of the original query. For this reason the *E* value cutoff used to choose the sequences to be employed at the next iteration of PSI-BLAST is chosen by default to be conservative, in order to minimize the risk of a false positive entering the search. A related problem that the internal workings of PSI-BLAST have to contend with is the domain structures of proteins. It is possible that an intermediate sequence (sequence B above) could be related to A by sharing a domain, and to C by sharing a different domain. In such a case there is no reason to infer that A is related to C. This is illustrated in *Fig. 12*.

A PSI-BLAST server can be found at the NCBI. *Table 2* shows an example of the use of PSI-BLAST. In this case the query sequence was the first PH domain from human pleckstrin and the objective of the search was to find other PH domains.

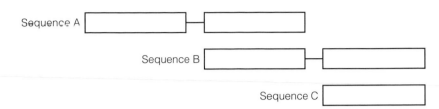

Fig. 12. Shared domains can mean that an intermediate sequence B does not relate sequence A to sequence C. Domains are indicated by boxes. Sequence B shares its first domain with sequence A and its second with sequence C, but nothing is shared between A and C.

These domains are known to be very common, but to have very divergent sequences. The parameters for the search considered a hit to be significant if its E value was below 0.01, and hits with E values below 0.005 were used to form the sequence profile at each iteration. While the first iteration (an ordinary BLAST search) found only 93 hits, subsequent iterations added more than 500 new sequences. The search converged (no more new sequences were found) at the fourth iteration. This test case shows conclusively the advantage of PSI-BLAST over an ordinary BLAST search in detecting distant sequence similarity.

Table 2. Detection of PH domain sequences at progressive PSI-BLAST iterations

Iteration number	Number of PH domain sequences found
1	93
2	607
3	622
4	622

J4 SEQUENCE FAMILIES, ALIGNMENT, AND PHYLOGENY

Key Notes

Multiple alignment

Multiple alignment illustrates relationships between two or more sequences. When the sequences involved are diverse the conserved residues are often key residues associated with maintenance of structural stability or biological function. Multiple alignments can reveal many clues about protein structure and function.

Software

The best-known software is the Clustal package, available by ftp from ftp://ftp.ebi.ac.uk/pub/software/clustalw2/

Progressive alignment

Most commonly used software uses the method of progressive alignment. This is a fast method, but frozen-in errors mean that it does not work perfectly always. Biological knowledge can provide information about likely alignments, and where automatically produced alignments turn out to be imperfect, software for manual alignment editing is required. A recent introduction of the program MUSCLE has improved on the performance of earlier progressive alignment methods.

Phylogenetics

Similarities and differences among species can be used to infer evolutionary relationships (phylogenies). This is because, if two species are very similar, they are likely to have shared a recent common ancestor. Many different types of character can be used in phylogenetic analysis, but nucleic acid and protein sequences are the most popular because they are common to all life forms (allowing both closely and distantly related taxa to be studied), and they can be compared objectively. However, caution must be exercised when inferring phylogenies from sequences because the rate of mutation may not be constant, and sequences may be subject to differential selection.

Graphs and trees

A graph is a diagram showing relationships between particular entities; for example, evolutionary relationships between species. Evolutionary relationships are generally represented by a special type of graph called a tree, which has n nodes and n-1 edges.

Phylogenetic trees and cladograms

A phylogenetic tree is a simple way to show evolutionary relationships, with species represented by nodes and lines of descent represented by links. Phylogenetic trees may be unrooted or rooted, the latter including the position of the last common ancestor of each tree member. A phylogenetic tree that shows the evolution of species as a series of bifurcations is a binary tree or cladogram. The edges in a cladogram may vary in length to convey a sense of evolutionary time.

Classification and ontologies

Classification systems are arbitrary in nature; that is, there is no standard measure of difference that defines a species, genus, family, or order. Rigorous definitions are difficult to apply and such static ontologies, which are common in biological classification systems, may be replaced in the future by dynamic ontologies that are more flexible.

Similarity and distance tables

These are tables that show the relatedness between species for a set of chosen characters as either the percentage of matches (similarity table) or percentage of differences (distance table). Either table may be converted into a matrix and used to construct a phylogenetic tree, but distance tables are used normally for the analysis of macromolecular sequence data.

Distance matrix methods

These methods work by selecting the two most closely related taxa in a distance matrix and clustering them. The operation is repeated until there is only one cluster left. Several variations of the method exist, which differ in the way the distance between a new cluster and other candidate taxa is calculated.

Maximum parsimony methods

In maximum parsimony methods, sequences are compared and clustered on the basis of the minimum number of mutations required to convert one sequence into another at any given position. The final phylogenetic tree is based on the overall number of changes required throughout the whole sequence.

Maximum likelihood methods

Maximum likelihood methods are similar to maximum parsimony methods but include a user-defined model that allows the probability of a given substitution occurring at any given position in the sequence to be built into the algorithm. The likelihood of a given sequence change is calculated for each position in the sequence and the most reliable tree is that with the maximum overall likelihood.

Adding a root

Some methods produce rooted trees (e.g., UPGMA) and others produce unrooted trees (e.g., neighbor-joining). A root can be added to an unrooted tree either by using an outgroup, or by assuming a molecular clock.

Limitations of phylogenetic algorithms

Limitations to phylogenetic algorithms include incorrect sequence alignments, variation of evolution rates within a sequence, and variation of evolution rates for the same sequence in different branches of the phylogeny. These problems can be addressed by generating robust alignments and using algorithm modifications that correct known biases. Exhaustive analysis of more than about 10 sequences is not possible because of the large number of possible trees. Heuristic and branch-and-bound methods are useful alternatives.

Phylogenetic software

Many programs are available for phylogenetic analysis. Two of the most versatile and popular are PAUP and PHYLIP. These are suitable for distance matrix, maximum parsimony, and maximum likelihood analysis methods.

How reliable are phylogenetic trees?

There is no guarantee that a given tree will accurately represent evolutionary history. However, the reliability of data can be assessed by resampling and constructing more trees, using approaches such as boot-strapping and jack-knifing.

Related sections

Graph theory and its applications (H2)

Sequence analysis (J3)

Domain families and databases (J5)

Structural bioinformatics (L4)

Structural classifications (L5)

Multiple alignment

Protein and nucleic sequences exist in families and their inter-relationships can be illustrated by multiple alignment of the sequences. These are like the pair-wise sequence alignments of Section J3, but generally involve more than two sequences. Multiple alignment often tells us more than pair-wise alignment because it is more informative about functional conservation. For example, two identical amino acid residues may be aligned between two protein sequences, but the fact that these have not mutated may just be down to chance. On the other hand, if a residue is conserved throughout a family of sequences that are otherwise quite diverse, then this indicates that the residue might play a key structural or functional role.

An example of a multiple alignment is shown in *Fig. 1*. This is a section of a multiple alignment of serine protease sequences, the uppermost of which is human thrombin (THRB_HUMAN). This alignment section illustrates that there are two main reasons for family-wide conservation of amino-acid residues or residue properties in proteins, first to preserve function, and second to preserve structure. The biochemical function of serine proteases is the cleavage of peptide bonds, as their name suggests. The cleavage is carried out by a serine (S) residue that is activated as a nucleophile by transfer of charge using neighboring histidine (H) and aspartic acid (D) residues. These three residues are known as the "catalytic triad," and are essential to the function of the enzymes. Because they are essential to function, they are conserved in each member of the family. The conserved histidine residue is the sixth residue from the right-hand side of the alignment in *Fig. 1*, and as expected, it is conserved in all the aligned sequences. Conservation of residues to preserve function is very common in proteins, but it is important to remember that there are examples of homologous proteins that have different functions. For example, as discussed, lysozyme and α-lactalbumin are homologous proteins with very similar sequences, but different functions (the former is an enzyme and the latter a mammalian regulatory protein).

The maintenance of a stable three-dimensional structure is another reason for residue conservation in protein families. This is discussed in detail in Section L4.

```
SecStructure    ....................bBBBBb...----.bBBBBBb.....bBBb.aaa.bba
THRB_HUMAN      LESYIDGRIVEGSDAEIGMSPWQVMLFRKSP----QELLCGASLISDRWVLTAAHCLLYP
THRB_BOVIN      FESYIEGRIVEGQDAEVGLSPWQVMLFRKSP----QELLCGASLISDRWVLTAAHCLLYP
THRB_MOUSE      LDSYIDGRIVEGWDAEKGIAPWQVMLFRKSP----QELLCGASLISDRWVLTAAHCILYP
THRB_RAT        LDSYIDGRIVEGWDAEKGIAPWQVMLFRKSP----QELLCGASLISDRWVLTAAHCILYP
LFC_TACTR       SDSPRSPFIWNGNSTEIGQWPWQAGISRWLADHNMWFLQCGGSLLNEKWIVTAAHCVTYS
FA9_RAT         EPINDFTRVVGGENAKPGQIPWQVILNGEIE------AFCGGAIINEKWIVTAAHCLK--
FA9_RABIT       QSSDDFTRIVGGENAKPGQFPWQVLLNGKVE------AFCGGSIINEKWVVTAAHCIK--
FA9_PIG         QSSDDFIRIVGGENAKPGQFPWQVLLNGKID------AFCGGSIINEKWVVTAAHCIEP-
FA7_BOVIN       NGSKPQGRIVGGHVCPKGECPWQAMLKLNGA------LLCGGTLVGPAWVVSAAHCFER-
FA7_MOUSE       NSSSRQGRIVGGNVCPKGECPWQAVLKINGL------LLCGAVLLDARWIVTAAHCFDN-
FA7_RABIT       GASNPQGRIVGGKVCPKGECPWQAALMNGST------LLCGGSLLDTHWVVSAAHCFDK-
PRTC_HUMAN      QEDQVDPRLIDGKMTRRGDSPWQVVLLDSKK-----KLACGAVLIHPSWVLTAAHCMDE-
PRTC_RAT        EELELGPRIVNGTLTKQGDSPWQAILLDSKK-----KLACGGVLIHTSWVLTAAHCLES-
PRTC_MOUSE      DELEPDPRIVNGTLTKQGDSPWQAILLDSKK-----KLACGGVLIHTSWVLTAAHCVEG-
PSS8_HUMAN      CGVAPQARITGGSSAVAGQWPWQVSITYEGV------HVCGGSLVSEQWVLSAAHCFPS-
                :  *        ***. :              **. ::    *:::****.
```

Fig. 1. Part of a multiple alignment of some serine protease sequences. The uppermost sequence is a section of the human thrombin sequence and its secondary structure is given in the line above (a,A = helix (α, 3₁₀ or Pi), b,B = β strand). Given in the line at the foot of the alignment are symbols indicating the degree of conservation in each column ('' indicates a completely conserved column (same residue in each sequence), ':' indicates a column containing only very conservative substitutions, and '.' indicates a column containing mostly conservative substitutions).*

There is therefore a very strong link between the conservation patterns seen in multiple sequence alignments and the underlying protein three-dimensional structure. Here we simply note some structural features in the alignment of *Fig. 1*, and refer the reader to Section L4 for a full discussion. The key conserved structural features in *Fig. 1* are the two conserved cysteine residues, which form a disulfide bond; the strong tendency is to conserve hydrophobic residues in β-strand secondary structure elements, and the positioning of insertions and deletions outside the secondary structure elements (e.g., the four residue insertion in the LFC_TACTR sequence).

Software

The most commonly used multiple alignment software is the Clustal package (Thompson J.D., Higgins D.G., Gibson T.J). This software is freely available by ftp from ftp://ftp.ebi.ac.uk/pub/software/clustalw2/ It works by the progressive alignment method outlined below.

Progressive alignment

The problem of multiple sequence alignment is much more difficult computationally than the pair-wise alignment described in Section J3. Most current programs use the method of **progressive alignment**, which has the advantage of being relatively fast. This involves making a preliminary assessment of how the sequences are related using pair-wise alignments, using this to form a **guide tree** and then using this guide tree to add sequences progressively to the alignment, beginning with the most closely related sequences and finishing with the most distant. This process is illustrated in *Fig. 2*.

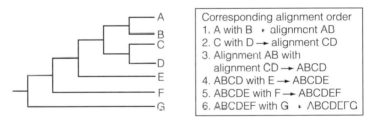

Fig. 2. Progressive alignment. The seven sequences A–G are related as shown in the guide tree on the left. This tree can be viewed as a possible way in which the sequences evolved. Lengths of the branches (horizontal lines) indicate the degree of difference between the sequences. Thus, A and B are closely related and have diverged from a near common ancestor, as are C and D. These four sequences have a common ancestor and are more closely related to each other than to sequences E, F, G, etc. This leads to the alignment order shown in the box on the right. First, the closely related sequences A and B are aligned, followed by C and D, and then these two alignments are aligned to form an alignment of four sequences. Then sequences E, F, and G are added sequentially to this alignment. All these alignments are performed with the dynamic programming algorithms described in Section J3.

Progressive alignment is usually very effective, but it suffers from the problem that alignment errors made early in the process can never be rectified. They are "frozen" into the alignment. Thus, in the example of *Fig. 1*, there might be important information in sequences C and D that could improve the alignment of A and B, but this can never be used, because with the progressive algorithm, A and B are aligned independently of C and D. Further, independent biochemical information can sometimes give information about the correct alignment of sequences. For instance, it may be known experimentally that certain residues are key structurally or functionally and should be in conserved columns. Most often alignment errors are very obvious, such as failure to align cysteine residues involved in disulfide bonds. For these reasons it is sometimes necessary or desirable to edit multiple alignments manually. Several software products are available for this purpose; for example, the Seaview or Cinema software.

There are several refinements that are applied often to the process of progressive alignment. For instance, in the Clustal suite of programs, gap penalties are varied so that gap insertion is more likely in hydrophilic loop regions, as would be expected from the discussion of the relationship of multiple alignment to protein structure above. Further, different amino acid substitution matrices (Section J3) are applied depending on the degree of relatedness of the sequences being aligned.

Recent advances　More recently new methods of multiple alignment have been introduced. These include MUSCLE and T-coffee. T-coffee uses both global and local alignments to improve accuracy. MUSCLE uses a progressive alignment strategy but uses refinement steps to improve accuracy compared to Clustal.

Phylogenetics　Living organisms are classified into groups based on observed similarities and differences. A general principle of classification systems is that the more closely related species a is to species b, the more likely they shared a recent common ancestor. In this way, similarities and differences between organisms can be used to infer phylogenies (evolutionary relationships). The branch of science that deals with resolving the evolutionary relationships among organisms is phylogenetics.

Phylogenetics can be studied in three ways. In phenetics, species are grouped with others they resemble phenotypically and all characters are taken into account. In cladistics, species are grouped only with those that share **derived** characters; that is, characters that were not present in their distant ancestors. The third approach, evolutionary systematics, incorporates both phenetic and cladistic principles. Cladistics is accepted as the best method available for phylogenetic analysis because it accepts and employs current evolutionary theory; that is, that speciation occurs by bifurcation (cladogenesis). Information on cladistics and a useful glossary of terms can be found at the following URL: http://www.cladistics.org

Many different criteria can be used for phylogenetic analysis, including morphological characteristics, biochemical properties and, most recently, the analysis of macromolecular sequences (nucleic acid and protein sequences). Macromolecular sequences are particularly useful for comparison because they provide a large and unbiased data set that extends across all known organisms, allowing the comparison of both closely related and distantly related taxa. Most importantly, however, the relatedness between sequences can be quantified objectively using sequence alignment algorithms. This is where bioinformatics plays an important role in phylogenetics.

The simple principle behind the phylogenetic analysis of sequences is that the greater the similarity between two sequences, the fewer mutations are required to convert one sequence into the other, and thus the more recently they shared a common ancestor. However, it is important to note that any evolutionary relationships inferred from such analysis sometimes assume a constant rate of mutation and the absence of differential selection in the sequences chosen for comparison. These conditions are seldom met!

Graphs and trees

The clearest way to visualize the evolutionary relationships among organisms is to use a graph. In mathematics, a graph is a simple diagram used to show relationships between entities, such as numbers, objects, or places. Entities are represented by nodes and relationships between them are shown as links or edges (connecting lines). Some simple graphs are shown in *Fig. 3*. These graphs can be used for a variety of purposes. For example, G1 might represent a cyclic chemical compound with atoms at the nodes and chemical bonds as the links, and G2 might represent a street map showing one-way streets. Note that in G2 the links have direction (i.e., they are represented by arrows rather than simple lines). Please refer to Section H2 for further details on graphical structures.

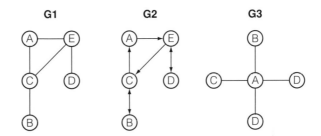

Fig. 3.   Three examples of simple graphs.

Phylogenetic trees and cladograms

Phylogenetic trees (also called dendrograms) are used to show evolutionary relationships. The nodes represent different organisms and edges are used to show lines of descent. As an example we consider the phylogenetic relationship between the entities C (chimpanzee), G (gorilla), H (human), and O (orangutan). Different trees representing this phylogeny are shown in *Fig. 4*.

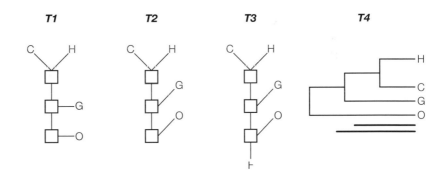

Fig. 4.   Trees showing the relationship between four species: C, G, H, O and F (chimpanzee, gorilla, human, orangutan and forebear). Ancestral nodes in T1, T2, and T3 are represented by boxes. The thick horizontal lines under T4 are discussed in the text.

The first point to note about these trees is that there are two types of node. The ancestral nodes (represented by boxes) give rise to branches. These may link to other ancestral nodes, or they may link to terminal nodes (shown as letters), which are also known as leaves or tips. Leaves represent known species and mark the end of the evolutionary pathway. Ancestral nodes may or may not correspond to a known species (e.g., the last common ancestor of humans and chimpanzees is unknown) but we can infer its existence.

The second point to note is that *T1* and *T2* are unrooted trees, whereas *T3* is a rooted tree. *T1* and *T2* are identical except that *T2* is drawn in a conventional style with angled branches to look more like a real tree. These are described as unrooted trees because neither of them show the position of the last common ancestor of all the species. In *T3* the position of this ancestor is indicated by the node F.

The third point to note is that each tree is binary; that is, no ancestral nodes have more than two branches. Thus, the evolution of species is represented as a series of bifurcations, which fits in with cladistic theory. For this reason, the trees also may be termed cladograms.

The fourth point to note is that the length of the branches may or may not be significant. In *T1*, *T2*, and *T3*, all the branches are of the same length, whereas in *T4* the branches are of different lengths. The lengths of the branches may be used to indicate the actual evolutionary distances between taxa. A cladogram that conveys a sense of evolutionary time using branch lengths may be called a phylogram. Note that, if *T4* represented differences between macromolecular sequences, there might be cases where the distance between H and C would be zero, thus H and C would appear on each end of a vertical line.

Finally, note that *T4* shows the same data as *T1* and *T2*, but it is presented as it would appear in the output format of multiple sequence alignment software such as ClustalW/X. In this format, the edges are still represented by lines, but the ancestral nodes are represented by vertical lines rather than boxes. Phylogenetic trees such as *T4* are useful for visualizing the concept of a **clade**, which is defined as a group of organisms descended from a particular common ancestor (i.e., an ancestor and all its descendents). The groups of organisms included within a clade are defined arbitrarily. If, for example, the distance represented by the upper of the two lines beneath *T4* was thought to be significantly close, H and C would be said to be in the same clade. If, however, the criterion was the length of the lower line, H, C, and G would be placed in the same clade. We explore this point in more detail below.

Classification and ontologies

For partly historical reasons, biological science abounds with examples of hierarchical or tree-like classifications. The current version of the Linnaean system of classification into species, genera, families, orders, classes, and kingdoms was the first to be developed, but the idea has been recruited into the nomenclature of the Enzyme Commission (EC numbers for enzymes), and protein structural classification systems such as SCOP and CATH (Section L5).

The phylograms represented by *Fig. 5* and *T4* in *Fig. 4* suggest there is a method for defining a classifier objectively. That is, by using some criterion or distance, species are represented by organisms that form a clade, genera by those that form a larger clade, families and orders by those that form still larger clades, etc. The problem that arises is that the criteria are not constant over all living organisms. For example, if the great apes in *Fig. 4* were bacteria, they would be regarded as minor variations within the same species, whereas they actually represent different families of mammals. Any attempt to overcome this problem by imposing extra rules leads to inconsistency. For example, one might say chimpanzees and humans are at least distinct species because they cannot interbreed. However, inter-species and even inter-generic crosses are commonplace among ornamental plants and among fish, so this rule cannot be applied rigorously. Among the bacteria (eubacteria), lateral gene transfer occurs widely, but does this mean that all bacteria belong to the same species?

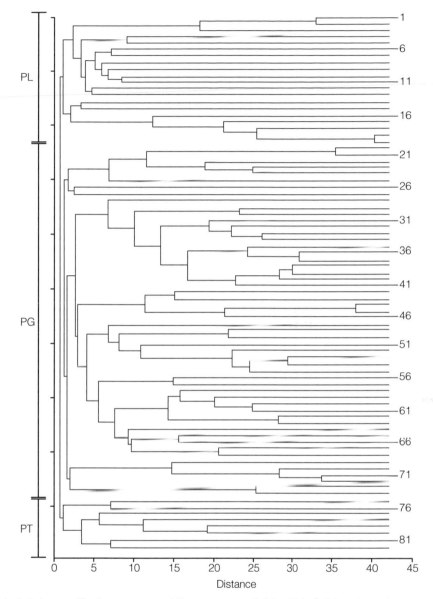

Fig. 5. *Example of an actual cladogram. The leaves represent the sequences of all the 12-helix integral membrane proteins known in 1996 (with redundant members removed). The sequences are numbered arbitrarily from 1 to 82. The key is not included here. PL, PG, and PT are three major clades identified by the authors. "P" refers to the PAM matrix (Section J3) and 'L,' 'G,' and 'T' refer to typical members of the clades (lactose transport, glucose transport, and tetracycline resistance, respectively). This is an unrooted tree and the point of trichotomy is about half way down the PG clade.*

An important point arising from the above is that classifications have to be not only objective and logical but also useful. It could be argued that all bacteria logically belong to the same species, but this is not useful when we want to find the causes of bubonic plague, leprosy, and typhoid, or when we want to know which organisms to use in the manufacture of cheese and vinegar. The above exemplifies the problems of a static ontology. Ontology is a term used in artificial intelligence systems to describe the relationships between entities within a given area of interest. In a static ontology, there is a formal and explicit specification of the relationships between entities, rather like the formal and inflexible classification systems discussed above. The future of classification may well lie in alternative ontologies, which are more flexible. A dynamic ontology, for example, allows the relationship between entities to be progressively refined and updated; that is, the rules can be relaxed to fit around the problem rather than forcing descriptions onto the entities themselves.

Similarity and distance tables

Phylogenetic trees can be constructed from either similarity tables or distance tables, which show the resemblance among organisms for a given set of characters. *Fig. 6* shows an example of each type of table, where *a* to *e* are five species (or taxa) that have been scored for resemblance. The characters chosen for comparison may be morphological or biochemical in nature, or may be DNA, RNA, or protein sequences. If sequences are used, they are compared initially using multiple sequence alignment tools such as the ClustalW/X programs.

(a)

	a	b	c	d	e
a	100	65	50	50	50
b	65	100	50	50	50
c	50	50	100	97	65
d	50	50	97	100	65
e	50	50	65	65	100

(b)

	a	b	c	d	e
a	0	6	11	11	11
b	6	0	11	11	11
c	11	11	0	2	6
d	11	11	2	0	6
e	11	11	6	6	0

Fig. 6.   Hypothetical (a) similarity table, and (b) distance table for five organisms, a–e.

The numbers in the similarity table (*Fig. 6a*) show the *percentage of matches*, thus the diagonal (in which each species *a* to *e* is compared to itself) consists of 100% values. Such data form the basis of Adamsonian analysis or numerical taxonomy. The numbers in the distance table (*Fig. 6b*) show *percentage differences or distances*, thus the diagonal consists of 0% values. Although either table is suitable for phylogenetic tree-building using essentially the same methods, macromolecular comparisons are recorded usually as differences, so we shall use distance tables in the discussion below. A common measure of difference between macromolecular sequences is 100-S, where S is the percentage of identical monomers when the sequences have been optimally aligned (Section J3).

Distance matrix methods

Some of the most commonly used methods for tree-building in phylogenetic analysis involve agglomerative hierarchical clustering based on distance matrices. The essential basis of this type of algorithm is that the taxa represented in a distance table (such as that shown in *Fig. 6b*) are transformed into a series of nested partitions by merging two taxa together in each step until only one cluster remains.

This process can be formally expressed as follows.

● Begin by creating *n* singleton clusters; that is, clusters containing one taxon from the distance table. For this purpose, a distance table is regarded as an *n* × *n* matrix, with *n* representing the number of taxa being compared. In *Fig. 6b*, *n* = 5, and the five singleton clusters are *a*, *b*, *c*, *d*, and *e*.

- Then, determine the differences between each possible pair of clusters. This is formally expressed as a distance function, dis(i,j), where i and j represent any given pair of clusters. These differences are the data in the difference table shown in *Fig. 6b*.
- Next, select the pair of clusters for which dis(i, j) is minimal; that is, select the two most similar clusters. In *Fig. 6b*, this would be c and d.
- Then merge these data into a new cluster (ij). In *Fig. 6b*, this would mean defining a new cluster (cd) as the union of c and d. In the resulting phylogenetic tree, a new ancestral node would be defined with c and d as branches.
- The number of clusters in the matrix is now reduced by one. Repeat the analysis until there is only one cluster left. This is a recursive process, which must be carried out $n - 1$ times.

There are a number of alternative distance matrix algorithms, which differ in the way that the distance between a new (merged) cluster and the remaining taxa in the matrix are calculated for the purpose of recursive searching. The four popular variations of the method are single linkage, complete linkage, average linkage, and the centroid method. In the single linkage method, the distance between the merged cluster (ij) and any candidate taxon k is minimized; that is, the smaller of dis(ik) and dis(jk) is chosen. In the complete linkage method, the distance is maximized; that is, the larger of dis(ik) and dis(jk) is chosen. There are two variations of the average linkage method, in which the distance between the merged cluster (ij) and candidate taxon k is taken as the arithmetic mean of dis(ik) and dis(jk). In the unweighted pair group method using arithmetic mean (UPGMA), the distance is calculated as a simple average because each candidate is weighted equally. In the weighted pair group method using arithmetic mean (WPGMA), the clusters are weighted according to their size; that is, so that the candidate taxon k is equivalent in weighting to all previous taxa in the cluster. The calculations for these algorithms are therefore slightly more complex than those for the single and complete linkage methods. Similarly, there are weighted and unweighted variations of the centroid method also, in which the centroid value is used rather than the arithmetic mean. These variations are known as the unweighted and weighted pair-group methods using centroid value (UPGMC, WPGMC).

The UPGMA method discussed above is popular because of its simplicity, but it makes trees with two important assumed properties that are particularly important when trees are made from macromolecular sequence data. First, it is assumed that evolution occurs at the same rate on all tree branches (this is known as the assumption of a molecular clock), and second it is assumed that distances in the trees are additive. The additive assumption is that the distance between any two leaves is the sum of distances on edges connecting them. As a consequence of these assumptions the method can create incorrect trees. For example, two sequences might be very similar not because they have a direct common ancestor, but simply because they are evolving very slowly by comparison with other sequences being analyzed. UPGMA would produce a tree in which they had a direct common ancestor.

Neighbor-joining (NJ) is a clustering method related to UPGMA that is able to solve some of the problems discussed above. In particular it does not make the assumption of additivity. It is quite fast computationally also, and so is almost always a better choice of method, and is used in the ClustalW/X programs to estimate trees from multiple sequence alignments. To start building an NJ tree all taxa are placed in a star; that is, individually joined to a single central node or hub through n spokes. From this star, the two taxa with the greatest similarity are chosen and are connected to a new internal node; these taxa are then known as neighbors. The process is repeated until the whole star is resolved into a tree. For an unrooted tree there will be $n - 3$ internal branches and the process must be repeated $n - 3$ times. ClustalW/X also offers bootstrapping to estimate the robustness of the generated trees (see below).

Maximum parsimony methods

In distance-matrix methods, all possible sequence alignments are carried out to determine the most closely related sequences, and phylogenetic trees are constructed on the basis of these distance measurements. As an alternative, maximum parsimony methods can be used, in which trees are constructed on the basis of the **minimum number of mutations** required to convert one sequence into another. In proteins this is achieved by multiple sequence alignment followed by the identification and analysis of corresponding positions in each sequence. For each aligned residue, the minimum number of base substitutions required to convert one amino acid into another is calculated. The final tree is generated by grouping those sequences that can be inter-converted with the smallest number of overall changes. This method is very attractive intellectually, but like the maximum likelihood method below can be expensive in computer time, so neighbor-joining (above) is often to be preferred.

Maximum likelihood methods

Maximum likelihood methods also involve multiple sequence alignment and the analysis of changes at each position of the sequence. However, the difference between maximum likelihood and maximum parsimony is that the former incorporates an **expected model** of sequence changes, which weights the probability of any residue being converted into any other. This model can be set by the experimenter. For each possible tree, the likelihood of different sequence changes at each position is calculated, and these values are multiplied to provide an overall likelihood for each tree. The most reliable tree is that with the maximum likelihood.

Adding a root

Some methods automatically generate rooted trees. For instance, the clustering methods described above place the root between the final two clusters to be joined before the algorithm terminates. On the other hand, neighbor-joining produces unrooted trees. There are two main ways in which roots can be added to unrooted trees. The easiest is to use an **outgroup**. For instance, if the tree had been generated from a mixture of mammalian and bacterial sequences then it is clear that the root should lie between these two groups so that the first divergence in the tree is to divide bacteria and mammals. In this case one of the distantly related groups (e.g., the bacterial sequences) is known as the outgroup. In the case where there is no obvious outgroup in the sequences, a root can be added half way between two most distantly related sequences, essentially assuming a molecular clock.

Limitations of phylogenetic algorithms

All clustering methods suffer from three major limitations: incorrect sequence alignments, failure to account for variation of evolution rates at different sites within a sequence, and failure to account for sequences evolving at different rates in different taxa. These limitations often generate incorrect trees through a process known as long-branch attraction, in which rapidly evolving sequences are grouped even if their relationship is very distant.

The problems can be addressed in several ways. The first is to make sure that sequence alignments are robust. If many diverse sequences are included in the analysis and the alignment contains many gaps, this could be a source of error. It is much better to eliminate outliers before commencing the analysis because each clustering step is definitive and cannot be undone later. Caution should be exercised in particular if a newly built tree disagrees with others generated through the analysis of different genes or proteins.

Improvements and modifications to the clustering algorithms have made tree-building much more accurate also. For example, the Farris transformed distance method is a useful preliminary to UPGMA and the Fitch-Margolish method is a robust modification of WPGMA. The paralinear distances algorithm (LogDet) addresses the problem of unequal evolution rates and is now incorporated into many phylogenetic software packages (see below).

It is important to understand the limitations of computational power in phylogenetic analysis. Essentially, the objective of any tree-building experiment is to select the correct tree from many incorrect trees. Assuming that all other limitations have been overcome, how many trees is it necessary to build to get the correct one? This depends on the number of data points being analyzed. For example, if there are three sequences, there are three possible rooted trees and one possible unrooted tree. If there are five sequences, there are 105 possible rooted trees and 15 possible unrooted trees. If there are seven sequences, there are 10 395 possible rooted trees and 954 possible unrooted trees. Exhaustive tree search methods, where all possible trees are created and tested, therefore, can only be used for up to about 10 sequences. If more sequences need to be compared, alternative methods must be used. For example, branch-and-bound analysis ignores families of trees that cannot possibly give a better answer than a tree that already been found. Heuristic analysis samples trees randomly and can be used for many sequences, but the best tree may be missed.

Phylogenetic software

A large number of software packages are available, some free over the internet, for the phylogenetic analysis of macromolecular sequences. Some popular programs, such as **PAUP** (phylogenetic analysis using parsimony) and **PHYLIP** (phylogenetic inference package) are versatile and allow distance matrix, parsimony, and maximum likelihood method analysis to be carried out. Such packages are frequently updated with the most recent modifications and corrections to the phylogenetic algorithms. Other packages, such as MacClade, are useful for tree manipulation. A comprehensive resource for phylogenetic software can be found at the following URL: http://evolution.genetics.washington.edu/phylip/software.html

How reliable are phylogenetic trees?

There is no guaranteed way to verify that a phylogenetic tree represents the true path of evolutionary change. However, there are ways in which to test the reliability of phylogenetic predictions. First, if different methods of tree construction give the same result, this is good evidence that the tree is reliable. Second, the data can be re-sampled to test their statistical significance. In a technique called bootstrapping, data are sampled randomly from any position within a multiple sequence alignment, and are built into new artificial alignments, which then are tested by tree-building. Since the sampling is random, some positions may be sampled more than once and others not at all. Ideally, the trees built by bootstrapping should match the original tree always, and this would be defined as "100% bootstrap support". In reality, bootstrap support of 70% or more for any given branch of a tree is taken to provide 95% confidence that the branch is correct. Jack-knifing is a similar process in which about 50% of the original data are re-sampled and used to make a new matrix, from which phylogenetic relationships are reconstructed.

J5 DOMAIN FAMILIES AND DATABASES

Key notes

Protein families

Assigning sequences to protein families is a very valuable way of predicting protein function. Many ways have been developed to represent protein family information and these have been stored in secondary protein family databases.

Consensus sequences

These condense the information from a multiple alignment into single sequence. Their main shortcoming is the inability to represent any probabilistic information apart from the most common residue at a particular position. Derivation of consensus sequences illustrates that any protein family representation is subject to bias if the set of sequences from which it was derived is biased.

PROSITE

The PROSITE database contains sequence patterns associated with protein family membership, specific protein functions and post-translational modifications. A special notation involving square brackets (e.g., [LIVM]), curly brackets (e.g., {FD}), and x(n) is used to express alternative residues at each position in the pattern. The database is curated manually and any known false positives or false negatives are reported. Some of the patterns, particularly short post-translational modification patterns, suffer from a lack of specificity and occur many times in some sequences. The database also contains some sequence profile entries.

PRINTS and BLOCKS

These represent protein families of multiply aligned un-gapped segments (motifs) derived from the most highly conserved regions of the sequences. By representing more of the sequence, they have the potential to be more sensitive than short PROSITE patterns. The ability to match in only a subset of the motifs associated with a particular family means that they have the ability to detect splice variants and sequence fragments and to represent sub-families. WWW-based search engines for the databases are available.

Domain families

Many proteins are built up from domains in a modular architecture. Study of protein families is best pursued as a study of protein domain families. Prodom is a database of protein domain sequences created by automatic means from the protein sequence databases. The resources described in this section can be viewed as protein domain family descriptions.

Sequence profiles

These represent complete domain sequences with scores for each amino acid at each position in a multiple alignment, and position specific measures of the likelihood of insertion and deletion. They are used as an alternative to sequence patterns in some PROSITE database entries.

Hidden Markov models

These are rigorous statistical models of protein domain family sequences. They can be viewed as a sequences of match, insert, and delete states that generate protein sequences according to probability distributions for each state and each transition between states. A model representing a protein domain family generates sequences from that family with high probability and sequences from other families with lower probabilities. Algorithms are available to approximate the probability that a new protein sequence was generated by a particular family model, and these can be used to assign new protein sequences to families.

Resources

Pfam and SMART can be used for protein domain family analysis. The integrated resource Interpro unites PROSITE, PRINTS, Pfam, Prodom, and SMART.

Related sections

Statistical approaches to artificial antelligence and machine learning (I2)
Sequence analysis (J3)

Sequence families, alignment, and phylogeny (J4)
Structural bioinformatics (L4)
Structural classifications (L5)

Protein families

In Section J4 we discussed how multiple alignments of sequences from the same protein family can be used to deduce much important information about the structure, function, and key amino acid residues of the family. For this reason it has been important to store information from multiple alignments of protein sequences in databases. There are many possible representations of the information within a multiple alignment, including the alignment itself, consensus sequences, conserved residues and residue patterns, sequence profiles, and other probabilistic models of the sequence family. These are all useful depending on the application in mind, and most have been developed and stored in databases for large numbers of different protein families. We refer to these databases as **secondary databases**, because the information within them is not raw experimental data, which would be stored in **primary databases**, but has been derived in some way from experimental data.

Consensus sequences

Perhaps the simplest and most intuitive way of condensing the information in a multiple alignment is to use a consensus sequence. These can be derived in several ways, depending on the software used, but the same general idea is used in all cases. This is to produce a single sequence in which each residue is the most common, or consensus, for the sequence family. A method of derivation of a consensus sequence is illustrated in *Fig. 1*, where the consensus residue at a particular position is the most common residue in the corresponding alignment column, as long as it is shared by more than 60% of the sequences. If no residue is shared by 60% of the sequences then the consensus residue is X (any residue). Consensus sequences can be useful summaries of protein families, but are less powerful than some of the methods we will discuss later, and so databases of consensus sequences are not commonly used.

Fig. 1 illustrates some important general principles of protein family representation. First it shows the essential weakness of the consensus sequence approach. This is that much information from the sequences that do not contain the consensus residue is ignored, even though these hold information about allowed substitutions at that position. It also illustrates an important

```
THRB_HUMAN    LESYIDGRIVEGSDAEIGMSPWQVMLFRKSPQELLCGASLISDRWVLTAAHCLLYP
THRB_BOVIN    FESYIEGRIVEGQDAEVGLSPWQVMLFRKSPQELLCGASLISDRWVLTAAHCLLYP
THRB_MOUSE    LDSYIDGRIVEGWDAEKGIAPWQVMLFRKSPQELLCGASLISDRWVLTAAHCILYP
THRB_RAT      LDSYIDGRIVEGWDAEKGIAPWQVMLFRKSPQELLCGASLISDRWVLTAAHCILYP
FA9_RAT       EPINDFTRVVGGENAKPGQIPWQVILNGEIE--AFCGGAIINEKWIVTAAHCLK--
FA9_RABIT     QSSDDFTRIVGGENAKPGQFPWQVLLNGKVE--AFCGGSIINEKWVVTAAHCIK--
Consensus     XXSYIXGRIVEGXDAEXGXXPWQVMLFRKSPQELLCGASLISDRWVLTAAHCXLYP
```

Fig. 1.   Deriving a consensus sequence from a multiple alignment using thrombin (THRB) and Factor 9 (FA9) sequences as an example. Each position in the consensus corresponds to a column in the alignment. The consensus residue is the most common residue in the column if it is shared by more than 60% of the sequences, or X otherwise. The 60% threshold is usually variable.

consideration for all protein family representation methods. The set of sequences used in *Fig. 1* contains four thrombin sequences and two Factor 9 sequences; it is therefore biased in favour of thrombin. The consensus sequence derived is biased also, being a better representation of thrombin than Factor 9, and not a general representation of these two serine protease families. For instance, because the requirement is that a residue be shared by 60% of the sequences in order to be the consensus residue, it is sufficient that it appear in all four thrombin sequences, irrespective of the corresponding residues in the Factor 9 sequences. This is an important point that must be considered in all family representation methods. It is very important to address the issue of possible bias in the set of sequences or multiple alignment that have been used in the derivation, otherwise the representation derived will be biased itself.

PROSITE

PROSITE is a database of sequence patterns associated with protein family membership. It is developed by a largely manual process of seeking the patterns that best fit particular protein families and functions. For instance two patterns are associated with the serine protease family. These are

$$[LIVM]-[ST]-A-[STAG]-H-C$$

and

$$[DNSTAGC]-[GSTAPIMVQH]-x(2)-G-\ [DE]-S-G-[GS]-[SAPHV]-$$
$$[LIVMFYWH]\ -PA-[LIVMFYSTANQH].$$

Within these patterns square brackets indicate sets of possible residues that can occur in a particular position in sequences from the associated family. The first pattern represents a sequence of six residues and should be read as 'one of the residues L, I, V, or M, followed by S or T, then A, then one of S, T, A, or G, then H, then C.' In this case the penultimate residue (H) is a key catalytic residue of the serine protease family, and so it makes sense that there should be no alternatives at this position. The second pattern above is the sequence pattern that occurs around another key catalytic residue (serine). This pattern is longer, representing a sequence of 14 residues, and it includes an extra feature written as x(2). This means two residues of any type, and not necessarily of the same type. In other patterns the x notation may be extended further; for instance, x(2,4) indicates that between two and four residues of any type may be present at a particular position in the family. Another extension to the notation is the use of curly brackets; for instance, {LT} means that the position can be occupied by any residue except L or T. PROSITE patterns are derived most commonly from multiple alignments, indeed the first pattern given above is present in the multiple alignment example of Section J4.

PROSITE patterns differ from consensus sequences in that they tend to be much shorter than the total sequence length, and that they give a means of describing a set of acceptable residues in a multiple alignment column. The patterns can be useful in assigning distant homologs to sequence families when, for instance, all that remains of the sequence similarity is limited to a few important residues around the key functional machinery of the protein. Therefore they can be indicative also of shared biological function. As well as patterns associated with family membership, PROSITE contains generic patterns associated with processes like post-translational modification (glycosylation, phosphorylation, etc.). For instance the pattern

$$N-\{P\}-[ST]-\{P\}$$

is associated with N-glycosylation of asparagine (N). While these patterns are useful in some circumstances, they tend to be very short, and this leads to a lack of **specificity**. The pattern is likely to occur in many sequences in positions that are not real glycosylation sites. Such occurrences are known as false positives.

PROSITE patterns have a number of weaknesses. First, their shortness tends to lead to false positive occurrences in unrelated sequences (as above), and this effect is not limited to the short patterns associated with post-translational modification. Second, while they allow the description of variation at a particular position, they have no way of attaching probabilities to the variation. For instance, [LIVM] says that a position might be L, I, V, or M, but it does not say that perhaps L occurs in 90% of sequences in the family and I, V, or M only in the other 10%. Where PROSITE patterns have known false positives (or false negatives) they are annotated in the database. In order to overcome some of the difficulties associated with PROSITE patterns, the PROSITE database now contains sequence **profiles** in some entries. These attempt to describe longer sequence segments (usually complete domains) than the patterns, and are discussed in detail in Section J4. There are also several other databases that are complementary tools to PROSITE in sequence analysis, including PRINTS and BLOCKS (below), and Pfam and SMART.

The PROSITE database and various PROSITE pattern search options are available from the EXPASY WWW site (http://ca.expasy.org/prosite).

PRINTS and BLOCKS

PRINTS and BLOCKS (Henikoff and Henikoff) are closely related. Each represents protein families in terms of multiply aligned un-gapped segments derived from the most highly conserved regions in a group of proteins or protein family. Such multiply aligned un-gapped segments are termed **blocks** (in BLOCKS) or **motifs** (in PRINTS). In PRINTS a set of such motifs representing a family is called a **fingerprint**. The PRINTS database is very high quality, having been created with a great deal of manual effort, and contains extensive annotation and description for the protein family and function concerned. The original version of BLOCKS was created by automatic means, but now many databases (including PRINTS) are available in BLOCKS format. An example PRINTS entry for the SH3 domain is shown in *Fig. 2*. PRINTS represents this domain by four motifs covering the most conserved areas in the multiple alignments of many SH3 domain sequences.

The motifs within these databases typically cover larger regions of the sequence than PROSITE patterns, and unlike PROSITE, matching of motifs in sequences usually takes account of amino acid substitution matrices (Section J3),

Motif 1	Motif 2	Motif 3	Motif 4
GYVSALYDYDA	DELSFDKDDIISVLGR	EYDWWEARSL	KDGFIPKNYIEMK
YTAVALYDYQA	GDLSFHAGDRIEVVSR	EGDWWLANSL	YKGLFPENFTRHL
RWARALYDFEA	EEISFRKGDTIAVLKL	DGDWWYARSL	YKGLFPENFTRRL
PSAKALYDFDA	DELSFDPDDVITDIEM	EGYWWLAHSL	YKGLFPENFTRRL
EKVVAIYDYTK	DELGFRSGEVVEVLDS	EGNWWLAHSV	VTGYFPSMYLQKS

Fig. 2.   Example sequences for the four conserved motifs used to represent the SH3 domain in the PRINTS database. These motifs represent the most conserved regions from the alignment of many SH3 domains. For brevity, we show only five example sequences for each motif, but there are many more examples in the PRINTS database. Nevertheless, the conservation patterns in each motif should be clear, particularly with reference to the preferred amino acid substitutions (Section J3).

thus not requiring exact matches to a fixed pattern. For these reasons, matches to PRINTS/BLOCKS patterns are potentially more sensitive (more distant relationships can be found) and more specific (fewer false positives occur) than matches to PROSITE patterns. An advantage of the representation as a set of motifs or blocks is a natural way of representing sequence fragments and splice variants that contain only a subset of the motifs. Such sequences present more problems for the profile and hidden Markov model methods discussed later in this section.

Search engines for the databases are available from http://bioinf.man.ac.uk/ dbbrowser/PRINTS and http://blocks.fhcrc.org/ An example search of the PRINTS database, using FingerPRINTScan, for a sequence fragment known to contain an SH3 domain is shown in *Fig. 3*. It can be seen that this sequence does indeed match the four motifs (*Fig. 2*) held in PRINTS for the SH3 domain with significant (low) p values. These p values can, as usual, be interpreted as the probability that a match scoring at least as well as the identified match would occur by chance in a random sequence.

Domain families
We have discussed simple representations of conserved features in protein families, including PROSITE patterns and fingerprints. We will move on now to discuss some perhaps more sophisticated ways of describing protein families. These attempt to describe complete sequence families, including position-specific insertion and deletion probabilities, rather than just the most conserved parts.

Query sequence: YEDEEAAVVQYNDPYADGDPAWAPKNYI**EKVVAIYDYTK**DKD
DELSFMEGAIIYVIKKNDD**GWYEGVCN**RV**TGLFPGNYVESI**MHYTD

Fingerprint	Motif number	Pval	Sequence
SH3 domain	1 of 4	3.02e-04	EKVVAIYDYTK
	2 of 4	1.45e-06	DELSFMEGAIIYVIKK
	3 of 4	6.39e-02	DDGWYEGVCN
	4 of 4	1.25e-05	VTGLFPGNYVESI

Fig. 3.   PRINTS database search. The sequence fragment above (top) was searched against the PRINTS database. A significant match to all four motifs in PRINTS, representing the SH3 domain, was found, and a simplified version of the output is shown. For each of the motifs the matching sequence and p value are given. The overall E value for these four motif matches was 9.0e-11 (9.0e-11 means 9.0×10^{-11}). Matching regions in the query sequence are shown in bold type.

The modular architecture of proteins must be emphasized at this point. Many proteins are constructed from more than one **domain**, and some domains are common to many protein families. For instance SH2 and SH3 domains appear in many proteins associated with signaling, and PH domains appear in many proteins that bind phospholipids. This modular architecture probably reflects the way proteins have evolved. Genetic events can result in domain swapping, domain duplication, and loss and gain of domains. Acquiring new domains with specific functions allows proteins to acquire new and more complex functions very quickly. For instance, an enzyme might gain a new domain associated with regulation of its activity, producing a protein that would be active in more specific circumstances.

In the above we have not defined precisely what we mean by a protein domain. This is because such a definition is difficult to construct. Domains might be defined as parts of a protein sequence with a single well-defined function (e.g., binding a particular ligand), or they might be parts of the sequence able to fold into a three-dimensional structure independent of the rest of the sequence. Equally they might be defined as just parts of the protein three-dimensional structure that appear to be geometrically distinct. However, one important aspect of the definition of a domain is that it must be an independent unit able to exist in many, otherwise unrelated, protein sequences. Because a large number of such domains exist, it is sensible that databases of protein family descriptions should describe *domain families*. For instance, PH domains from many functionally distinct and diverse proteins may be described as a single domain family, and the structure and evolution of the domain studied independently.

The **Prodom** database (http://prodom.prabi.fr) is a database of protein domain family sequences. It is created automatically from the database of known protein sequences, using pair-wise sequence comparison (as described in Section J3) with the BLAST tools, followed by clustering together of related sequence segments from different proteins into domain families. This procedure follows the domain definition above, in that it attempts to identify independent sequence segments with similarity detectable at the sequence level, which are shared by many proteins. These are identified with domains. The server allows a protein sequence to be compared to the database of domain sequences, in order to identify shared domains.

Sequence profiles
Sequence profiles (alternatively known as weight matrices) are a way of describing related sequences from a protein domain family. They have been adopted by the PROSITE database to supplement the PROSITE pattern type entries. Their principal advantage is that they describe a complete domain sequence, including the likelihood of observing each amino acid, along with the likelihood of insertion and deletion, at each position in the sequence. An example of a PROSITE sequence profile is shown in *Fig. 4*. Each multiple alignment column in the aligned family sequences is associated with a set of 20 scores, one for the appearance of each amino acid at that position. Observed amino acids tend to score highly, but the profile is derived in such a way that amino acids that substitute well for the observed amino acids score reasonably highly as well. A new sequence can be aligned to a profile so that its amino acid residues match with columns where they obtain high scores. A high total alignment score indicates that the sequence is likely to belong to the family.

	F	K	L	L	S	H	C	L	L	V
	F	K	A	F	G	Q	T	M	F	Q
	Y	P	I	V	G	Q	E	L	L	G
	F	P	V	V	K	E	A	I	L	K
	F	K	V	L	A	A	V	I	A	D
	L	E	F	I	S	E	C	I	I	Q
	F	K	L	L	G	N	V	L	V	C

A	-18	-10	-1	-8	8	-3	3	-10	-2	-8
C	-22	-33	-18	-18	-22	-26	22	-24	-19	-7
D	-35	0	-32	-33	-7	6	-17	-34	-31	0
E	-27	15	-25	-26	-9	23	-9	-24	-23	-1
F	60	-30	12	14	-26	-29	-15	4	12	-29
G	-30	-20	-28	-32	28	-14	-23	-33	-27	-5
H	-13	-12	-25	-25	-16	14	-22	-22	-23	-10
I	3	-27	21	25	-29	-23	-8	33	19	-23
K	-26	25	-25	-27	-6	4	-15	-27	-26	0
L	14	-28	19	27	-27	-20	-9	33	26	-21
M	3	-15	10	14	-17	-10	-9	25	12	-11
N	-22	-6	-24	-27	1	8	-15	-24	-24	-4
P	-30	24	-26	-28	-14	-10	-22	-24	-26	-18
Q	-32	5	-25	-26	-9	24	-16	-17	-23	7
R	-18	9	-22	-22	-10	0	18	-23	-22	-4
S	-22	-8	-16	-21	11	2	-1	-24	-19	-4
T	-10	-10	-6	-7	-5	-8	2	-10	-7	-11
V	0	-25	22	25	-19	-26	6	19	16	-16
W	9	-25	-18	-19	-25	-27	-34	-20	-17	-28
Y	34	-18	-1	1	-23	-12	-19	0	0	-18

Fig. 4. A PROSITE sequence profile (taken with permission from the PROSITE user manual, copyright by Amos Bairoch). A multiple alignment segment is shown above the profile. Within the profile, numbers represent scores for each of the amino acids at the corresponding position in the multiple alignment. In the first column, for instance, scores are high for the observed amino acids (F, Y, L), and also for amino acids that substitute well for these (W, V, I, M).

Hidden Markov models

Hidden Markov models (HMMs, outlined in Section I2) are the most statistically sophisticated way of representing a family of protein-domain sequences. An HMM represents a protein domain family by generating sequences from that family with very high probability, and other sequences with much lower probabilities. The name hidden Markov model is used because the sequences are generated by a Markov process, which is defined as a process in which the probability of a particular state depends only on the state immediately preceding it in a sequence. The term hidden is used because what you observe for a member of a protein family is its sequence, but hidden from you is the sequence of states (path through the model) that generated it.

You might ask what the use is of a model that generates protein sequences. Bioinformatics has enough of those already! What you really want is a way of deciding whether a particular real sequence, perhaps from a genome sequencing program, contains a domain from a particular family. The answer is that there are sophisticated algorithms that can find, for a given sequence, the most likely path through the model to generate it, and can use the probability of this path to estimate how likely it is that the sequence belongs to the family.

There is an obvious correspondence between the structure of the HMM in *Fig. 5* of Section I2 and multiple sequence alignments. The match states correspond to conserved columns in the alignment, generating amino acids similar to those observed in these columns with high probabilities. Delete states are used for sequences from the family in which the amino acid from such a column has been deleted, and insert states represent sequences with one or more inserted amino acids between the columns. The model thus describes the process of insertion and deletion of amino acids in a position-dependent way, and even gives probabilities for each possible inserted amino acid.

Resources

The Pfam database (http://pfam.sanger.ac.uk) is a collection of protein-domain family multiple-alignments and HMMs. This is maintained at the Sanger Center and can be used via a WWW interface to analyze new sequences. An example of such an analysis for the sequence of human pleckstrin is shown in *Fig. 5*. In this case Pfam has elucidated the full domain structure of the protein, including two PH domains and a single DEP domain. Another useful resource that will carry out this type of analysis is **SMART** (Simple Modular Architecture Research Tool) (http://smart.embl-heidelberg.de/).

A recent and very valuable development is the integration of several of the protein family resources into a resource called **Interpro** (http://www.ebi. ac.uk/interpro/). This unites information from many of the resources we have discussed, including PROSITE, PRINTS, Pfam, Prodom, and SMART, allowing integrated search and sequence analysis.

Fig. 5. Pfam analysis of the human pleckstrin sequence, showing matches to the HMMs

representing the N- and C- terminal PH (pleckstrin homology) domains and the central DEP domain. The PH domains match the pleckstrin sequences in amino acids 5-101 and 245-347, and the DEP domain matches amino acids 136-221.

K1 TRANSCRIPT PROFILING

Key Notes

Global expression analysis and DNA microarrays

RNA expression can be measured by hybridizing the RNA to other oligonucleotides. Analyzing signal intensities under different conditions can identify the levels of differential gene expression.

DNA microarrays

There are many types of microarray that are appropriate for different types of analysis. DNA microarray technology uses a grid of oligonucleotides on a chip that hybridize to the complementary target RNA. The level of transcript is determined by the hybridization signal of either Affymetrix- or Agilent-based microarrays. Affymetrix microarrays comprise 25-mer oligonucleotides and there are 11–20 probe-pairs per probe-set per gene. Agilent microarrays have longer oligonucleotide probes, which are more specific, but this method has only one probe per gene.

Tiling arrays

Tiling arrays have probes that target sections of the genome at a very high resolution. The tiling probes can be overlapping or have a short gap between them. Tiling arrays are used in ChIP-chip, MeDIP-chip, and DNase-chip studies.

SNP microarrays

SNP microarrays measure single nucleotide polymorphisms that occur in different diseases, or a comparison of tissues or treatments. They are the key technology in human disease genome-wide association studies and drug development studies.

Next generation sequencing

Next generation sequencing encompasses new technologies that provide fast and accurate sequence reads. The number of sequencing reads can be converted into levels of differential gene expression.

Related sections

Statistical issues for transcriptome analysis (K2)	Multivariate techniques and network inference (K4)
Analyzing differential gene expression (K3)	Data standards and experimental design (K5)

Global expression analysis

Traditionally, expression has been studied on a gene-by-gene basis using techniques such as northern blots. More recently, methods have been developed for **global expression analysis**; that is, the study of all genes simultaneously. Such studies generate very large amounts of data, which must be mined for relevant information.

For analysis at the RNA level, **direct sequence sampling** from RNA population cDNA libraries, or even from sequence databases, can be useful. A simple approach is to sequence say 5000 randomly picked clones from a cDNA library. Abundant mRNAs would appear at a higher frequency among the sampled sequences than rare ones, and statistical analysis of these data would allow relative expression levels to be determined. Although simple in concept, this type of experiment is expensive because of all the sequencing reactions that must be carried out.

A more sophisticated technique is **serial analysis of gene expression (SAGE).** In this method, very short sequence tags (usually 8–15 nucleotides) are generated from each cDNA sequence and hundreds of these are joined together to form a concatemer prior to sequencing. Thus, in one sequencing reaction, information on the abundance of hundreds of mRNAs can be gathered. Each **SAGE tag** uniquely identifies a particular gene, and by counting the tags the relative expression level of each gene can be determined. The advantage of all direct sampling methods is that expression levels are automatically converted into numbers; that is, the data are digital. Although direct sampling is a powerful approach to expression analysis, it is labor intensive.

The technology that has had the widest impact on global RNA expression profiling is **DNA array hybridization (DNA chips).** As discussed in more detail below, the principal advantage of DNA arrays is that the expression levels of all genes can be monitored in parallel in a single experiment. Expression data are obtained as signal intensities, and these are clustered to identify similarly expressed genes. Array-based assay formats have been developed for proteins also but these 'protein chips' are not yet as versatile as the equivalent DNA chips.

DNA microarrays   Nucleic acids (DNA and RNA) can form double-stranded molecules by **hybridization**; that is, complementary base pairing. The specificity of nucleic acid hybridization is such that a particular DNA or RNA molecule can be labeled (e.g., with a radioactive or fluorescent tag) to generate a **probe**, and can be used to isolate a complementary molecule from a very complex mixture, such as whole genomic DNA or whole cellular RNA. This specificity also allows thousands of hybridization reactions to be carried out simultaneously in the same experiment.

A **DNA microarray** or **DNA chip** is a dense grid of DNA elements (often called **features** or **cells**) arranged on a miniature support, such as a nylon filter or glass slide. Each feature represents a different gene. The array is usually hybridized with a **complex RNA probe**; that is, a probe generated by labeling a complex mixture of RNA molecules derived from a particular cell type. The composition of such a probe reflects the levels of individual RNA molecules present in its source. If non-saturating hybridization is carried out, the intensity of the signal for each feature on the microarray represents the level of the corresponding RNA in the probe, thus allowing the relative expression levels of thousands of genes to be visualized simultaneously.

Two major technologies are used in microarray manufacture. The cheaper and more widely accessible method involves the robotic spotting of individual DNA clones onto a coated-glass slide. Such **spotted DNA arrays** can have a density of up to 5000 features per square cm. The features comprise double-stranded DNA molecules (genomic clones or cDNAs) up to 400 bp in length, and must be denatured prior to hybridization. Prefabricated arrays can be purchased from a number of companies and many laboratories have their own in-house production facilities. The alternative method is **on-chip photolithographic synthesis**, in which short oligonucleotides are synthesized in situ during chip manufacture. These arrays are known as **GeneChips** and are manufactured exclusively by the US company Affymetrix Inc. They have a density of up to 1 000 000 features per square cm and each feature comprises single-stranded oligonucleotides of 25 nucleotides in length. Owing to the reduced hybridization specificity of a short oligonucleotide, each gene on a GeneChip is represented by 11–20 features (20 non-overlapping oligos), and 11–20 mismatching controls.

Both technologies can be used to look at differential gene expression between samples (e.g., healthy tissue vs. disease tissue). Spotted DNA arrays use fluorescent probes, since different fluorophores can be used to label different RNA populations. These can be hybridized simultaneously to the same array, allowing differential gene expression to be monitored directly. Typically, Cy3 is used to label one probe and Cy5 to label the other. Cy3 fluoresces bright red, and Cy5 fluoresces bright green. If a particular RNA is present only in the Cy3-labeled probe, the corresponding feature on the array appears red. If another RNA is present only in the Cy5-labeled probe, that spot appears green. RNAs found in both probes would hybridize in equivalent amounts, and these features would appear yellow. Dual labeling is not used on GeneChips. Instead hybridization is carried out with separate probes on two identical chips and the signal intensities are measured and compared by the accompanying analysis software.

Tiling microarrays

Tiling arrays are microarrays with a far higher resolution of probes spanning the whole genome of the species. The aim of tiling arrays is to use probes to measure transcription at regular intervals of regularly spaced probes that either overlap or are very close together. The step between each probe can be complementary to one strand or both strands. The tiling probes can span across exonic regions (coding) and intronic regions (non-coding), and this approach can give a huge insight into the post-translational processes of a species. Tiling arrays are one of the most powerful tools in genome-wide association studies.

Human Affymetrix tiling arrays contain 45 million 25-mer probes across the whole genome and this offers the highest resolution with the smallest cost per probe. The Affymetrix probes are interspersed every 35 base pairs, and there are other chips for *Arabidopsis*, *C. elegans*, *Drosophila*, and mouse. NimbleGen is another tiling-array platform that has 2.1 million probes across the human genome. Nimblegen provides other chips for chicken, cow, *C. elegans*, dog, *Drosophila*, *Plasmodium*, rat, mouse, Rhesus monkey, yeast, and zebrafish.

The most common use of a tiling microarray is in **ChIP-chip** (chromatin immunoprecipitation on chip experiments), a technique that is used to investigate protein interactions and binding sites of DNA-binding proteins. This technique identifies genomic functional elements involved in transcription-factor binding, histone modification, chromatin structure, and polymerases in a target genome.

DNA methylation controls transcription and the genome-wide methylation sites can be identified by using a **MeDIP-chip** (**methyl DNA immunoprecipitation-chip**). Additionally, **DNase chip technology** identifies transcriptionally regulated regions by digesting the chromatin with a DNase, and then using the DNase chip owing to the chromatin being cleaved when open for transcription.

SNP microarrays

SNP arrays are used to find which **single nucleotide polymorphisms (SNPs)** are present in an individual within a species. An SNP is a change in the genomic sequence at a single site in the genome and this is the most common cause of variation within a species. There are predicted to be to be millions of SNPs in the human genome, and these slight changes in sequence are being studied as potential markers of disease and variation.

SNPs are important in **drug targeting**, the study of drug efficiency and specificity, which is very important in medicine with adverse reactions to drug treatments. The drug discovery pipeline has five main parts: disease characterization, target selection, pharmacokinetics (absorption, disposition, metabolism, and excretion – ADME), and clinical trials to determine adverse drug reactions (ADR) and drug efficacy. The drug discovery process incurs huge costs for pharmaceutical companies, especially with regard to failures in the clinical research.

The **genome-wide association studies** are involved in finding the SNPs that have been found to be involved in the diseases cancer, heart disease, diabetes (type I and II), and bipolar disorder, and many more. The finding of ways to predetermine diseases as a preventative measure will change current approaches to medicine and tailor treatments to the individual, which are so-called personalized medicines.

Next generation sequencing

Next generation sequencing technologies are revolutionizing many areas of biology. These highly efficient parallel sequencing techniques allow many billions of bases of sequence to be generated per day on each machine. In principle, this allows individual human genomes to be sequenced in 1 day. There are several **next generation sequencing platforms** including Ilumina Solexa, Roche 454, and Applied Biosystems SOLiD System Sequencing. Different systems have different trade-offs between sequence read-length and the total amount of sequence generated. The technology is advancing rapidly with huge increases in read-length, and subsequently data, along with a reduction in costs.

These advances open up diverse areas of traditional biology to sequencing technologies previously unavailable to them, which includes transcriptomics. The total RNA population is reverse transcribed to cDNA, which is then sequenced. Transcript levels are determined by the number of sequences that align against one region of the genome. Advantages over traditional microarray approaches include: base pair accuracy in the determination of transcriptional start and stop sites, splice variants with greater accuracy in determination of transcript level, and application to organisms of which the genome sequence is undetermined.

K2 STATISTICAL ISSUES FOR TRANSCRIPTOME ANALYSIS

Key Notes

Raw data from microarrays

Primary microarray data comprise images from hybridized arrays representing the hybridization signal intensities at individual spots. These may be generated by single-fluorescent, dual-fluorescent, radioactive, or colorimetric labels.

Data quality

It is essential to record the signal intensities of individual spots accurately as errors in data recording cannot be detected or corrected at a later stage. Software for reading microarrays is generally provided with the recording equipment (scanner or phosphorimager) but manual adjustment is necessary to compensate for variations in array manufacture. The signal must be corrected for background (non-specific hybridization, autofluorescence, contamination), and hybridization controls must be used when comparing results across different arrays.

Normalization methods

Microarray normalization is carried out to remove bias and variation across the arrays. There are several normalization methods including Mas-5, RMA, GC-RMA, and Li-Wong. The choice of normalization approach is important with subsequent statistical techniques in microarray analysis.

Related sections

Raw data from microarrays

The previous section describes how microarrays are miniature devices comprising a large number of DNA sequences immobilized on a substrate slide. The intensity of the hybridization signal for each feature corresponds to the amount of a particular molecule in the probe, and this normalized intensity is directly proportional to the level of gene expression in the cell type or tissue from which the probe was prepared. In this way, microarrays can report the relative expression levels of thousands or tens of thousands of genes.

The raw data from microarray experiments thus consist of images from hybridized arrays. The exact nature of the image, however, depends on the **array platform** (the type of array used). Most array experiments now are carried out using glass **spotted microarrays** or **high-density oligonucleotide chips**. In both cases, the substrate has minimal autofluorescence so a **fluorescent probe** can be used. Data are acquired by confocal laser scanning of the hybridized array at the appropriate **excitation wavelength** and recording at the appropriate **emission wavelength** (or **channel**). A single label is used for oligonucleotide chips; so differential gene expression is detected by hybridizing different probes to duplicate arrays. However, in the case of spotted arrays, two probes can be labeled with different fluorophores and hybridized simultaneously to the same array, allowing differential gene expression to be monitored directly.

Data quality

DNA arrays may contain many thousands of features. Therefore, data acquisition and analysis must be automated. The software for initial image processing normally is provided with the scanner (or phosphorimager). This allows the boundaries of individual spots to be determined and the total signal intensity to be measured over the whole spot (this is termed the **signal volume**). Locating spots precisely can be a problem, particularly if there is distortion on the array surface. Therefore, it often is necessary to align the grid manually. This is very important because signal intensities can vary across individual spots and the shape and size of different spots may not be uniform. Most importantly, the signal intensity has to be corrected for **background** (noise), which may be generated by non-specific hybridization, autofluorescence, dust and other contaminants, or poor hybridization technique (e.g., partial dehydration). The background can vary over the array surface, so signal intensities must be normalized for local background values. Correction for background is difficult when the signal intensity for a particular spot is itself very low.

Usually control features are included on the array to measure non-specific hybridization and variable hybridization across arrays. For example, Affymetrix GeneChips incorporate a set of **mismatching oligonucleotides** for each perfect match set to determine non-specific hybridization. Controls are particularly important where duplicate arrays are being used to study differential gene expression, since variation in array manufacture or experimental protocol can influence the signal intensities on different arrays. The bottom line is that errors and artifacts introduced before or during data acquisition cannot be detected or corrected at a later stage.

Microarray normalization

Microarray normalization is essential in microarray analysis as it removes bias and variation across each array. There are several normalization methods including **Mas-5**, **RMA**, and **GC-RMA,** whose details fall outside the scope of this work so references will be supplied. **Mas-5** normalization utilizes both the perfect-match and mis-match probe level, normalized using **Tukey biweight**. RMA (**Robust Multi-array Average**) uses just the perfect-match in the analysis and normalizes across all the chips, and then normalizes by a procedure known as **Median Polish**. GC-RMA is the same as RMA normalization with an additional background correction based on GC content. There is also the **Li-Wong method** of normalization that uses the invariant set and multiplicative model-fitting summarization. A summary of these normalization methods can be found in *Table 1*.

Table 1. Summary of four normalization methods from Lim et al., *2007*

Procedure	Background correction	Normalization	Summarization	Reference
Mas5	Ideal (full or partial) MM subtraction	Constant	Tukey biweight	Hubbell *et al.*, 2002
RMA	Signal and noise close-form transformation	Quantile	Median Polish	Irizarry *et al.*, 2005
GC-RMA	Optical noise, probe affinity	Quantile	Median Polish	Wu *et al.*, 2004
Li-Wong	None	Invariant set	Multiplicative model fitting	Li & Wong, 2001

K3 ANALYZING DIFFERENTIAL GENE EXPRESSION

Key Notes

Gene expression matrices
The raw data from microarray experiments is converted into tables known as gene expression matrices. The rows represent genes and the columns represent experimental conditions. The values in the matrices are measurements of signal intensities, representing relative levels of gene expression.

Grouping expression data
Each gene in a gene-expression matrix has an expression profile, relative to the changes in expression measurement over a range of conditions. The analysis of microarray data involves grouping these data on the basis of similar expression profiles. If a pre-defined classification system is used to group the genes, the analysis is described as supervised. If there is no pre-defined classification, the analysis is described as unsupervised.

Tools for microarray data analysis
Many software applications are available for the analysis of microarray data and these can be downloaded and installed on local computers. Two examples of microarray analysis software platforms include GeneSpring and Bioconductor. There are also several resources available for the analysis of microarray data over the internet; Expression Profiler is the most widely used. Several gene expression databases have been constructed for the storage and dissemination of microarray data. These include the NCBI Gene Expression Omnibus and the EBI ArrayExpress database.

Differential expression
Differential expression refers to the up- and down-regulated genes a microarray experiment. The levels of expression are commonly determined by a fold-change at a set cutoff value. Volcano plots are a common representation of genes that are selected by fold-change and p value.

Mapping of expression data onto networks
Gene expression data can be mapped onto a network, which can be protein interaction, gene regulatory, or metabolic networks. Cytoscape is the main network visualization and analysis tool.

Related sections
Transcript profiling (K1)
Statistical issues for transcriptome analysis (K2)

Multivariate techniques and network inference (K4)
Data standards and experimental design (K5)

Gene expression matrices
Whichever platform is used, the aim of data processing is to convert the hybridization signals into numbers, which can be used to build a **gene-expression matrix**. Essentially this matrix can be regarded as a table in which the

rows almost always represent genes (the different features on the array) and the columns represent treatments or conditions used in the experiment. For a dual-hybridization experiment using a glass microarray, each of the probes represents a different experimental condition. In other cases, a whole series of conditions or treatments may be used; for example, representing a series of concentrations of a particular drug, or a series of developmental time points.

Grouping expression data

The interpretation of microarray experiments is carried out by grouping the data according to similar expression profiles (Section K4). An **expression profile** can be defined, in this context, as the expression measurements of a given gene over a set of conditions. Essentially this means reading along a row of data in the gene-expression matrix. In this case, the intensity of shading is used to represent expression levels (*Fig. 1*). If we concentrate on experimental conditions C1 and C2, we can conclude that genes G1 and G2 are functionally similar and G3 appears to be different. However if we include C3, this suggests a functional link between G1 and G3.

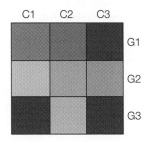

Fig. 1.   Schematic of an idealized expression array, in which the results from three experiments are combined. Three genes (G1, G2, and G3) are labeled on the vertical axis and three experimental conditions (C1, C2, and C3) are labeled on the horizontal axis, giving a total of nine data points. The shading of each data point represents the level of gene expression, with darker colors representing higher expression levels.

Analysis methods can be described as either supervised or unsupervised. **Supervised methods** are essentially **classification systems**. That is, they incorporate some form of classifier so that expression profiles are assigned to one or more **pre-defined categories**. For example, the supervised analysis of gene-expression profiles from different leukemias allows the samples to be divided into two distinct subtypes: acute myeloid leukemia and acute lymphoblastoid leukemia. **Unsupervised methods** have no in-built classifiers, so the number and nature of groups depends on the algorithm used and the nature of the data only. This type of analysis is known as **clustering** and is discussed in more detail in Section K4.

Tools for microarray data analysis

Many software applications are available for implementing these methods, and a list of useful resources is shown in *Table 1*. In many cases, installing and running software locally is used to carry out the analysis. There are several microarray analysis software tools available including GeneSpring and Bioconductor.

GeneSpring is a very powerful microarray analysis tool with an attractive user-interface and integrated Bioconductor R libraries (*Table 1*). GeneSpring analyzes patterns in gene-profile differential-expression and provides statistical tests to find relevant gene-sets. Pattern finding within the data can be performed using principal components analysis (PCA), hierarchical clustering analysis (HCA), and K-means clustering tools (Section K4). GeneSpring also performs functional enrichment of gene-sets by use of gene ontology, gene enrichment analysis, and pathway-based analysis (Section K4). The advantage to GeneSpring is the sheer volume of analysis that can be carried out with a microarray data set; the disadvantages are that GeneSpring is expensive and requires an annual license.

Another tool is **Bioconductor**, which is an open-source microarray-analysis tool built on the statistical programming language R. It is popular with both biologists and bioinformaticians as it has an extensive set of R libraries that offer a wide range of analysis (*Table 1*). The major strength of Bioconductor is the building of microarray-analysis pipelines, which is essential with huge numbers of microarrays. Another strength is that the software is community-driven, which facilitates the development of new algorithms for new types of analysis. The disadvantages with Bioconductor are poor documentation and limited testing of algorithms. The main libraries, however, are very well documented and user tested (e.g., Affy package for Affymetrix microarray analysis). A list of the Bioconductor packages available can be found at: http://www.bioconductor.org/packages/release/Software.html

The European Bioinformatics Institute (EBI) site provides one of the most user-friendly WWW-based microarray-analysis tools, which is called **Expression Profiler**. This is a suite of programs designed for the analysis and integration of sequence and expression data. Another example of a relevant program for microarray analysis is **EPCLUST**, which allows data to be uploaded from source or accessed from publicly available files. Data can be grouped using a variety of alternative methods for distance measurement and clustering. Another tool is **Genevestigator**, which is a Java-based tool that starts in the browser, and it has access to publicly available microarray data for six organisms: human, mouse, rat, *Arabidopsis*, barley, and rice. The main advantage of Genevestigator is that the microarray data can be analyzed with clustering and pathway-based analysis.

Microarray databases are community-driven public repositories. The reason for creating them is so that microarray data can be further analyzed by other scientists with new software, or methods either confirm the analysis or to find new ways to extract meaningful results. Microarray repositories have the additional advantage of ensuring that microarray data are of an acceptable quality and in accordance with the MIAME standard (Section K5).

The **NCBI Gene Expression Omnibus** is the main curated microarray repository and allows the user to browse and retrieve the data. The other main microarray repository is **ArrayExpress**, which is maintained by the EBI. The **Stanford Microarray Database** (SMD) is a web-interface to access raw and normalized microarray experiments, and also used as a public resource to connect expression data to real biology. **Gene Pattern** is integrated with the SMD and is a web-based workflow platform of over 90 analysis tools for clustering and marker selection. **NASCArrays** is a repository containing all microarray data generated by the National Arabidopsis Stock Centre (NASC). It contains mainly plant data, but does also include data for some other organisms.

Differential expression

The level of **differential expression** is most commonly determined by **fold-change**. The analyst can choose a cutoff of their choice; an example of a typical cutoff is two-fold change. However, it is an arbitrary cutoff, some genes will be only slightly up-regulated but can have huge effects on the organism (e.g., a transcription factor), but setting a two-fold cutoff will not find that important gene. *Fig. 2* shows a two-fold cutoff of up- and down-regulated genes, which has been compared with the treatment divided by the control expression values.

Statistical reliability of a change in expression is very important, and this is determined by **volcano plots**. A volcano plot is a graphical representation that sets a log fold-change cutoff at the x-axis, and a statistical p value cutoff on the y-axis (*Fig. 3*). The statistically relevant up- and down-regulated genes are found in the top left- and right-hand corners of the graph. The analyst can choose which fold-change and p value cutoffs to extract interesting genes involved in the microarray treatments.

Table 1.   Microarray analysis software tools and internet resources for microarray expression analysis

Product	Features	URL
Microarray analysis software tools		
GeneSpring GX	Very popular and powerful tool for biologists. Full support for most microarray platforms, Affymetrix and Agilent are examples. Extensive analysis including clustering, PCA, and pathway analysis. License required	http://www.chem.agilent.com/ en-US/products/software/ lifesciencesinformatics/pages/ gp35082.aspx
Bioconductor	R-based tool with many libraries for microarray analysis including extensive Affy and Agilent support. Free	http://www.bioconductor.org/
DChip	Analysis and visualization of gene expression and SNP arrays	http://biosun1.harvard.edu/complab/ dchip/
Examples of sites with extensive links to microarray analysis software and resources		
Gene Pattern	Extensive list of software resources from Stanford University and other sources, both downloadable and WWW-based	http://www.broad.mit.edu/cancer/ software/genepattern/
Examples of WWW-based microarray data analysis		
Expression profiler	Very powerful suite of programs from the EBI for analysis and clustering of expression data	http://www.ebi.ac.uk/ expressionprofiler/index.html
EPClust	Generic data clustering, visualization, and analysis tool	http://www.bioinf.ebc.ee/EP/EP/ EPCLUST/
Genevestigator	Provides gene expression meta-profiles for animals and plants (e.g., human, mouse, rat, and *arabidopsis*)	https://www.genevestigator.ethz.ch/gv/ index.jsp
The major microarray databases		
NCBI GEO (Gene Expression Omnibus)	GEO is a gene expression and hybridization array database, which can be searched by accession number, through the contents page, or through the Entrez ProbeSet search interface	http://www.ncbi.nlm.nih.gov/geo/
ArrayExpress	EBI microarray gene expression database. Developed by MGED and supports MIAME	http://www.ebi.ac.uk/microarray-as/ae/
Stanford Microarray Database	Microarray database that provides a resource for the scientific community. Many tools to explore and analyze the data	http://genome-www5.stanford.edu/
NASCArrays	Standard microarray database consisting of plant and other species data. Data mining tools and experiment search functions	http://affymetrix.arabidopsis.info/ narrays/experimentbrowse.pl

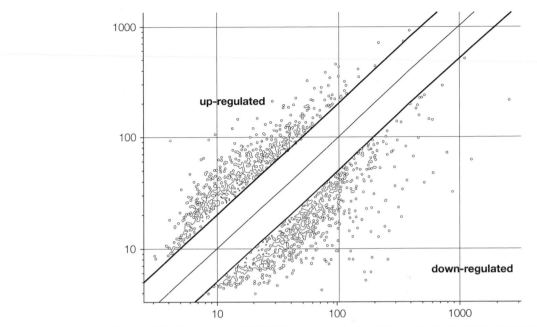

Fig. 2.   This figure shows differential expression at a two-fold cutoff of a treatment compared to a control microarray. The down-regulated genes were found beneath the cutoff line, and the up-regulated genes were found above. (Image produced using GeneSpring.)

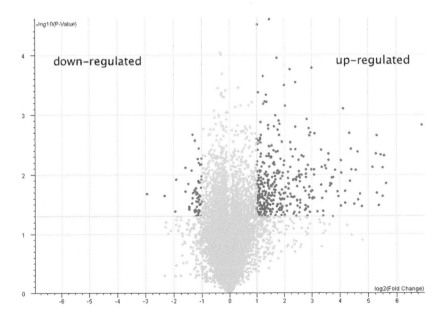

Fig. 3.   A volcano plot example showing log2-fold as vertical lines, and a statistically reliable p value as horizontal lines. The outliers (darker spots) are statistically relevant up-regulated and down-regulated genes. (Image produced by Genespring.)

Mapping of expression data onto networks

The fold-changes or levels of expression of a set of genes can be mapped onto networks and this can show relationships between genes, proteins, and metabolites. **Network biology** consists of **protein interaction**, **gene-regulatory**, and **metabolic** networks. Additionally, the interplay between these (so-called integrated networks) can show new relationships.

In these networks, a node can be a protein in a protein-interaction network, a gene in a gene-regulatory network, or a metabolite in a metabolic network. An edge is the relationship between the nodes. The main network visualization and analysis tool is called **Cytoscape**, which is Java-based with the added functionality of plugins (http://cytoscape.org/). Cytoscape is an open-source bioinformatics tool, and integrates large datasets for complex-systems analysis and has a good online tutorial (http://cytoscape.org/cgi-bin/moin.cgi/Presentations).

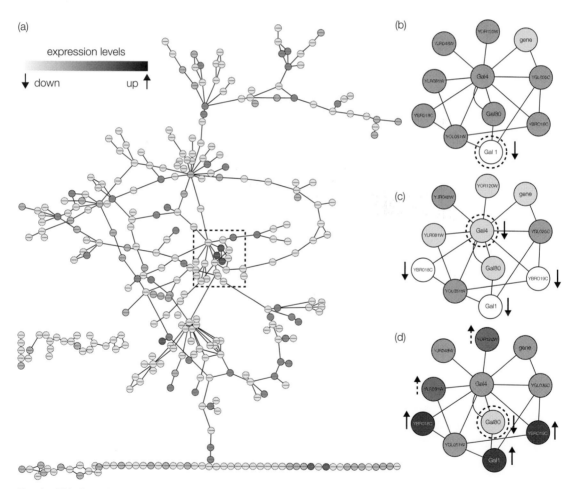

Fig. 4.   This figure is based upon a yeast gene-regulatory network, where each node is a gene or protein, color-coded in the case of genes on the basis of their transcript levels, with darker shades representing higher expression levels. (a) The overall network of 331 nodes and 362 edges with key genes denoted in the black rectangle; (b), (c), and (d) show the same subnetwork when Gal1, Gal4, and Gal80 are knocked out, respectively. The dashed circle shows which gene has been knocked out, and the arrows indicate whether the result is a slight (dashed) or major (solid) increase (upward arrow) or decrease (downward arrow) in gene expression compared to the wild-type. (Permission to use the Cytoscape tutorial to produce this figure was kindly given by Trey Ideker, UCSD.)

Expression data can be mapped onto a network and the nodes colored based on the levels of expression. *Fig. 4a* shows the overall yeast protein-protein and protein-DNA interactions involved in galactose metabolism. By mapping the gene-expression data from gene-knockout experiments onto the subnetworks, it was possible to elucidate key regulatory genes. Gal1, Gal4, and Gal80 are the yeast transcription factors, which control the levels of transcription of genes involved in galactose metabolism. The perturbed Gal1 only shows down-regulation of itself, so it does not have an effect on the rest of the subnetwork (*Fig. 4b*). The Gal4 knockout caused down-regulation in three genes, which shows that it is an **activator** (*Fig. 4c*). Gal80 is a **repressor** of Gal4, as shown by the Gal80 knockout causing up expression of Gal4, which causes up-regulation of the three highly expressed (black color) nodes (*Fig. 4d*).

K4 MULTIVARIATE TECHNIQUES AND NETWORK INFERENCE

Key Notes

Clustering methods

Clustering first involves converting the gene-expression matrix into a distance matrix, so that genes with similar expression profiles can be grouped together. This generally involves calculating either the Euclidean distance, or more often distances based on correlation measures or the Pearson linear correlation for each pair of values. Several clustering methods exist including hierarchical or k-means clustering and the derivation of self-organizing maps.

Feature reduction

A characteristic of microarray data analysis is the large number of features (data points). Clustering and classification algorithms can run more quickly if feature reduction is applied, to remove or amalgamate redundant and non-informative data.

From expression data to networks

Reconstructing molecular networks from expression data is a difficult task. One approach is to simulate networks using a variety of mathematical models, and then choose the model that best fits the data. Reverse engineering is a less demanding approach in which models are built on the basis of the observed behavior of molecular pathways. Models using systems of Bayesian statistics, differential equations, or Boolean networks each suffer from disadvantages, so hybrid models, such as the finite linear-state model, are preferred.

Related sections

Artificial intelligence and machine learning (I)	Statistical issues for transcriptome analysis (K2)
Sequence families, alignment, and phylogeny (J4)	Analyzing differential gene expression (K3)
Transcript profiling (K1)	Data standards and experimental design (K5)

Clustering methods

Clustering is a way of simplifying large data sets by partitioning similar data into specific groups (**clusters**). The successive stages of the analysis require a measurement of the distances between the genes in terms of their expression profiles followed by the use of some clustering algorithm. The first step is to convert a gene expression matrix into a **distance matrix**, so that the similarities and differences between data points can be determined. This is achieved often by calculating a value known as the **Euclidean distance**, which is the square root of the sum of the squared differences between any two data points. Alternatively, the **correlation measure-based distance** or the **Pearson linear correlation-based distance** may be used, because the Euclidean distance produces misleading clusters if two genes have similar expression profiles but different amplitudes or if two genes have different but very low expression profiles.

Having acquired a distance matrix, the next task is to reorder the data and generate a dendrogram (*Fig. 1*). The same process essentially is used to build phylogenetic trees (Section J4). However, the specific methods used in phylogenetic analysis are not readily applied to expression data, partly because the size of the data sets is much larger, and partly because the concept of an 'ancestor' is meaningless, in such cases a rooted tree is not required. Gene expression clusters are unrooted trees.

There are several different clustering strategies that can be applied to expression data, each with their own advantages. **Hierarchical clustering** methods, similar to those applied in phylogenetic analysis, are the most widely used. Initially each gene defines its own singleton profile and the algorithm searches for the two most similar profiles. These **neighbors** then are merged into a single cluster and the process repeated. The distance between a new (merged) cluster and any other given gene can be defined as the distance to the nearest of the neighbors in the merged cluster (single linkage), the distance to the farthest of the neighbors (complete linkage) or the distance to the average value of the neighbors (average linkage, with or without weighting). These methods generate dendrograms with different topologies (*Fig. 1*).

Another, non-hierarchical clustering method is **k-means clustering** in which the expected number of clusters is specified at the outset and defined as the parameter k (*Fig. 1b*). Initially, the center of each cluster (calculated as the **centroid value**, the weighted equivalent of a center of gravity) is randomly specified. Expression profiles are assigned to a particular cluster according to a distance matrix, and the centroid value is recalculated based on the incorporation of new profiles. Reiteration of this process eventually generates a tree in which groups of genes are assigned to clusters based on similar expression profiles. The advantage of this method over hierarchical clustering is that the boundaries between one cluster and another are not defined arbitrarily, but are recalculated in each iteration. A similar process, refined by the use of neural networks, involves the generation of **Kohonen self-organizing maps (SOMs)**, in which centroid values for the clusters are recalculated using not only information from profiles within each cluster, but also from profiles in adjacent clusters.

(a)   (b)

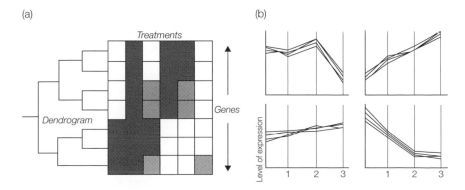

Fig. 1. *(a) A heat map is a graphical representation of microarray data in a data matrix that shows genes, which are up-regulated (white), down-regulated (black), and no change (gray), clustered together to form a dendrogram. A dendrogram represents the cluster arrangement of the genes. (b) A time-course of four k-means clusters showing similar profiles, the gray vertical lines represent the profile of a time-course of 1, 2, and 3 hours.*

Feature reduction Since microarray datasets are so large, classification and clustering can be laborious and demanding in terms of computer resources. Sometimes it is possible to use **feature reduction**, where non-informative or redundant data points are removed from the dataset, to make the algorithms run more quickly. For example, if two conditions have exactly the same effect on gene expression, the data for these conditions are redundant and one entire column of the matrix can be eliminated. Similarly, if the expression of a particular gene is the same over a range of conditions, it is neither necessary nor beneficial to use this gene in further analysis because it provides no useful information on differential gene expression. An entire row of the matrix can be removed.

Several approaches can be used to select such redundant or non-informative datasets automatically. A popular method is **principal component analysis** (PCA, also called **single value decomposition**) in which redundant data are combined to form a single, composite dataset, thus reducing the dimensions of the gene expression matrix and simplifying the analysis. Feature reduction can be used also in supervised analysis methods to reduce the number of features required to classify gene expression profiles correctly (sometimes this is called **cherry-picking**). In one method, this can be achieved simply by weighting classification features according to their usefulness and eliminating those that are least informative. *Fig. 2* shows how PCA can be used to group replicates in a set of microarrays. Clear clustering and separation of the controls from each treatment illustrate the reliability of the data.

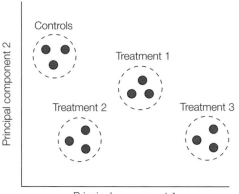

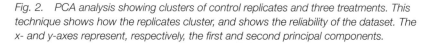

Fig. 2.   PCA analysis showing clusters of control replicates and three treatments. This technique shows how the replicates cluster, and shows the reliability of the dataset. The x- and y-axes represent, respectively, the first and second principal components.

From expression data to networks Often functional information about genes can help in the reconstruction of molecular pathways and networks, such as metabolic pathways, signal transduction cascades, and regulatory hierarchies. For example, the mutant phenotypes of genes acting in the same molecular pathway are very similar often, and pathways can be deduced also from information about protein-protein interactions. Genes in the same pathway may also have similar expression profiles. However, reconstructing molecular networks from expression data is difficult and a single correct method is yet to be discovered.

Bayesian statistics is the most widely used approach for inferring regulatory networks. **Bayesian (belief) networks** are a probabilistic graphical representation.

The variables are associated with a directed acyclic graph. There are conditional distributions for each variable. The nodes are the genes within the network, and the edges are inferred by using learning techniques on a gene expression dataset. Bayesian network inference features are computationally intensive and do not allow for cycles, which are important in genetic networks for homeostasis.

There are many **Bayesian-inference** resources in a variety of languages. There are two examples of Java-based tools: the Bayesian Network tools in Java (BNJ) (http://bnj.sourceforge.net/), and Bayesian Network Inference with Java Objects (http://www.cs.duke.edu/~amink/software/banjo/). There are sets of R packages that use Bayesian inference methods that have been contributed to the Comprehensive R Archive Network (CRAN) (http://cran.r-project.org/web/views/Bayesian.html). There are also lots of packages for specific model fitting. There is a Python-based Dynamic Bayesian Network toolkit called Mocapy (http://sourceforge.net/projects/mocapy/). Additionally, there is a MatLab toolbox called Bayes Net Toolbox (http://www.bnt.sourceforge.net/). Heuristic approaches (Section I) have been used to find optimal solutions in shorter times, for example artificial neural networks.

An alternative representation of gene regulation is as a **Boolean network**. This is a system in which the state of the network depends on its previous state, and any change is dependent on an explicit set of rules. As an example, consider three gene products – m, n, and p that interact with each other (Fig. 3a, b, c). We can implement a set of rules by which the original state of the system (the input, m, n, and p) can be changed into a new state (the output, m', n', and p'). The

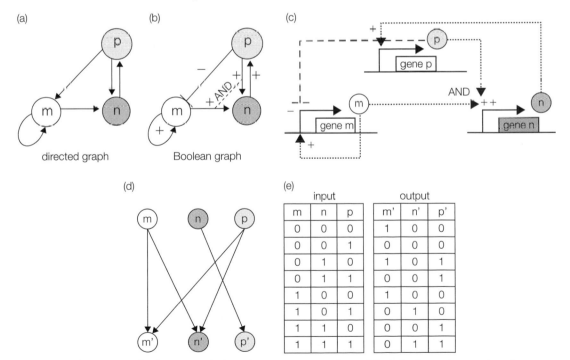

Fig. 3. *This figure shows Boolean networks based on genes m, n, and p, and the induction and repression of a gene regulatory network. Gene m induces itself and gene n, while gene n induces p, and p induces n. For gene n to be induced, p and m have to co-express. Gene p represses gene m. (a) Directed graph, and (b) a Boolean graph showing positive and negative regulation. (c) Biology-based gene regulatory diagram of genes m, n, and p. (d) Input- and output-based wiring diagram of the interactions. (e) Permitted output values (m', n', and p') given all possible input values (m, n, and p).*

rules are stated below and *Fig. 3e* shows the permitted output values for any given input. Input and output values can be either zero or one. A refinement of this representation has been to add probabilities to each edge. In this way, the networks can either take into account uncertainty or approximate the strength of the regulatory interaction.

Fig. 4 shows how a model can be used to describe a real regulatory network, namely the choice between lysis and lysogeny faced by bacteriophage γ when it infects *E. coli*. Initially the bacteriophage is committed to neither pathway, but a series of molecular events, depending on the growth conditions of the host cell, lead to the expression of either *cro* or *CI*. Each of these proteins can bind to the operators (O_R1, O_R2, and O_R3), which are shown as inputs in the model. If *cro* is expressed, the cro protein binds to the operator sites and prevents transcription from the promoters PR and PM so that the *CI* gene is not transcribed. This establishes lysis. If *CI* is expressed, the CI protein binds to the operator sites, preventing transcription from promoter PR (which prevents the transcription of *cro*) but allowing continued transcription of *CI* through the maintenance promoter PM. This establishes lysogeny. Hybrid methods are adequate for the modeling of such pathways whose behavior is well understood. However, models of unknown pathways based on expression data alone must be rigorously tested.

A final approach to inferring regulatory networks is to consider the dynamics of the process. Genes are regulated by other gene products and these can be represented by a matrix of ordinary differential equations. Using time-series data as a reference, there are various techniques (including PCA) to infer which genes are being influenced by which other genes. This approach is used in time-series network identification. The matrix can be made far sparser by assuming a maxiumum number of regulators of a given gene and also excluding influences below some arbitrary value.

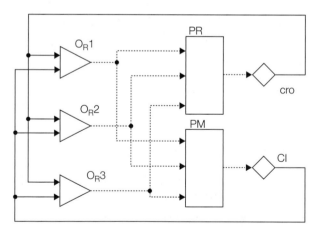

Fig. 4.   Model of a real regulatory network, the choice between lysis and lysogeny in bacteriophage λ, represented by inputs, control functions, and outputs.

K5 DATA STANDARDS AND EXPERIMENTAL DESIGN

Key Notes

Data standards	Data standards are essential with publication and the storage of microarray data in repositories. Microarray data repositories include NCBI GEO, ArrayExpress, Stanford Microarray Database, and NASCarrays.
Microarray data format	Unlike sequence and structural data, there is no international convention for the representation of data from microarray experiments. This is because of the wide variation in experimental design, assay platforms, and methodologies. An initiative to develop a common language for the representation and communication of microarray data has been proposed. Experiments are described in a standard format called MIAME (minimum information about a microarray experiment), and communicated using a standardized data-exchange model and a microarray mark-up language based on XML.
Experimental design	One major role of experimental design is to remove noise and variation in the samples. Well-planned microarray experiments provide a huge amount of information about the differential gene expression of a system.
	Technical issues are important when comparing microarray experimental data. The experiments must be robust, reliable, and reproducible. The same and cross-platform studies have shown a relative correlation between different laboratories and three major microarray platforms. However, there was a difference in same-platform comparisons across different laboratories.
Gene ontology	Gene ontology is a set of controlled vocabularies that determine a gene's biological process, cellular compartment, and molecular function.

Related sections	Data and databases (D)	Analyzing differential gene
	Transcript profiling (K1)	expression (K3)
	Statistical issues for transcriptome	Multivariate techniques and
	analysis (K2)	network inference (K4)

Data standards Data standards are essential in the storing of microarray data in data repositories. These standards are essential for validating publications and future cross analysis. This maintains a consistent standard that is statistically robust and reliable, which gives the community trust in the research. The sharing of data between scientific groups requires the development of mark-up languages (Section D) specific to bioinformatics analysis. Data repositories that require such high standards include NCBI GEO (Gene Expression Omnibus), ArrayExpress (an EBI expression database), Stanford Microarray Database, and NASCArrays (Section K3).

Microarray data format

Traditionally, bioinformatics has dealt with the analysis of sequence and structural data. There is a standard convention for the presentation of nucleic acid and protein sequences, and atomic coordinates in protein structures, allowing these data to be interpreted unambiguously by scientists around the world. More recently, the scope of bioinformatics has widened to include the analysis of gene and protein expression data. A standard format has been adopted for the representation of 2-DE protein gels but there is no similar convention for microarrays, even though microarray experiments produce some of the larger datasets bioinformatics has to deal with. This reflects the diversity of the different array platforms available (i.e., nylon macroarrays, spotted glass microarrays, high-density oligonucleotide chips), and also represents the large amount of variation in experimental design, hybridization protocols, and data gathering techniques.

Recently, there has been an international effort to develop a common language for the communication of microarray data. Essentially, the requirements for this language are that it should be minimal (i.e., no unnecessary embellishment of the experimental design and protocol) but convey enough information to enable the experiment to be repeated if necessary. The convention is known as **MIAME (minimum information about a microarray experiment)**, and incorporates six elements: overall experimental design, array design (i.e., identification of each spot on each array), probe source and labeling method, hybridization procedures and parameters, measurement procedure (including normalization methods), and control types, values, and specifications.

The language for expression data communication has been devised by the **MAGE group (Microarray And Gene Expression group)**. Currently, this involves a data exchange model (**MAGE-OM, Microarray And Gene Expression Object Model**) and a data exchange format (**MAGE-ML, Microarray And Gene Expression Mark-up Language**). MAGE-OM is modeled using the **Unified Modeling Language (UML),** and MAGE-ML uses **XML** (Section D). Information on these standards can be found in the **Microarray Gene Expression Database (MGED)** (http://www.mged.org).

Experimental design

The **experimental design** of a microarray experiment is essential to ensure the statistical reliability of the findings. Often control samples are as important as those relating to treatments. The number of replicates should depend upon how much variability there is within the experiment. This should be increased until the standard error is below an acceptable limit. However, microarray chips are expensive and so it is common practice to use three replicates and occasionally a small number more. Microarray manufacturers are helping in this regard by making chips more consistent.

There are differences in how a replicate is defined. Biological replicates are sourced from different experimental samples and therefore the RNA source is different. If the dataset is noisy, more than the standard three replicates will be required.

Technical replicates are from the same experimental RNA and are used to show variation in the laboratory procedure. Biological replicates are considered more important than technical replicates as biological replicates will find genuine biological phenomena and remove noise. To obtain the best results each stage of the process from experimental design to interpretation must be carried out (*Fig. 1*).

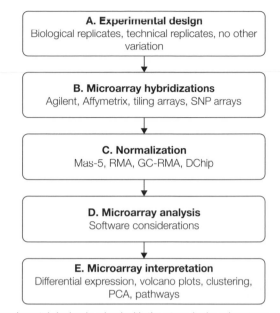

Fig. 1.   The experimental design involved with the steps in the microarray process.

Cross-platform comparisons were carried out with three leading microarray platforms: Affymetrix GeneChip, two-color cDNA arrays, and two-color long oligonucleotide microarrays. Cross-platform comparisons were relatively correlated when comparing the results from the same RNA samples. This study found that the major difference with same-platform studies using the same RNA samples was the differences between laboratories. Technical replicates are not required for every single experiment for an experienced microarray laboratory that is using a reliable and robust microarray platform. However, it is essential for technical replicates to be performed regularly as a quality control to show that the experiments are being conducted accurately within the laboratory.

An example of extensive statistical experimental design with large-scale microarray analysis is the genome-wide association studies using the human Affymetrix 500K mapping GeneChip, with comparisons in seven common diseases including type 1 and 2 diabetes, rheumatoid arthritis, hypertension, coronary heart disease, bipolar disease, and Crohn's disease. There were 14 000 cases for the 7 diseases with 3000 shared controls. The microarray design consisted of careful selection of subjects and extensive quality control.

Gene ontology

The **gene ontology project** focuses on creating controlled vocabularies, which describe a gene and its product (http://www.geneontology.org), and linking these using ontological concepts. These generally include the gene's involvement in one or more **biological processes**, its **molecular function**(s), and **cellular compartment**(s) in which the gene product is found. For every biological process (e.g,. carbohydrate metabolic process GO:0005975), cellular compartment (e.g., chloroplast GO:0009507), and molecular function (e.g., structural constituent of cell wall GO:0005199), there is a seven-digit reference GO identifier in the gene ontology database. Gene ontologies assist biologists in making rapid determination of gene function, and are useful in bioinformatics techniques such as text mining, and searching for gene products.

L1 PROTEOMICS TECHNIQUES

Key Notes

Raw data from 2-DE gels

Two-dimensional electrophoresis (2-DE) is a protein separation technique that allows the resolution of thousands of proteins on a single gel, on the basis of charge and mass. Separated proteins appear as spots, the nature and distribution of which constitute a protein fingerprint of any sample.

Data processing

Data extraction from 2-DE gels involves staining (to reveal the position of individual protein spots), scanning (to obtain a digital image), and then spot detection and quantitation. The quality of the image, in terms of spatial and densitometric resolution, is an important factor in accurate spot measurement. A number of algorithms are used to resolve complex overlapping spots and assemble a final spot list.

Gel matching

To study differential protein expression, a series of 2-DE gels must be compared. However, minute inconsistencies in gel structure and electrophoretic conditions make it impossible to replicate any experiment exactly. Sophisticated algorithms are required to follow individual spots through a series of gels, a process known as gel matching. MELANIE II is a widely used gel matching software application.

2-DE databases

Data from 2-DE experiments are deposited in dedicated 2-DE databases containing digital gel images with links from individual protein spots to useful annotations. Internet 2-DE databases are indexed at the ExPASy WORLD-2DPAGE. These allow 2-DE data to be shared with scientists around the world, and comparisons between gels can be carried out using Java applets such as Flicker or CAROL.

Raw data from mass spectrometry

The raw data from mass spectrometry (MS) experiments are the mass/charge (m/z) ratios of ions in a vacuum. These are used to determine accurate molecular masses. The masses can be used in peptide mass fingerprinting or fragment ion searching to find correlations in protein databases. Alternatively, peptide ladders can be generated and used to determine protein sequences *de novo*.

Protein expression profiles and databases

Differential protein expression data are assembled into a protein expression profile or matrix. This can be used to find distances between particular proteins or treatments, leading to classification or clustering of proteins according to similar expression profiles. There is no central database for protein expression profile data, though now there are common standards for sharing such data, MIAPE and PEML being the prime examples.

Virtual digests

Virtual digests are theoretical protein cleavage reactions performed by computers based on known protein sequences and the known specificity of a cleavage agent such as an endoproteinase. Although many different polypeptides can generate the same peptide digest pattern, in practice a correlation between the masses of two or more peptides produced from the same protein and the theoretical peptides produced in a virtual digest provides very strong evidence for a database match.

Dual digests

Dual digests, carried out on the same protein either separately or sequentially, can provide extra data to correlate experimentally determined molecular masses with less robust data resources such as dbEST. Alternatively, single digests can be carried out before and after protein modification, or ragged termini can be generated from proteins with clustered arginine and lysine residues, providing the masses of multiple fragments to use as database search terms.

Database search tools

Algorithms for database searching may attempt to match the experimentally determined mass of a peptide or peptide fragment to masses predicted from sequence database entries. The programs SEQUEST, X!Tandem, and Mascot work on this principle. Alternatively, the amino acid composition of a particular peptide or peptide fragment can be predicted from its mass. The order of amino acids cannot be predicted, so all permitted permutations are used as a database search query. The program Lutefisk works on this principle. Results from the programs SEQUEST and Mascot can be enhanced using new technology called Scaffold, which uses Bayesian statistics to provide more reliable identifications and reduce false positives.

Limitations of MS analysis

The failure of MS data to elicit a high-confidence hit on a sequence database may not always reflect the absence of that protein from the database. In some cases, it may reflect the presence of unknown or unanticipated post-translational modifications, or it may be caused by non-specific proteolysis or contaminating proteins. Imperfect matches may be generated if the experimental protein itself is absent from the database but a close homolog, with a related sequence, is present.

Related sections

Databases and data sources (J1)
Genome annotation (J2)
Transcript profiling (K1)
Analyzing differential gene
    expression (K3)

Multivariate techniques and
    network inference (K4)
Data standards and experimental
    design (K5)
Image analysis (Q)

Raw data from 2-DE gels

Two dimensional gel electrophoresis (2-DE) is a method used to separate proteins according to charge and mass (Section K3). The resolution of the technique is such that thousands of proteins can be distinguished, providing a diagnostic **protein fingerprint** of any particular sample. After the gel has been run, it is stained to reveal the position of individual proteins. These appear as **spots** of varying size, shape, and intensity. The role of bioinformatics in protein gel analysis is the extraction of useful information from the positions and intensities of the protein spots.

2-DE data can be used to derive general information on protein expression profiles without any knowledge of which specific proteins actually are present on the gel. However, the most powerful approach to 2-DE experiments is to couple the expression analysis with protein annotation by mass spectrometry.

Data processing   Data are extracted from 2-DE gels in several stages. First, the stained protein gel is scanned to obtain a digital image. Individual protein spots are then detected and quantified, and the intensity of the signal for each spot is corrected for local background.

The quality of the digital image is important since its **spatial resolution** determines how accurately protein spot sizes are recorded and its **densitometric resolution** determines how accurately the intensity of each spot is recorded. Protein spot measurement involves a special set of problems that are not found with microarray data. The features on a microarray are arranged in a precise grid, and the signals tend to be regular, discrete and non-overlapping. Conversely, the signals in a protein gel are irregularly distributed, the spots vary widely in morphology, and it is often the case that spots join together in clumps or lines that are difficult to resolve. Several algorithms are available which address these issues and generally these are based on either **Gaussian fitting** or **Laplacian of Gaussian spot detection (LOG).** Spots whose morphology deviates from a single Gaussian shape can thus be interpreted using a model of overlapping shapes. A simpler approach is **line-and-chain analysis**, in which columns of pixels from the digital image are scanned for peaks in signal density. This process is repeated for adjacent pixel columns allowing the algorithm to identify the centers of spots and their overall signal intensity (**signal volume**). A further approach is known as **watershed transformation (WST).** In this method, pixel intensities are viewed as a topographical map so that hills and valleys can be identified. This is useful for separating clusters, chains, and shouldered spots (small spots overlapping with larger ones), and also for merging regions of a single spot. The output of each method is a **spot list**, in which each individual protein spot is identified by the x and y co-ordinates of its center.

Gel matching   An important application of 2-DE is the analysis of **differential protein expression**. This can be used, for example, to look for proteins that are induced or repressed by particular treatments or drugs, to look for proteins associated with disease states, or to look at changes in protein expression during development. Differential protein expression can be studied only by running a series of 2-DE gels with alternative samples, and searching for novel spots or spots whose intensity changes significantly in different gels. Note that any observed changes; for example, the presence of a novel protein spot, may not reflect the *synthesis* of a new protein. Instead, the new spot may reflect the **post-translational modification** of an existing protein; for example, phosphorylation or glycosylation. Such modifications can radically alter the mass or charge of a protein, leading to different migration behavior during electrophoresis.

Owing to minute variations and inconsistencies in the chemical and physical properties of electrophoretic gels, it is impossible to reproduce exactly the conditions of any one electrophoresis experiment. This means that, even in a series of gels using exactly the same sample and electrophoresis parameters, the positions of individual protein spots are never the same. In order to compare serial gels and identify novel protein spots or spots that vary in intensity, it is necessary to use **gel matching**. Generally this involves establishing the positions of several unambiguous **landmark spots** and then using algorithms to match the positions of the remaining spots (*Fig. 1*). To bring the spots from two gels into register, simple image manipulations such as stretching and rotating can be carried out. This may be assisted by incorporating spot intensities and the (physical) distances between neighboring spots as variables in the algorithm. For example, an approach known as **propagation** involves determining the distance between a given landmark spot and all neighbors. If matches are found on other gels, the neighboring spots can be used as new landmarks, and the process reiterated. An implementation of such algorithms is available as the program **MELANIE II**. This is available to download for use on Windows-based computers (www.expasy.ch/melanie) but cannot be used online.

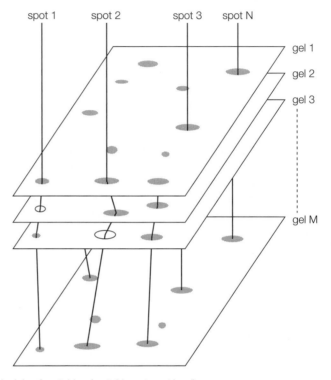

Fig. 1.   Principle of matching (matching at spot level).

2-DE databases   The results from 2-DE experiments are generally stored in **2-DE databases**, which use gel images as a basis for protein annotation. Digital images from 2-DE gels are presented, and each protein spot acts as a link to further information such as protein name, molecular mass, pI value, annotations from the SWISS-PROT database, bibliographic references and, if appropriate, graphs showing how spot intensities vary over a series of gels. Some 2-DE databases are free-standing and may come packaged with gel analysis software. For example, the analysis suite **PDQUEST** incorporates its own database, which may be installed locally and used to store the 2-DE experimental data from a single laboratory. There are free software packages also available that can be used to set up 2-DE databases, such as **Make2ddb**.

The construction of 2-DE databases on the WWW allows 2-DE data to be shared and compared over the internet. There are over 60 such databases currently available, many showing 2-DE data either for specific organisms or specific systems. For example, the HEART-2DPAGE contains 2-DE data related to heart development, physiology, and disease. All these pages are indexed by the ExPASy server on the **WORLD-2DPAGE** (http://www.expasy.ch/ch2d/2d-index.html). The format of data presentation is standardized. Images of 2-DE gels are presented overlain with a grid showing pI values and molecular masses. This allows each protein spot to be identified on the basis of its physical and chemical properties. Individual spots can be clicked, linking the user to an annotation page. On many gels, most of the spots are not annotated, and those with annotations are highlighted in some manner. For example, on the SWISS-2DE database each annotated protein spot is marked with a red cross. Annotations can be viewed as an original SWISS-PROT file, an example of which is shown in *Fig. 2*. Most of the fields are self-explanatory (see Section J1), but note that the 2D field may include data on both 2-DE electrophoresis and peptide mass fingerprinting, where such experiments have been carried out.

Another WWW resource, **2D Web Gel (2DWG)** can be found at the URL: http://www-lecb.ncifcrf.gov/2dwgDB/. This is a catalogue of some of the 2-D gel images found on the WWW, with associated search facilities so that images can be abstracted by key words. 2DWG represents a very good introduction to the world of 2-D gels. The images are stored in formats such as GIF that internet browsers can handle and there is an associated Java applet, called **Flicker**, to facilitate the comparison of two gels. The basis of flickering is that the images of two gels can be rapidly alternated to identify matching spots. Another useful applet is **CAROL** (http://gelmatching.inf.fu-berlin.de/Carol.html), which uses point pattern matching to compare any two gel images over the internet.

```
ID   ACTB_HUMAN; STANDARD; 2DG.
AC   P02570;
DT   01-AUG-1993 (Rel. 00, Created)
DT   01-DEC-2000 (Rel. 13, Last update)
DE   Actin, cytoplasmic 1 (Beta-actin).
GN   ACTB.
OS   Homo sapiens (Human).
OC   Eukaryota; Metazoa; Chordata, Craniata, Vertebrata; Euteleostomi;
OC   Mammalia; Eutheria; Primates; Catarrhini; Hominidae; Homo.
OX   NCBI_TaxID=9606;
MT   CSF_HUMAN, ELC_HUMAN, HEPG2_HUMAN, HEPG2SP_HUMAN, LIVER_HUMAN,
MT   LYMPHOMA_HUMAN, PLASMA_HUMAN, PLATELET_HUMAN, RBC_HUMAN, U937_HUMAN,
MT   KIDNEY_HUMAN, HL60_HUMAN, CEC_HUMAN, DLD1_HUMAN.
IM   CSF_HUMAN, ELC_HUMAN, HEPG2_HUMAN, HEPG2SP_HUMAN, LIVER_HUMAN,
IM   LYMPHOMA_HUMAN, PLASMA_HUMAN, PLATELET_HUMAN, RBC_HUMAN, U937_HUMAN,
IM   KIDNEY_HUMAN, HL60_HUMAN, CEC_HUMAN, DLD1_HUMAN.
RN   [1]
RP   MAPPING ON GEL.
RX   MEDLINE=93162045; PubMed=1286669; [NCBI, ExPASy, EBI, Israel, Japan]
RA   Hochstrasser D.F., Frutiger S., Paquet N., Bairoch A., Ravier F.,
RA   Pasquali C., Sanchez J.-C., Tissot J.-D., Bjellqvist B., Vargas R.,
RA   Appel R.D., Hughes G.J.;
RT   "Human liver protein map: a reference database established by
RT   microsequencing and gel comparison.";
RL   Electrophoresis 13:992-1001(1992).
RN   [2]
```

.....Several more literature references follow.....

```
CC   -!- SUBUNIT: SINGLE CHAIN WHICH CAN BIND UP TO 4 OTHER CHAINS.
2D   -!- MASTER: CSF_HUMAN;
2D   -!-  PI/MW: SPOT 2D-000C1S=5.24/44747;
2D   -!-  MAPPING: MATCHING WITH THE PLASMA MASTER GEL [2].
2D   -!- MASTER: ELC_HUMAN;
2D   -!-  PI/MW: SPOT 2D-000ED0=5.21/41208;
2D   -!-  PI/MW: SPOT 2D-000ED7=5.12/41300;
2D   -!-  MAPPING: MATCHING WITH THE LIVER MASTER GEL [2].
2D   -!- MASTER: HEPG2_HUMAN;
2D   -!-  PI/MW: SPOT 2D-00030B=5.15/41700;
2D   -!-  PI/MW: SPOT 2D-00030Z=5.09/41700;
2D   -!-  PI/MW: SPOT 2D-00031Z=5.23/41272;
2D   -!-  MAPPING: MATCHING WITH THE LIVER MASTER GEL [2].
```

Fig. 2. continued on next page.

2D   -!- MASTER: HEPG2SP_HUMAN;
2D   -!- PI/MW: SPOT 2D-000952=5.27/38525;
2D   -!- PI/MW: SPOT 2D-000954=5.16/38400;
2D   -!- PI/MW: SPOT 2D-000955=5.22/38223;
2D   -!- PI/MW: SPOT 2D-000959=5.11/38223;
2D   -!- MAPPING: MATCHING WITH THE PLASMA MASTER GEL [2].
2D   -!- MASTER: LIVER_HUMAN;
2D   -!- PI/MW: SPOT 2D-0000WF=5.26/41839;
2D   -!- PI/MW: SPOT 2D-0000WN=5.19/41722;
2D   -!- PI/MW: SPOT 2D-0000WO=5.22/41605;
2D   -!- MAPPING: MATCHING WITH A PLASMA GEL [1].
2D   -!- MASTER: LYMPHOMA_HUMAN;
2D   -!- PI/MW: SPOT 2D-0007PE=5.26/41898;
2D   -!- PI/MW: SPOT 2D-0007PQ=5.15/42194;
2D   -!- MAPPING: MATCHING WITH THE LIVER MASTER GEL [2].
2D   -!- MASTER: PLASMA_HUMAN;
2D   -!- PI/MW: SPOT 2D-00050N=5.28/43590;
2D   -!- PI/MW: SPOT 2D-00050Q=5.24/43244;
2D   -!- MAPPING: IMMUNOBLOTTING [3].
2D   -!- NORMAL LEVEL: PLATELET CONTAMINATION.
2D   -!- MASTER: PLATELET_HUMAN;
2D   -!- PI/MW: SPOT 2D-000FWX=5.27/41946;
2D   -!- PI/MW: SPOT 2D-000FZ5=5.18/41400;
2D   -!- PI/MW: SPOT 2D-000FZM=5.06/41400;
2D   -!- PI/MW: SPOT 2D-000FZN=5.30/41400;
2D   -!- MAPPING: MATCHING WITH RBC AND LIVER MASTERS [5].
2D   -!- MASTER: RBC_HUMAN;
2D   -!- PI/MW: SPOT 2D-00064D=5.20/42104;
2D   -!- PI/MW: SPOT 2D-00064O=5.14/42209;
2D   -!- PI/MW: SPOT 2D-00064W=5.27/42104;
2D   -!- MAPPING: IMMUNOBLOTTING [3] AND MATCHING [4].
2D   -!- MASTER: U937_HUMAN;
2D   -!- PI/MW: SPOT 2D-000CXH=5.23/41807;
2D   -!- MAPPING: MATCHING WITH THE LIVER MASTER GEL [2].
2D   -!- MASTER: KIDNEY_HUMAN;
2D   -!- PI/MW: SPOT 2D-000N5T=5.18/41212;
2D   -!- PI/MW: SPOT 2D-000N5U=5.25/41503;
2D   -!- PI/MW: SPOT 2D-000N67=5.14/41406;
2D   -!- PI/MW: SPOT 2D-000N6J=5.12/41309;
2D   -!- MAPPING: MATCHING WITH THE LIVER MASTER GEL AND IMMUNODETECTION
2D   [6].
2D   -!- MASTER: HL60_HUMAN;
2D   -!- PI/MW: SPOT 2D-000YZK=5.25/41925;
2D   -!- PI/MW: SPOT 2D-000Z0A=5.15/41390;
2D   -!- PI/MW: SPOT 2D-000Z0G=5.08/41497;
2D   -!- MAPPING: MATCHING WITH THE LIVER MASTER GEL [7][8].
2D   -!- MASTER: CEC_HUMAN;
2D   -!- PI/MW: SPOT 2D-000TWW=5.05/41396;
2D   -!- PI/MW: SPOT 2D-000TWX=5.11/41497;
2D   -!- MAPPING: MATCHING WITH THE LIVER MASTER GEL [9].
2D   -!- MASTER: DLD1_HUMAN;
2D   -!- PI/MW: SPOT 2D-001E5J=5.15/41545;
2D   -!- PI/MW: SPOT 2D-001E5O=5.19/41545;
2D   -!- PI/MW: SPOT 2D-001E6E=5.10/41391;
2D   -!- PEPTIDE MASSES: SPOT 2D-001E5J: 976.504; 1132.57; 1198.7;
2D   1516.72; 1790.89; 1954.05; 2231.02; TRYPSIN.
2D   -!- PEPTIDE MASSES: SPOT 2D-001E5O: 976.464; 1132.53; 1198.7;
2D   1516.7; 1790.87; 1954.05; 2231.07; TRYPSIN.
2D   -!- PEPTIDE MASSES: SPOT 2D-001E6E: 976.513; 1132.57; 1516.73;
2D   1790.89; 1954.04; 2231.05; TRYPSIN.
2D   -!- MAPPING: MASS FINGERPRINTING [10].
CC ---
CC This SWISS-2DPAGE entry is copyright the Swiss Institute of Bioinformatics.
CC There are no restrictions on its use by non-profit institutions as long as
CC its content is in no way modified and this statement is not removed. Usage
CC by and for commercial entities requires a license agreement (See
CC http://www.isb-sib.ch/announce/ or send an email to license@isb-sib.ch).
CC ---
DR   SWISS-PROT; P02570; ACTB_HUMAN.
DR   Siena-2DPAGE; P02570; ACTB_HUMAN.
//

Fig. 2.   Example SWISS-2DPAGE entry (for human actin B). Some material has been deleted for brevity, as shown. Note the extensive 2-D data, which include information from 2-DE experiments and mass spectrometry.

Raw data from mass spectrometry

Mass spectrometry (MS) is a method for accurately determining the **mass/charge ratio (m/z)** of ions in a vacuum, thus allowing the precise determination of **molecular masses**. These raw data can be used in three different approaches for the identification of proteins. In **peptide-mass fingerprinting**, a protein is digested with a specific **cleavage agent** (usually the enzyme trypsin) and the masses of each of the resulting peptides are determined. Then these masses are used in **correlative database searching** to identify the protein. In **fragment ion searching**, peptide fragments are generated as above and then fragmented in a **collision cell** between two quadrupole mass analyzers (this is called **tandem MS**, often abbreviated to **MS/MS**) or by a process known as **post-source decay**, which occurs when MALDI-TOF MS is used at a higher acceleration voltage than usual. The resulting **fragment ions** are short peptide fragments. The molecular masses of such fragments can be used to search not only protein databases, but also other sequence repositories including dbEST (Section J1). Both these approaches require that the protein in question has been identified and its sequence deposited. Where this is not the case, *de novo* **sequencing** can be carried out. In this method, peptide ladders are generated either by sequential chemical degradation of terminal amino acid residues or by separating the fragment ions generated as above into nested sets. Mass differences between sequential fragments correspond to the known masses of individual amino acids, thus allowing protein sequences to be deduced without any correlative information from sequence databases.

Protein expression profiles and databases

Once protein expression data have been recorded they are built into a **protein expression matrix** (or **profile**). This is similar in principle to the gene expression matrix discussed in Section K3. Spots, representing proteins, are arranged in rows, while experimental treatments are listed in columns. The data points in the matrix represent signal intensities for each spot. Comparison of values along a given row can identify proteins whose expression levels change according to different treatments. As with gene expression data, multivariate statistical analysis can group data by similar expression profiles, which can be used for classification purposes or clustering (Section K3).

In addition, proteomics is following the lead of transcriptomics in setting data standards for the sharing of such expression profile data. The commonest standards are minimal information about a proteomics experiment (MIAPE – very analogous to MIAME), and Protein Expression Mark-up Language (PEML).

Table 1. Protein expression profile databases

Resource	URL	Features and comments
Integrative Gene and Protein Expression Database (InGaP)	https://webcreate.kazusa.or.jp/create/servlet/create.LoginServlet	Stores data generated from mKIAA constructs
Lung Cancer Protein Expression Database		Details in Oh, JMC *et al.*, (2001)
Lymphoid Neoplasia Database		Fujii, K *et al.*, (2006)
Pancreatic Expression Database	http://www.pancreasexpression.org/	Gene and protein expression profiles relevant to the study of pancreatic cancers
Proteome Experimental Data Repository (PEDRo)		Garwood, K *et al.*, (2004)
Yale Protein Expression Database (YPED)	http://info.med.yale.edu/proteome/	A collaboration between Yale and the Inst. for Systems Biology (ISB) in Seattle

Unlike the gene expression databases, which are now coalescing into a small number of principal resources, the protein expression databases are scattered still, often system-specific, and may contain a mixture of gene-expression and protein-expression profile data (see *Table 1* for examples).

Virtual digests

It would be impossible to link peptide mass data to proteins in sequence databases such as SWISS-PROT without first knowing the *expected* peptides from such proteins. This information can be obtained by carrying out **virtual digests**; that is, theoretical digests based on the known protein sequence and the known specificity of the cleavage agent used. Cleavage agents with high specificity are most suitable and the endoproteinase **trypsin**, which cleaves a polypeptide chain after each basic amino acid (lysine or arginine), providing the next residue is not a proline, is the most widely used. Given a protein of known sequence, the **tryptic peptides** can be predicted. For example, a protein with the sequence shown below

```
MCLTAKGAATCSATFRYLIFALSLATKPACALLASALLARACATTAVA
```

would generate the following tryptic peptides

```
MCLTAK GAATCSATFR YLIFALSLATKPACALLASALLAR ACATTAVA
```

The four peptides provide four theoretical molecular masses. Correlation between these theoretical masses and the actual masses obtained in a mass spectrometry experiment would provide very convincing evidence for a database match. Of course the same molecular masses could be obtained in many other ways. Each of the four peptides has a potentially very large number of anagrams (same amino acids in a different order), which would all have the same molecular mass. Theoretically, this places limitations on the usefulness of the technique, but in practical terms the chances of a series of peptides from the same protein generating spurious matches to the same deposited protein sequence because of permutations in the order of amino acids are very small indeed. Generally, correlation between two or more peptides is taken to be unambiguous confirmation of a database match. Another theoretical limitation is that the amino acids leucine and isoleucine have the same molecular masses. This provides a small technical problem in *de novo* sequencing, but for peptide mass fingerprinting and fragment ion fingerprinting it does not have a practical impact.

Dual digests

Peptide mass fingerprinting allows the rapid identification of proteins if they are already represented in databases such as SWISS-PROT. Where this is not the case, both peptide mass fingerprinting and fragment ion searching can be used to match mass spectrometry data to other sequence databases, including the EST database dbEST. The problem with this approach is that the databases contain a large amount of irrelevant data (e.g., non-coding sequence and other 'noise'), which can reduce the efficiency of the search. More confidence can be placed in any results if **dual digests** are carried out (i.e., combining the information from two protease digests using enzymes with different specificities, such as trypsin and endoproteinase LysC). Another approach is to carry out a single digest with the protein in the native state and then carry out the same digest after modifying the protein, for instance by methylation. Furthermore, since lysine (K) and arginine (R) are two of the most common amino acids in proteins, there is a relatively large number of doublets and triplets (e.g., RR, KKR, KRK). Trypsin cleaves randomly at such sites generating peptide fragments with **ragged termini**. More confidence can be placed in database hits that are compatible with such peptides because this is effectively the same as searching with a larger number of peptides.

Database search tools

A number of algorithms have been developed for sequence database searching using MS data. Among these, the most commonly used is **SEQUEST**, which works by searching for all peptides in the specified database(s) with the same mass as a given peptide ion. Then a virtual digest is performed on the matched protein and a theoretical mass spectrum generated. The data from the theoretical mass spectrum are compared then to the experimental data and the best matches are scored. Programs **X!Tandem** and **Mascot** also work on this principle. In a different approach, peptide mass data are used to generate a collection of possible sequences, and this profile is used as a query in a modified BLAST or FASTA search. A program called **Lutefisk** has been developed for this purpose. Many software resources for the analysis of MS data are available over the internet and some are listed in *Table 2*.

Table 2. Internet resources for MS-based protein identification

Resource	URL	Features and comments
CBRG, ETH-Zurich	www.cbrg.inf.ethz.ch/services/MassSearch	Peptide mass search
EMBL, Heidelberg	www.mann.embl-heidelberg.de/Group Pages/PageLink/peptidesearchpage.html	Peptide mass and fragment ion search
ExPASy	www.expasy.ch/tools/	Peptide mass and fragment ion search
Mascot	www.matrixscience.com/search_form_select.html	Peptide mass and fragment ion search
Rockefeller University, New York	prowl.rockefeller.edu/	Peptide mass and fragment ion search
X!Tandem	www.thegpm.org/TANDEM/	Peptide mass and fragment ion search
Scaffold	www.proteomesoftware.com/Proteome_prod_Scaffold.html	Peptide mass and fragment ion search validation
OMSSA	Pubchem.ncbi.nlm.nih.gov	Peptide mass and fragment ion search
University of California	prospector.ucsf.edu	Peptide mass (MS-Fit) and fragment ion (MS-Tag) search
	donatello.ucsf.edu	As above, on-line access
University of Washington	http://fields.scripps.edu/sequest/	Instruction on how to get the SEQUEST fragmention search program

Limitations of MS analysis

Although a powerful technique for protein annotation and sequencing, there are some limitations to MS which need to be taken into account when interpreting and analyzing experimental data. One of the most important factors to take into account is that MS data may not match any database entry because of the presence of an unknown **post-translational modification**. Where such modifications are known, exact mass differences between unmodified and modified amino acids can be predicted. Indeed several algorithms, including **SEQUEST** and **Scaffold**, have built-in parameters for detecting such modifications. However, the presence of a modified residue should always be confirmed experimentally. Another potential problem is the occurrence of non-specific proteolysis. This depends on the purity of the cleavage agent used. Many algorithms will carry out peptide mass searches without a specified cleavage agent to take non-specific proteolysis into consideration. A common problem is that protein spots isolated from 2-DE gels often contain a mixture of proteins, and these contaminants may be difficult to identify. Finally, imperfect matches may result because the actual protein does not exist in the database, but a close homolog from the same species or a different species, which may have a related sequence, does exist.

L2 INTERACTION PROTEOMICS

Key Notes

Interactions and pathways

Proteins that physically interact with each other may be involved in the same molecular pathway or network, or may form part of a multi-subunit complex. Using this principle, pathways can be reconstructed based on evidence of protein interactions. However, information from other sources; for example, gene expression patterns and mutant phenotypes, may be useful as well.

Yeast two-hybrid analysis

Yeast two-hybrid screens produce large amounts of protein interaction data, but there is a relatively high level of spurious results (false positives and false negatives). This problem can be addressed by scoring interactions for reliability, based either on the repeatability of interactions over multiple experiments, or by the number of times a given bait will trap independent clones representing the same prey. Even so, similar large-scale screens tend to identify different (although overlapping) sets of interactions.

Multi-subunit and organelle protein composition

There are techniques for tagging proteins that form part of a complex. Purifying the complex then enables the other members of the complex to be identified. These techniques complement yeast two-hybrid analysis. LOPIT is a technique for determining compartment-specific proteins.

Related sections   Transcript profiling (K1)   Proteomics techniques (L1)

Interactions and pathways

The function of individual proteins can only be fully elucidated by studying them in the context of pathways and other higher-order systems. Pathways can be reconstructed from different types of functional data, including the study of mutants and gene expression patterns (see Section K for methods). Protein interaction data, however, provide not only a functional basis but also a physical basis for pathway reconstruction. Protein interaction data may demonstrate the existence of multi-subunit complexes also. In this section, we consider the role of bioinformatics in the handling and analysis of protein interaction data and show how this helps in the definition of molecular pathways, networks, and complexes.

Most of the techniques used to study protein interactions are low-throughput methods; that is, they are most suitable for studying the interactions of a single protein, or interactions among a small group of proteins such as a multi-subunit complex. For these methods, the challenge of bioinformatics is to find ways of assimilating the data from diverse individual experiments into a single resource (see below). Conversely, library-based methods – particularly the yeast two-hybrid (Y2H) system – are geared for high-throughput screening and thus generate large datasets. The challenge for bioinformatics in this context is to mine the data and extract meaningful results.

Yeast two-hybrid (Y2H) analysis

This type of analysis is a medium-throughput genetic approach to the detection of protein–protein interactions. In yeast there is a transcription factor that comprises domains for DNA-binding and recruitment of RNA polymerase. These domains can be separated. The first stage of the analysis consists of making a library of protein-coding sequences cloned into a gene construct that encodes a hybrid protein with one of these domains. This library constitutes 'the prey'. The other half of the analysis consists of making a hybrid construct that encodes the other domain and the protein for which protein interactions are to be determined. This is 'the bait.' The latter is cloned into a yeast cell line, which is then transformed (in the genetic manipulation sense) with the library of prey molecules and plated onto selective media. The principle of the technique is that the yeast will only express certain essential genes when two hybrid proteins (the bait and prey) interact for long enough to recruit RNA polymerase (see *Fig. 1*), hence the name of the technique. DNA for the prey proteins are sequenced then to discover which gene/protein interacts with the bait.

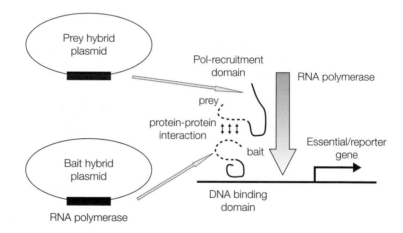

Fig. 1.   Yeast two-hybrid rationale. Inside a yeast cell are two plasmids containing hybrid genes. Each expresses a fusion protein. Recruitment of RNA polymerase to allow the expression of the gene is dependent upon there being a protein-protein interaction between the bait and prey protein domains. The gene encodes a protein that is either essential for cell survival and/or marks the yeast colony in some way to 'report' that the gene is being expressed.

The greatest problem with Y2H screens is the relatively high proportion of spurious results. These comprise **false positives**, where the reporter gene is activated in the absence of any specific interaction between the bait and prey, and **false negatives**, when the reporter gene is not activated even if the bait and prey normally interact, as shown by other experimental techniques. False positives occur for many reasons; for example, if the prey is **sticky** (interacts nonspecifically with many proteins) or capable of **autoactivation** (activation of transcription without interacting with the bait). False negatives often occur if the fusion protein does not fold in the same manner as the native protein.

The analysis of Y2H data thus requires each result to be assessed for reliability. Before considering how confidence is assessed, an appreciation of the Y2H assay format is required. Essentially, large-scale Y2H screens come in two types. In a **matrix assay**, each bait construct and each potential prey construct is individually produced; for example, using the polymerase chain reaction, and haploid yeast strains are arranged as a matrix in a series of microtiter plates, allowing systematic and exhaustive crossings to be carried out, therefore testing each possible interaction. Conversely, in a **random library assay**, the bait constructs are produced individually but the prey are represented by a library of random DNA clones. Libraries of high complexity are required to cover an entire genome. In a matrix assay, reliability is generally judged by the repeatability of a particular interaction. That is, the crossings are carried out a number of times, and only those interactions that are highly repeatable are taken to be genuine results. In a random library assay, each prey is represented generally by a number of overlapping clones. Reliability can be judged by the number of independent clones, representing the same prey that are trapped by the bait. A further advantage of this approach is that the domain of the prey protein responsible for the interaction can be narrowed down (*Fig. 2*). Despite these precautions, it is apparent that similar Y2H screens tend to identify different sets of interacting proteins. For example, between 1999 and 2001 two research groups carried out large-scale interaction screens covering the entire yeast genome (see *Further Reading*). In total, over 1500 high-confidence interactions were cataloged, but less than 10% of these were identified in both screens.

Multi-subunit and organelle protein composition

Cellular machinery often consists of complexes involving multiple different protein subunits. Where one component of a complex is known, it can be tagged in a variety of ways, for example through a genetic construct to make a fusion protein that includes an N- or C-terminal peptide (e.g., Protein A) recognized by an affinity purification agent. Following gentle cell lysis, the tagged complex can be affinity purified, its proteins separated by gel electrophoresis and identified using mass spectroscopic methods (see Section L1). This approach (called TAP-tag – tandem affinity purification tags) was originally developed by Gavin *et al.* (2002), but various refinements have been made in the interim, most notably iTAP in which RNA interference is used to suppress endogenous expression of a given gene while its tagged counterpart is expressed unhindered. This has the effect of increasing the yield of tagged complexes.

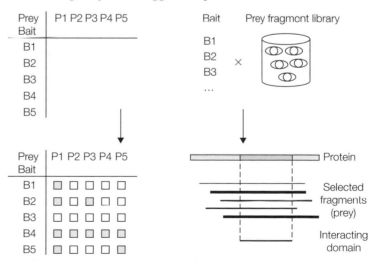

Fig. 2.   Principle of the matrix and random library formats for Y2H screens. In the matrix format, different bait–prey combinations are tested systematically. In the library format, coverage of the genome is dependent on library complexity but the domains of interacting proteins can be identified by the analysis of overlapping prey clones.

Some protein may act in concert by virtue of being localized in the same cellular compartment, and techniques now exist to determine these in a comprehensive manner. The technique is known as localization of organelle proteins by isotope tagging (LOPIT). A sample of cells is first carefully disrupted leaving the organelles largely intact. The lysate is then subject to differential centrifugation, which causes different organelles and compartments to sediment at different times. The centrifuged mixtures are then aliquoted and the amounts of the proteins in sequential samples determined by protein separation and mass spectroscopic techniques. The spectra of proteins are compared to those which are known to be confined to particular compartments. Proteins with matching distributions are highly likely to be restricted to the same compartment.

L3 INTERACTION DATABASES AND NETWORKS

Key Notes

Higher-order systems

Although genes and proteins can be studied individually, more insight into their functions can be gained by studying higher-order systems; that is, molecular pathways and networks, cells, tissues, organs, and whole organisms. This allows their physical and functional interactions to be determined in the widest possible context.

Protein interaction databases

Several databases have been set up to store the interaction data arising from large-scale Y2H screens. However, much more information on protein interactions is available in the scientific literature, and a current challenge in bioinformatics is the assimilation of these interaction data from diverse sources.

The interactome

The interactome is the sum of all protein interactions in a cell. The simplest way to represent protein interactions is a graph with proteins as nodes and interactions as edges. However, when large numbers of proteins are considered, the graphs become too complex to interpret all the interactions potentially taking place.

Network simplification

Interaction maps can be simplified in several ways including collapsing sets of proteins to single nodes by various criteria (such as whether or not they are part of a multi-protein complex or belong to some specific aspect of cellular physiology), or taking a subset of the total map on the basis of which proteins are thought, on the basis of biomolecular profile data, to be present in the cell/compartment/physiological state of interest.

Related sections

Graph theory and its applications (H2)	Interaction proteomics (L2) Biochemical dynamics (O)

Higher-order systems

Genes and proteins often have been studied as individual entities. However, to gain a fuller insight into the function of proteins and their role in physiological processes, it is necessary to assimilate these data into **higher-order systems**, which include pathways, networks, cells, tissues, organs, and whole organisms. This section will focus specifically on protein interaction networks, which incorporate signaling pathways, multi-protein complexes, and trafficking. Bioinformatics has a crucial role to play in the representation, reconstruction, and modeling of pathways and networks, and in the creation of database resources. There are analogous approaches for metabolic networks (see Section O), gene regulatory networks (Section K) and integrated networks (which combine all three).

Protein interaction databases

Protein interaction data are very useful for functional annotation and the reconstruction of molecular pathways and networks. One role of bioinformatics is to provide **protein interaction databases** that allow interaction data to be stored, queried, assessed for confidence, and used for pathway reconstruction. Dozens of databases have been set up to store protein interaction data, with a selected subset shown in *Table 1*. Readers are recommended to go in the first instance to PathGuide (http://pathguide.org/) or BioGRID (http://www.thebiogrid.org/), which include databases of all types of molecular interaction. Protein–protein interaction databases currently fall into five categories.

1. Manually collated sets of known interactions that make up some aspect of the physiology of cells or organisms.
2. High-throughput datasets usually generated specifically for one organism, and often developed as part of a functional genomics project looking at that specific organism.
3. Datasets inferred in some way from other sources. These include sophisticated data mining of literature databases, and the combination of sequence similarity and known interactions in other organisms.
4. Databases that attempt to capture all the previous three categories.
5. Databases containing all types of biomolecular interaction, which includes those in this category.

Table 1. A selection of protein interaction databases available over the internet

Database	URL	Category/comments
STKE	http://stke.sciencemag.org/	1) Signal Transduction Knowledge Environment
Reactome	http://www.reactome.org/	1) Manually curated knowledge base of biological pathways; includes some metabolism and gene regulation
Hybrigenics	http://pim.hybrigenics.com/ pimriderext/common/	2) Lists protein interactions in the bacterium *Helicobacter pylori*. Includes graphical interface for the visualization of interactions
OPHID	http://ophid.utoronto.ca/	3) Online predicted human interactions database
AtPID	http://atpid.biosino.org/	3) Predicted protein interactions of *Arabidopsis thaliana*
IntAct	http://www.ebi.ac.uk/intact	4) Protein interaction database hosted at the European Bioinformatics Institute
DIP	http://dip.doe-mbi.ucla.edu	4) Database of Interacting Proteins. Comprehensive resource for protein-protein interactions
BIND	http://bind.ca	5) Biomolecular Interaction Network Database. Lists protein interactions with a variety of molecules (proteins, nucleic acids, small ligands), and includes resources for protein complexes and molecular pathways

The interactome

The ultimate challenge for bioinformatics in the field of protein interaction technology is the reconstruction of the **interactome**, defined as the sum of all protein interactions in the cell. The representation of these data in an accessible and user-friendly format is very difficult. The simplest way to represent protein interactions is a mathematical graph called an **interaction map**, with proteins at the nodes and interactions represented by edges (see Section H2). For a small number of proteins, interaction maps are very useful because it is possible to read the names of the protein nodes. However, larger numbers of proteins yield graphs of incredible complexity, which can be used only to give an impression of the overall structure/ topology of the network. *Fig. 1* gives examples of these extremes.

(a) (b)

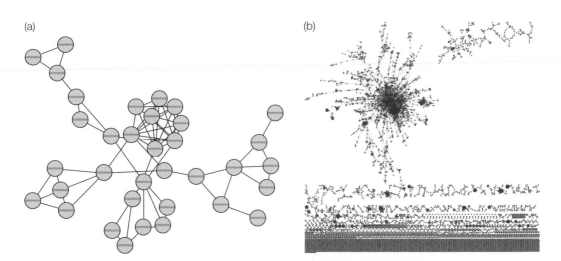

Fig. 1.   Examples of protein-protein interaction networks. (a) A comparatively small network of 35 proteins (nodes), which is one of the smaller disconnected sub-graphs in (b), which represents the entire inferred Arabidopsis thaliana interactome, known as AtPID, and consists of 8024 proteins and 14 837 interactions. The latter shows that there are many disconnected sub-graphs/networks in this database. The main cluster is in the top left quadrant of the picture, with a few larger sub-clusters on the top right. Then underneath in rows are the progressively smaller interaction subsets until there are over 10 rows of pair-wise interactions. In Arabidopsis, there are almost certainly more interactions than have been captured here, but this is a useful start. Both layouts were generated in Cytoscape, courtesy of A. Marshall.

Network simplification

Although these dense pictures can provide a general overview, they make it impossible to focus on subsets of the whole that are of particular interest. There are various ways to reduce this complexity, which will be considered in turn.

The first is to decrease the crowding around hubs. For example, most proteins link to ubiquitin as part of the protein-degradation process, but pulling protein interactions around this process is a serious distraction from the many varied signaling mechanisms taking place in cells. Reducing such crowding can be achieved either by removing the hubs or having as many nodes for that hub molecule as there are edges. This results in looser networks that have no loss of detail, making it less difficult to identify both pathways/complexes of interest and their context.

In cells, multi-protein complexes act as a unit, so an obvious second way to decrease complexity is through the identification of such complexes, by using the computer to look for 'cliques,' and then collapsing these down to a single node per clique. When searches for cliques have been carried out on holistic protein interaction networks, some of the most dense cliques do not correspond to complexes, but rather to interactions of hubs in stress response and cell cycle regulation. Collapsing these physiological units to individual nodes also can be beneficial for understanding how cellular processes fit together. This type of functional summarization can be carried out in a more formal manner by taking advantage of gene ontology assignments for the nodes in the network.

The final way to reduce network complexity is by removing the nodes and edges that are irrelevant. This approach can be based on a broad range of criteria. The obvious first is to include in the sub-graph only the specified nodes. The edges from the larger network will show whether the specified nodes form a pathway, cycle, or some form of lattice. The locality of the node(s) of interest can be explored by adding to them the nodes that are some defined number of edges away. Given the scale-free topology of biological networks, then one is not advised to look more than three to four edges away, as this will bring back a large proportion of the total network because this value is close to the network diameter.

A more interesting variant on this approach involves using sets of proteins, say from proteomics or transcriptomics profile analysis, to find the sub-graph that connects them. This would include nodes (proteins) that are not in the input list. Such analysis may reveal interactions of which the user was unaware, or components of a process that had not been previously recognized. However, this can still lead to large sub-graphs; for example, when a cell's response to an environmental signal results in multiple physiological processes being triggered. This can lead to a network in which there are long (>4 node) paths that connect clusters of input nodes. These connecting paths can be restricted in length, say to two or three nodes, which results in a set of unconnected but more physiologically relevant sub-graphs. Other ways to trim down large networks include restricting protein sets to those that are specific to some specified subcellular compartment or tissue type. Also the gene ontology may be used to study the interactions in some specified subset of biochemical or physiological process.

This and the previous sections of this chapter have focused on the identification of proteins and their interactions. The following sections pay attention to protein structure.

L4 STRUCTURAL BIOINFORMATICS

Key Notes

Structural types and conceptual models

Globular proteins are soluble in predominantly aqueous solvents such as the cytosol and extra-cellular fluids, and integral membrane proteins exist within the lipid-dominated environment of biological membranes. Conceptual models of protein structure are valuable aids to understanding protein bioinformatics.

Globular proteins

In globular proteins the linear amino acid polymer forms a three-dimensional structure by folding into a globular compact shape. Globular proteins tend to be soluble in aqueous solvents and folding is dominated by the hydrophobic effect, which directs hydrophobic amino acid side chains to the structural core of the protein, away from the solvent.

Secondary structure

Globular proteins usually contain elements of regular secondary structure, including α-helices and β-strands. These are stabilized by hydrogen bonding and contribute most of the amino acids to globular protein cores. Residues in regular secondary structures are given the symbols H meaning helix, or E (or B) meaning extended or β-strand.

Tertiary structure

The tertiary structure is the full three-dimensional atomic structure of a single peptide chain. It can be viewed as the packing together of secondary structure elements, which are connected by irregular loops that lie predominantly on the protein surface. Loop residues are given the symbol C to distinguish them from residues in helices or strands.

Quarternary structure

Several tertiary structures may pack together to form the biologically functional quarternary structure.

Integral membrane proteins

These exist within biological lipid membranes and obey different structural principles compared with globular proteins. They contain runs of generally hydrophobic amino acids, associated with membrane spanning segments (often but not exclusively helices), connected by more hydrophilic loops that lie in aqueous environments outside the membrane. Membrane proteins are very important components of cellular signaling and transport systems.

Domains

Proteins tend to have modular architecture and many proteins contain a number of domains, often with mixed types; for example, mixed integral membrane and globular domains.

Evolution

In globular proteins, surface residues in loops evolve (change) more quickly than residues in the hydrophobic core. In integral membrane proteins the most slowly evolving residues are those in the membrane spanning regions.

Structure and function	Proteins rely upon the shapes and properties of key functional areas of their three-dimensional structures to carry out biological functions. Knowledge of protein structure is key to understanding protein function and this is one reason for its importance in bioinformatics.
Structural and functional constraints	Evolution accepts changes to amino acid residues in proteins where they have a neutral or advantageous effect on protein structural stability or protein function. Residues can be conserved for structural or functional reasons. Amino acids are conserved where they are uniquely able to fulfill particular structural roles. This often occurs with cysteine, glycine, and proline.
Multiple sequence alignment	Understanding how structures evolve can help us understand multiple sequence alignments. Key structural and functional residues often are observed to be conserved. Insertions and deletions are seen to occur preferentially in hydrophilic surface loops by comparison with regular secondary structure elements, which are also subject to faster mutational change. Conservation of hydrophobic core residues in secondary structure elements is also common, as are conservation patterns associated with amphipathic helices.
Evolution of the overall protein fold	If two naturally occurring protein sequences can be aligned to show more than 25% sequence similarity over an alignment of 80 or more residues then they will share the same basic structure. The Sander-Schneider formula gives the higher threshold percentage identities necessary to guarantee structural similarity from shorter alignments.
Conservation of structure	Protein structures tend to be conserved even when evolution has changed the sequence almost beyond recognition. Structural knowledge is therefore a key factor in understanding protein evolution.
Evolution of function	While structure tends to be conserved by evolution, function is observed to change. There are many examples of proteins whose sequence and structure are very similar yet which have different functions. When function has changed, key functional residues change as well, and this is often clear in multiple sequence alignments.
Software and WWW sites	A large variety of software for structure visualization and analysis is available on the WWW.
Obtaining data	All published protein structures are submitted to a public database. Data search and download can be performed at various WWW sites.
Visualization of structures	RasMol, Chime, and Cn3d are commonly used programs for viewing structural data.
General structural analysis	There is an enormous amount of software available for structural data analysis, and also several WWW sites hold pre-prepared analyses.
Analysis of functional sites	Functional sites in protein structures typically contain a few residues in defined spatial positions. Software and databases have been developed to locate and search for similarity in such sites.

Structural alignment	It can be very difficult to find correct biologically meaningful alignments of very distantly related protein sequences because they contain only a very small proportion of identical monomers. In such cases structural information can help because evolution tends to change structure less. Superimposing the backbones of similar structures implies structurally equivalent residues, and this process is known as structural alignment.
Software	A variety of software is available for structural alignment.
Structural similarity	Structural alignment methods often produce measures of structural similarity. The most common of these is the **root mean square deviation** (RMSD), which is reported by most programs. This is the root mean square difference in position between the alpha-carbon atoms of aligned residues in optimal structural superposition.
Structural similarity searches	Similarity searches of the structural database are available from several WWW sites.

Related sections

Protein structure in 'Instant Notes in Biochemistry,' Section B3
Genomes and other sequences (J)
Structural classifications (L5)

Structure prediction and modeling (L6)
Molecular dynamics and drug design (L7)

Structural types and conceptual models

Conceptual models can add enormously to our understanding of protein sequence, structure, and function. It is the purpose of this section to introduce some key conceptual ideas that are essential to understand protein bioinformatics. It is useful to distinguish three different protein structural types: **fibrous proteins** (e.g., collagen); **globular proteins**, which tend to exist in aqueous solvents like the cytosol and extra-cellular fluids; and **integral membrane proteins**, which exist within the lipid environment of biological membranes. In this text we will be concerned mainly with globular proteins and integral membrane proteins.

Globular proteins

Proteins with a wide variety of functions fall within this broad class, including enzymes, antibodies, and a variety of molecules associated with signaling and transport. Overall they tend to have globular (spheroidal) shapes. Structure formation in globular proteins is dominated by the **hydrophobic effect**. Non-polar chemical compounds with few charged atoms are **hydrophobic**: they do not dissolve easily in water. In contrast, polar compounds with charged atoms form electrostatic and hydrogen bonding interactions with water molecules and therefore tend to dissolve easily. Some of the amino acid side chains are hydrophobic (valine, leucine, etc.), while other amino acids are hydrophilic (aspartic acid, lysine, etc.). When a protein folds it is able to minimize its free energy by placing hydrophilic amino acids on the surface of the globule, in contact with the aqueous solvent, and hydrophobic amino acids in the central core of the globule, away from the solvent. It is thought that this effect is the strongest force driving the linear polymer of amino acid residues to fold into globular shape in water. Although most core side chains are hydrophobic, there are sometimes some hydrophilic side chains in the core, very often making hydrogen bonds or ionic interactions with other core side chains, and equally, there is often a small proportion of hydrophobic side chains on the protein surface. An important feature of protein cores is that they are efficiently packed, so that most of the available space is filled with atoms.

Secondary structure

Globular proteins usually contain elements of regular secondary structure. The best known example is the α-helix, where four or more consecutive amino acid residues in the polymer adopt the same conformation, resulting in the adoption of a regular helical shape by the polypeptide backbone. This helix is stabilized (held together) by hydrogen bonds between the main chain C=O group of each amino acid and the H-N group of the amino acid four residues further along the helix. This forms a helix with 3.6 amino acid residues per helical turn (see *Fig. 1*). Although helices with slightly different hydrogen bonding patterns (3_{10} and π-helices) do occur in protein structures, α-helices are by far the most common. The illustration in *Fig. 1* shows that the amino acid side chains point away from the helical axis, forming a surface for the helix. The notation H is used to indicate that a particular residue is a member of an α-helix.

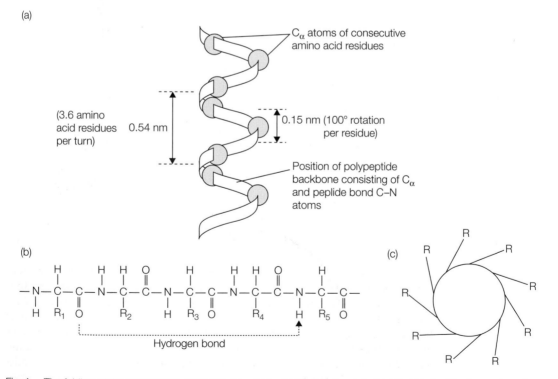

Fig. 1. The folding of the polypeptide chain into an α-helix. (a) Model of an α-helix with only the C_α atoms along the backbone shown; (b) in the α-helix the CO group of residue n is hydrogen bonded to the NH group on residue (n + 4); (c) cross-sectional view of an α-helix showing the positions of the side-chains (R groups) of the amino acids on the outside of the helix.

The other common type of regular secondary structure is the β-strand (see *Fig. 2*). This secondary structure element is formed by consecutive amino acids in their most extended conformation, and for this reason the letter E (Extended) is used to indicate that a residue adopts this structure. In this case hydrogen bonds between main chain C=O and N-H groups form not to residues in the same strand, but to residues in strands formed by other parts of the polypeptide. These hydrogen bonds mean that single beta-strands do not exist in isolation, but are always spatially adjacent to at least one other strand. The twisted, pleated structure formed by consecutive, spatially adjacent, hydrogen-bonded strands is known as β-pleated sheet. If such a sheet is curved round so that the strands that would have been on the edges of the sheet are spatially adjacent and hydrogen bonded, the structure is known as a β-barrel. The illustration in *Fig. 2* shows that the amino acid side chains point alternately above and below the sheet.

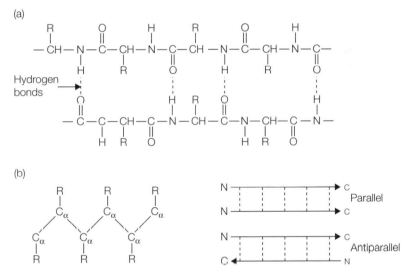

Fig. 2. *The folding of the polypeptide chain in a β-pleated sheet. (a) Hydrogen bonding between two sections of a polypeptide chain forming a β-pleated sheet; (b) a side view of one of the polypeptide chains in a β-pleated sheet showing the side chains (R groups) attached to the C_α atoms protruding above and below the sheet; (c) because the polypeptide chain has polarity, either parallel or antiparallel β-pleated sheets can form.*

Tertiary structure

The tertiary structure or fold of the protein is the position of every protein atom in three-dimensional space. This can be considered conceptually as the product of a process in which the secondary structure elements of the polypeptide chain pack together with their inward-facing sides contributing most of the residues to the hydrophobic core. These secondary structure elements are connected by sections of polypeptide chain called **loops**. For our purposes we can consider the loops to be of irregular structure (sometimes called **coil** or **random coil**, and denoted by the letter C), although a more detailed treatment would reveal some regular structures. The loops tend to lie on the protein surface and to contain mainly hydrophilic residues.

Quarternary structure

Many proteins exist as multi-meric molecules formed as several polypeptide chains bind together to form a complex. This complex is known as the quarternary structure of the protein, and it is the biologically functional unit. Sometimes complexes have a single biochemical function, but there are many examples of multi-functional complexes; for instance, enzymes which catalyze consecutive steps in a metabolic pathway.

Integral membrane proteins

Integral membrane proteins are key elements of biological signaling and trans-membrane transport systems. Examples are the G-protein-coupled receptors (involved in signaling), and channel proteins responsible for rapid transport of ions across membranes. The fact that significant parts of these proteins exist in the lipid environment of biological membranes, in contrast to globular proteins that exist in aqueous solvents, means that different structural principles apply to them. Typically integral membrane proteins have one or more segments that actually cross the membrane. Amino acids in these segments tend to be of hydrophobic nature, compatible with the lipid structure of the membrane, while portions of the protein that exist in the more aqueous environment on either side of the membrane tend to be more polar. Very often the membrane spanning segments adopt an α-helical structure, as in G-protein-coupled receptors and many ion channels, but there are other examples; for instance, the bacterial porins, where the membrane spanning segments are β-strands. A schematic illustration of a membrane protein is shown in *Fig. 3*. The number of membrane spanning sections within a typical membrane protein varies from one (often a membrane 'anchor' for the protein) to more than ten. The well known G-protein-coupled receptors, like rhodopsin, contain seven membrane spanning segments.

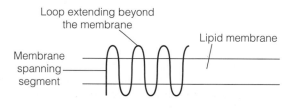

Fig. 3.   An integral membrane domain with seven membrane spanning segments.

Domains

From the knowledge of protein sequence and structure we have now it is clear that many proteins have modular architecture. Nature creates proteins with complex functions by combination of quasi-independent modular units or domains, typically with much simpler functions. For instance, there are many eukaryotic proteins that contain the homeobox domain whose function is to bind to DNA, but these proteins also contain other domains that are responsible for other aspects of their overall function. This is illustrated in *Fig. 4*. The precise definition of a protein domain depends on your point of view. Some would define it as a protein sub-sequence or sub-structure with a recognized function, as above. Others view domains as protein sub-sequences that are able to fold independently of the rest of the protein, and others regard domains as geometrically distinct parts of the protein structure. Perhaps the most useful definition of a domain, however, is that it is a protein unit observed to occur in many otherwise unrelated proteins.

Whatever your preferred definition of a protein domain, it is very important to bear in mind the modular architecture of proteins in many bioinformatics analyses. It is also important to appreciate that proteins can have domains of mixed types; for instance, integral membrane domains often are found in combination with globular domains. This is the case for many receptors, which are anchored to a biological membrane by an integral membrane domain, and have extra-cellular globular domains responsible for recognizing biochemical signals.

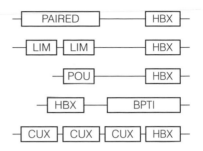

Fig. 4. Multi-domain proteins containing a homeobox domain. The sequences are shown from amino terminus (left) to carboxy terminus (right) with recognized domains shown as boxes. From top to bottom the proteins are: human PAX4, human LHX2, human OTF2, C. elegans C02F12.5, and mouse CUX2. HBX – homeobox domain, PAIRED – paired box domain, LIM – LIM domain, POU – POU domain, BPTI – Kunitz BPTI domain, CUX – CUX domain.

Evolution

In earlier sections we explained that changes in the amino acid sequences of proteins occurring during evolution are accepted if they have a neutral effect on protein structure and function (i.e., they have little effect on the stability of the protein and its ability to perform its normal function), or if they have a positive effect (i.e., they enhance stability or functional efficacy). When a change has a negative effect it is likely to be rejected by natural selection, except in the cases of the changes occurring when a duplicated gene is in the process of evolving a new function.

When we observe the evolution of protein structures we see that evolutionary change is very slow for residues in the structural core, and significantly faster for residues on the protein surface. This can be understood in the light of the conceptual models introduced above. Within the structural core the introduction of changes can affect the tight atomic packing, resulting in reduction of protein stability, and therefore is less likely. On the other hand, residues on the protein surface are not subject to this constraint (essentially they need only to be hydrophilic) and therefore are much more susceptible to evolutionary change. Because secondary structure elements tend to contribute most of the core residues, while loop residues lie mainly on the surface, it is observed often that loops evolve much more quickly than secondary structure elements. A related effect is observed in integral membrane proteins, where the membrane spanning segments (most often helical) evolve more slowly than the loops which connect them.

Structure and function

The ability of proteins to perform biological functions depends on the formation of a three-dimensional structure (fold) that is stable in the normal environment of the protein. For example, enzymes often catalyze reactions using an **active site**. Typically this is a cavity in the three dimensional structure of the enzyme that is accessible to the reactants from the protein surface. Active sites are multi-functional. They contain the key catalytic machinery of the protein, which is typically one or more residues that are actively involved in the chemical reaction catalyzed, and stabilize transition states. They exclude solvent from the reaction, and their shape and physico-chemical properties are such that they bind the intended reactants much more strongly than any alternatives, thereby creating catalysis specific to certain molecules or molecular classes. All this depends on the active site adopting a specific three-dimensional shape, and ultimately on the fold of the peptide chain itself. The importance of the three-dimensional form of the active site in serine protease enzymes is illustrated in *Fig. 5*.

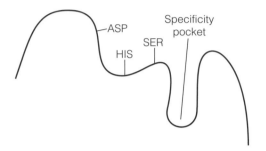

Fig. 5.   A schematic view of a serine protease active site. The key catalytic triad comprising residues serine (SER), histidine (HIS), and aspartic acid (ASP) is shown. During catalysis, the serine residue is activated as a nucleophile by transfer of charge to the other key residues in order to attack and cleave peptide bonds. Also shown is the specificity pocket responsible for recognizing specific types of molecules for catalysis. While all serine proteases share the catalytic triad, the shape and properties of the specificity pocket vary to produce enzymes of different specificities. In trypsin, the pocket recognizes peptides with basic side chains, while in chymotrypsin, it recognizes large hydrophobic side chains. The shape and properties of the active site are created by residues from distal parts of the sequence, which are brought close together in space by the three-dimensional fold of the peptide chain.

Not all proteins are enzymes, but most rely on their three-dimensional structures in order to perform biological functions. Molecular recognition without catalysis is a function of many proteins. Transport proteins need to recognize the molecules they carry. Protein–protein recognition is important in the recognition of foreign proteins by antibodies, the interactions between components of signaling pathways, and the formation of multi-functional complexes. Similarly the recognition of other macromolecules by proteins is important in the regulation of gene expression (by DNA binding proteins), and the formation of mixed protein–RNA complexes like the ribosome. Finally many receptor proteins need to recognize a molecular signal specifically (for example, the recognition of steroid hormones by receptors in the cell nucleus).

For molecular recognition to take place it is necessary that the molecules concerned are able to bind together in an energetically favorable conformation. This depends on the **complementarity.** They must be able to form complementary shapes, so that they fit snugly together, and where their surface atoms are in contact there must be complementarity of physico-chemical properties. This means that negatively charged areas of one interacting partner must contact positively charged areas of the other, and hydrophobic areas must be in contact with other hydrophobic areas. All this depends on the formation of a stable three-dimensional structure by the protein.

Structural and functional constraints

The evolution of protein sequences was discussed in Section J3 where amino acid substitutions were understood in terms of physico-chemical relationships between the different amino acids. Amino acids with similar size and physico-chemical properties are likely to make reasonable replacements for each other, and to form accepted substitutions during protein evolution. With structural knowledge we can be more precise. Fundamentally, amino acid substitutions are accepted in evolution if the change is either neutral or advantageous to protein structural stability and/or function. Although matrices such as PAM250 (Section J3) can tell us about average rates of substitution between amino acids, when the three-dimensional structure is known we are able to understand which substitutions are acceptable at particular positions in the protein sequence.

Some amino acid residues have key roles to play in the stability of particular structures (see *Table 1*). When a residue plays a key role it is often the case that no other residue can substitute for it while maintaining the stability of the structure, and the residue is conserved throughout the evolutionary family. In a similar way, key functional residues are often conserved also. Many enzymes rely on the chemical properties of certain amino acids to effect catalysis; for instance, serine, histidine, and aspartic acid residues in the serine protease active site. If these properties are unique to the residue involved they will tend to be conserved to preserve function.

Table 1.   Unique structural roles played by some amino acid residues in protein structures

Residue	Structural role
Cysteine	The formation of disulfide bridges (S-S bonds) with other cysteine residues, making covalent links between sequence separated parts of the polypeptide chain. These bonds can make a major contribution to structural stability
Glycine	The side chain of glycine is very small (a single hydrogen atom). This means that the structure of glycine is more flexible than that of the other amino acids. It can adopt some conformations that are accessible to no other residues, and is often found where the peptide chain goes through a very tight turn
Proline	Amino acids tend to prefer to adopt *trans* conformations around the planar peptide bond; that is, with the two alpha carbons located on opposite sides of the bond. *Cis* conformations (with the alpha carbons on the same side of the bond) occur occasionally, and often involve a proline residue, which is better able to form the *cis* conformation than the other amino acids

When these roles are adopted, the residue concerned often is found to be conserved throughout the protein evolutionary family because loss of the role, which no other residue could fulfill, would incur a loss in structural stability.

Multiple sequence alignment

The relationship between multiple sequence alignments (Section J4) and protein structure and function is very important. Many of the features of the alignment can be understood in the light of a known structure for one member of the family. *Fig. 6* shows a multiple sequence alignment of lysozyme and α-lactalbumin sequences with key conserved structural (disulfide forming cysteines) and functional (key catalytic residues) features annotated.

Earlier we explained that residues on the protein surface evolve more quickly than residues buried in the structural core because they are not subject to the strong constraints involved in the maintenance of efficient packing of the structural core. We also described a conceptual model of globular protein

```
KEY RES          $                                 $   !                        !              $
LYC_CHICK   GRCELAAAMKRHGLDNYRGYSLGNWVCAAKFESNFNTQATNRN-TDGSTDYGILQINSRWWCND
LYC_MELGA   GRCELAAAMKRLGLDNYRGYSLGNWVCAAKFESNFNTHATNRN-TDGSTDYGILQINSRWWCND
LYC_PAVCR   GRCELAAAMKRLGLDNYRGYSLGNWVCAAKFESNFNTHATNRN-TDGSTDYGILQINSRWWCND
LYC_CERAE   ERCELARTLKRLGLDGYRGISLANWVCLAKWESGYNTQATNYNPGDQSTDYGIFQINSHYWCNN
LYC_GORGO   ERCELARTLKRLGMDGYRGISLANWMCLAKWESGYNTRATNYNAGDRSTDYGIFQINSRYWCND
LYC_HUMAN   ERCELARTLKRLGMDGYRGISLANWMCLAKWESGYNTRATNYNAGDRSTDYGIFQINSRYWCND
LYC_SHEEP   ERCELARTLKELGLDGYKGVSLANWLCLTKWESSYNTKATNYNPGSESTDYGIFQINSKWWCND
LYC_BOVIN   ERCELARTLKKLGLDGYKGVSLANWLCLTKWESSYNTKATNYNPSSESTDYGIFQINSKWWCND
LYC_AXIAX   ERCELARTLKELGLDGYKGVSLANWLCLTKWESSYNTKATNYNPGSESTDYGIFQINSKWWCDD
LCA_CAPHI   TKCEVFQKLKDL--KDYGGVSLPEWVCTAFHTSGYDTQAIVQN--NDSTEYGLFQINNKIWCKD
LCA_SHEEP   TKCEAFQKLKDL--KDYGGVSLPEWVCTAFHTSGYDTQAIVQN--NDSTEYGLFQINNKIWCKD
LCA_BOVIN   TKCEVFRELKDL--KGYGGVSLPEWVCTTFHTSGYDTQAIVQN--NDSTEYGLFQINNKIWCKD
LCA_HUMAN   TKCELSQLLKDI--DGYGGIALPELICTMFHTSGYDTQAIVEN--NESTEYGLFQISNKLWCKS
LCA_MOUSE   TKCKVSHAIKDI--DGYQGISLLEWACVLFHTSGYDTQAVVND--NGSTEYGLFQISDRFWCKS
LCA_RAT     TKCEVSHAIEDM--DGYQGISLLEWTCVLFHTSGYDSQAIVKN--NGSTEYGLFQISNRNWCKS
CONSERV          :*:       ::  :    ..* * :* :   *       *.::::*    :  . **:**::**..: **..
SS               HHHHHHHHHH         HHHHHHHHHHH       EEEE       EEEEE     EEEE
```

Fig. 6. *A multiple alignment of lysozyme and α-lactalbumin sequences. Sequence names starting LYC are lysozymes and those starting with LCA are α-lactalbumins. The KEY RES lines indicates key structural residues (disulfide forming cysteines) with the symbol '$', and key lysozyme catalytic residues with the symbol '!'. Key structural residues are conserved in all sequences, but the lysozyme functional residues are not conserved in the α-lactalbumin sequences. The CONSERV line indicates the degree of conservation in a particular column ('*'= identically conserved, ':' contains only very conservative substitutions, '.' contains conservative substitutions). The SS line shows secondary structure (taken from one sequence of known 3-D structure).*

structure in which secondary structure elements (helices and strands) contribute most residues to the hydrophobic core of the protein and alternate with loops whose residues lie predominantly on the surface. Loops are therefore observed to evolve more quickly than secondary structure elements. This effect is observed in multiple-sequence alignments, where it is often clear that the alignment consists of alternating blocks of well-conserved residues corresponding to secondary structure elements and less well-conserved residues from surface loops. This is evident in the alignment of *Fig. 6*, where it is clear that insertions and deletions in the sequences fall in loops (outside the labeled secondary structure elements), and that strong conservation of hydrophobic residues is observed in some secondary structure elements. For example, the second helix contains a group of three resides of hydrophobic character, which are W, V, and C in the first (top) sequence.

As we commented in Section J4, multiple alignments made on the basis of sequence alone by most standard software methods are often not perfect. Structural information can be used to make manual improvements. For instance, if it is known that a particular cysteine is involved in a disulfide bridge it should be verified that it is aligned with a cysteine in all sequences in the family. Similarly if the alignment involves insertions and deletions it should be borne in mind that these are much more likely in loops than in secondary structure elements.

In the case where there is no known three-dimensional structure, a multiple alignment can serve as a first stage in structure prediction. Conserved cysteines can be predicted to be involved in disulfide bonds (although this is not the only reason for conservation of this residue, which also is found often as a metal ligand). Similarly the alternation between blocks of conserved residues and blocks where the sequences are more variable, which is apparent in many multiple alignments, can be interpreted in terms of probable secondary structure elements alternating with surface loops.

Another structural feature that is often visible in multiple sequence alignments is the amphipathic helix. The amino acid side chains point away from the helical axis and form a surface for the helix. Very often, α-helices are positioned in protein structures so that one side of the helix is part of the hydrophobic structural core and the other side lies on the protein surface. Hydrophobic side chains therefore dominate on one side and hydrophilic ones on the other. Since an α-helix has 3.6 residues per turn, an amphipathic helix tends to exhibit an alternation between hydrophilic and hydrophobic residues with a periodicity of three to four residues. This is often visible in multiple sequence alignments as conserved hydrophobic residues occurring alternately every three or four residues in the sequence. This is illustrated in *Fig. 7*. Helical wheel illustrations of the type shown in *Fig. 7* are produced by various software programs.

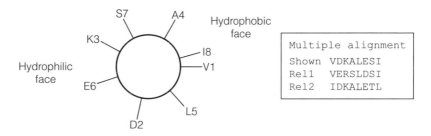

Fig. 7. An amphipathic α-helix. The helical wheel (left) shows the positions of the side chains for two turns of the α-helix. The helical wheel is a projection of the helix down its own axis onto a plane, showing the positions of the side chain. For instance, in this case residue I8 is about two helical turns from V1, on the same side of the helix but almost directly above. The helix has a hydrophobic face populated by A4, I8, V1, and L5, and a hydrophilic one populated by D2, E6, K3, and S7. The multiple alignment of the sequence shown and two relatives illustrates the conservation of hydrophobic residues alternately four (V1 and L5) and three (L5 and I8) residues apart in the sequence

Evolution of the overall protein fold

So far we have considered the evolution of parts of the protein structure in isolation, now we consider how the overall structure evolves. In 1991, Sander and Schneider published a study of naturally occurring proteins of known structure showing how much similarity in protein sequence is needed to guarantee similarity in structure. They found that the threshold of similarity depended on the lengths of the sequences involved, with more similarity being required for shorter sequences. They showed that if two protein sequences can be aligned to show more than 25% identical residues over an alignment length of more than 80 residues, then they will share the same basic three-dimensional structure. This result only applies to the naturally occurring proteins currently of known structure. There are now examples of artificially engineered proteins with sequence similarity of as much as 50% yet with different structures. Other potential exceptions to Sander and Schneider's rule are proteins like the prion protein that appear to have two distinct possible structures (one normal and the other disease associated). The Sander-Schneider rule was derived from a database in which the proteins were in their normal biologically active conformations, and these are the conformations to which it applies.

Sander and Schneider's threshold for structural similarity was

$$t(L) = 290.15L^{-0.562}$$

where L is the length of the alignment and t is the percentage identity threshold required to guarantee structural similarity. For alignments with L equal and greater than 80 the value of t approaches the 25% mentioned above, but it is larger for smaller values of L. For instance a similarity of $t = 43\%$ is required for structural similarity in an alignment of $L = 30$ residues.

Sander and Schneider's result was very important because it provided the theoretical basis for a method of structure prediction known as comparative modeling (see Section L6 for more details). Using the fact that sequences with a sufficiently high level of similarity can be assumed to have the same structure, sequences of known structure can be used in structure predictions for sufficiently similar sequences of unknown structure.

Conservation of structure

Sequences with sufficient similarity share the same structure, but observations made on known sequences and structures have shown us that divergent evolution of proteins with a common ancestor can operate until their sequence similarity is almost undetectable. The best known example of this comes from the globin family, including the hemoglobin chains, myoglobin, and plant leghemoglobin. These sequences adopt the same overall structure, and carry out similar functions (oxygen transport) by the same mechanism, but often exhibit sequence identities that are below 20% of identical residues. This is a level of sequence similarity that is found often between the sequences of completely unrelated proteins. This observation that sequence evolves much faster than structure means that structure is very important to bioinformatics. Many potential evolutionary relationships that were not apparent on the basis of sequence analysis have been discovered when three-dimensional structures have become available.

Evolution of function

We have commented in previous sections that protein function evolves, so that proteins that are clearly homologous (i.e., have closely related sequences and/ or structures) sometimes have different functions, and we have identified gene duplication as a possible mechanism for the development of new functions. When this happens, residues that have a key role in the performance of one function that are conserved throughout the members of the family performing that function are often not the same or even conserved in members of the family that perform a different function. Lysozyme and α-lactalbumin are examples of homologous proteins that perform very different functions (the former is an enzyme and the latter a regulatory protein). The multiple alignment in *Fig. 6* shows that the key lysozyme functional residues are not conserved in α-lactalbumin, while key structural residues are (in particular the disulfide bond forming cysteine residues).

Gene duplication is not the only way in which protein function evolves. Multidomain proteins are proteins that consist of more than one domain. Often the different domains have different functions that are combined to produce another, much more complex function. For example, some enzymes have domains associated with regulation of their activity. It seems that nature evolves many new proteins by swapping and recombination of modular units (domains), and that this is a major route for the evolution of new and more complex protein functions.

Software and WWW sites

It would be impossible to produce a comprehensive list of available software and WWW sites for the display and analysis of protein and other macromolecular structures. All we provide here are pointers to the most commonly used and freely available utilities. It should be noted that this book is not intended to serve as a software manual for commonly used programs; that function is much better provided by the documentation that is distributed with the software itself. Such documentation will be much more extensive than the space available here would allow, and should remain up-to-date with changes to programs as they are made.

Obtaining data We have already discussed the structural database (Section J1) and the experimental methods that produce the data contained within it. Like the sequence databases discussed in earlier sections, the structural database can be searched by many of the standard search tools, including SRS (Section J1) and tools available at the NCBI WWW site (Section J1). There are also specialized search engines maintained by the macromolecular structural database group at the EBI (http://www.ebi.ac.uk/msd/), and the RCSB (http://www.rcsb.org/pdb). Both these tools allow easy downloading of structural data in the standard PDB format which is input to much of the software discussed below.

Table 2. Summary of software and WWW sites for protein structure visualization and analysis

Resource	Type	URL	Function
RasMol	Computer program for most computer operating systems	www.openrasmol.org	Visualization of protein structures in three dimensions
Cn3D	Helper application to permit viewing of 3D structures in a WWW browser, for most computer operating systems	http://www.ncbi.nlm.nih.gov/Structure/CN3D/cn3d.shtml	Visualization of protein structures in three dimensions, which can be linked to sequence alignments
Chime	Helper application to permit viewing of chemical structures in a WWW browser	http://www.mdli.com/	Visualization of protein structures in three dimensions
Molscript	Computer program for UNIX operating systems	http://www.avatar.se/molscript/	Visualization of protein structures in three dimensions
JMol	Computer program for most operating systems	http://jmol.sourceforge.net/	Visualization of protein structures in three dimensions
SWISS-PDB viewer	Computer program for Windows and Macintosh	http://expasy.org/spdbv/	Visualization of protein structures in three dimensions
TOPS	Computer program for UNIX-related operating systems and WWW server	http://www.tops.leeds.ac.uk	Program for visualization of protein folding topologies
DSSP	Computer program for most computer operating systems	http://swift.cmbi.ru.nl/gv/dssp/	Finds secondary structure elements in an input protein structure. Also calculates relative solvent accessibility
MSMS	Computer program for UNIX-related operating systems	http://www.scripps.edu/~sanner/html/msms_home.html	Protein surface calculation
Surfnet	Computer program for UNIX-related operating systems	http://www.biochem.ucl.ac.uk/~roman/surfnet/surfnet.html	Visualization of protein surfaces

Table 2.   continued.

HBPLUS	Computer program for UNIX-related operating systems	http://www.biochem.ucl.ac.uk/bsm/hbplus/home.html	Finds internal hydrogen bonds and non-bonded interactions
NACCESS	Computer program for UNIX-related operating systems	http://www.bioinf.manchester.ac.uk/nacccess/	Calculates atomic and residue solvent accessibilities
PROCHECK	Computer program for UNIX-related operating systems	http://www.biochem.ucl.ac.uk/~roman/procheck/procheck.html	Checks stereochemical quality of protein structures
PROMOTIF	Computer program for UNIX-related operating systems	http://www.biochem.ucl.ac.uk/bsm/promotif/promotif.html	Analyzes protein structural motifs
LIGPLOT	Computer program for UNIX-related operating systems	http://www.biochem.ucl.ac.uk/bsm/ligplot/ligplot.html	Produces graphical displays of ligands and their binding sites
PDBSum	WWW database	http://www.ebi.ac.uk/pdbsum/	A WWW resource containing detailed summaries of structural analyses carried out on entries in the public structural database
Catalytic Site Atlas	WWW database	http://www.ebi.ac.uk/thornton-srv/databases/CSA/	A database of three-dimensional atomic templates defining particular enzymic activities, taken from active sites in known enzyme structures
SITES	WWW database and search server	http://www.bioinformatics.leeds.ac.uk/siteCompare	An automatically curated database of protein active sites and ligand binding sites with geometrical search facilities
Relibase	WWW database and search server	http://relibase.ebi.ac.uk	A database of protein ligand complexes with geometrical search facilities.
SPASM	WWW search server	http://xray.bmc.uu.se/usf/spasm.html	Geometrical searches for structural motifs
ASSAM	Computer program	–	Geometrical searches for structural motifs
PINTS	WWW search server	http://www.russell.embl-heidelberg.de/pints/	Database of protein-active sites and ligand-binding sites with geometrical search facilities
EF-site	WWW search server	http://ef-site.hgc.jp/eF-site/	Database of protein-active sites and ligand-binding sites with geometrical search facilities
CavBase	WWW search server	http://relibase.ebi.ac.uk	Database of protein-active sites and ligand-binding sites with geometrical search facilities
SitesBase	WWW search server	http://www.modelling.leeds.ac.uk/sb/	Database of protein-ligand binding sites with geometrical search facilities

Visualization of structures

The program RasMol (*Table 2*), written by Roger Sayle, is perhaps the best known viewer for macromolecular structures. Given a structure in the standard structural database format this software displays a three-dimensional image of the structure. The image can be rotated by use of the mouse to produce different views, and displayed in various formats, including 'wireframe' (bonds displayed as lines with atoms implied at junctions and ends), 'space filling' (each atom represented by a sphere of proportionate size), and 'ball-and-stick' (atoms represented by small spheres and chemical bonds by sticks). There are also special 'cartoon' formats that give clear displays of secondary structure elements. The user can choose between various color schemes and even use customized colors. Finally, there are flexible ways of selecting parts of structures to enable highlighting with a different display format. For instance, key residues or sub-structures can be highlighted by variation in the display mode. *Fig. 8* shows a small protein structure displayed in RasMol in two different display formats, the first showing all atoms, and the second abstracted to show only secondary structure elements, with helices as helical ribbons and strands as extended ribbons. A more complex example of an enzyme with its inhibitor displayed is shown in *Fig. 9*.

Some other related programs are Chime (a plug-in for use in a WWW browser like netscape) which has similar functionality to RasMol, Jmol (a viewer compatible with various file formats), Cn3d which is also able to link the structure display directly to a multiple-sequence alignment, and SWISS-PDB viewer which is a sophisticated viewer capable of viewing several structures at once. The visualization of protein structures in three dimensions can be very difficult, and for this reason there are several tools that produce schematic or summary displays of protein folds. One example of this is the RasMol cartoons format (*Fig. 8*), and similar more sophisticated displays can be produced with Molscript and Ribbons. Finally, TOPS is a utility that produces two-dimensional summaries of three-dimensional protein folds, an example of which is shown in *Fig. 10*.

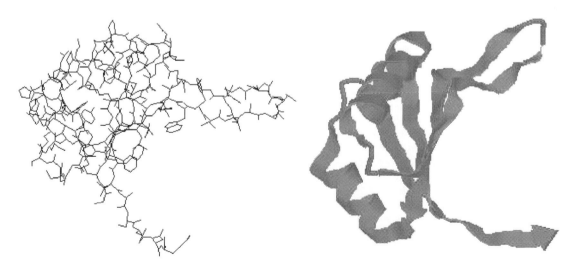

Fig. 8.   Two images of protein structures displayed with the RasMol program. On the left all atoms are displayed in a format known as 'wireframe.' The image on the right displays the secondary structure elements of the protein fold in a format known as 'cartoon.' The fold on the right comprises a four-stranded anti-parallel β-sheet packed against two α-helices.

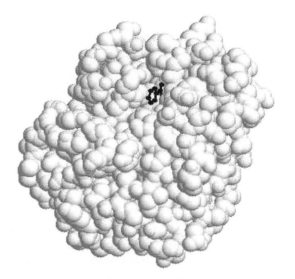

Fig. 9.   The structure of the serine protease enzyme trypsin in 'space filling' format with an inhibitor (benzamidine) shown in black, bound in the main specificity pocket of the enzyme, adjacent to the active site.

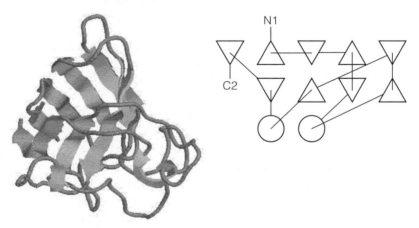

Fig. 10. A TOPS protein-folding topology diagram (right) for a superoxide dismutase, and the full structure in RasMol cartoon format (left). The RasMol diagram shows β-strands as arrows and α- or 3_{10}-helices as helical ribbons. The core of the protein comprises two anti-parallel β-sheets, one having four strands and the other five. The sheets pack face to face to form a sandwich structure, and two short helices lie in loops in front of the four-stranded sheet. This is difficult to see from the three-dimensional RasMol diagram, but is much clearer in the TOPS cartoon. In the TOPS cartoon, the peptide chain runs from N terminus N1 to C terminus C2 and can be traced by following the connecting lines from symbol to symbol. The triangular symbols represent β-strands and the circular ones, helices. The symbols should be thought of as representing secondary structure elements which are perpendicular to the plane of the diagram. They have a direction (N to C), which is either 'up' (out of the plane of the diagram) or down (into the plane of the diagram). 'Up' strands are represented by upward pointing triangles and 'down' ones by downward pointing triangles. The five-stranded sheet is represented by the upper horizontal row of five triangles and the four-stranded sheet by the lower one. The alternation of strand directions along these rows shows that the sheets are anti-parallel. The observant reader may have noticed that the connecting lines between the triangular strand symbols are drawn sometimes to the edge of the symbol and sometimes to the center. This is in fact another way of showing the direction of the secondary structure element: if the amino- (N-)terminal connection is drawn to the edge of the symbol and the carboxy- (C-) terminal connection to the centre, then the direction is up; otherwise the N-terminal connection is drawn to the center and the C-terminal one to the edge, and the direction is down. So there are two ways in which the direction of strands is shown, but the direction of helices must be deduced by looking at the connecting lines.

General structural analysis

Protein structures offer almost unlimited possibility for analysis programs, including the automatic annotation of secondary structure elements, the analysis of residue solvent accessibility to find surface and core residues, the creation of protein surfaces, checking the stereochemical quality of structures, and analysis and display of functional sites. Some examples of such programs are listed in *Table 2*. One WWW resource that is worthy of special mention is PDBSum, provided by the bioinformatics group at University College London. This resource contains a well-presented summary of structural data and analyses for every entry in the structural database, accessed by the accession code for that database (Section J1). The information presented includes secondary structure, disulfide bonds, ligand binding sites, active sites, key residues, plots of intermolecular interactions, folding topologies, EC numbers for enzymes, and much more.

Analysis of functional sites

Probably the most important parts of any protein structure are its functional sites; that is, active sites of enzymes, ligand binding sites, metal ion binding sites, sites where other proteins or DNA might bind, etc. Locations of such sites in the database of known structures is surprisingly difficult. Some sites are annotated by those who determined the structures, and others can be located because they contain bound ligands, and most of these are part of PDBSum (described above). Nevertheless, many potentially interesting sites in known protein structures cannot be found in this way, and there is a need for methods that can locate such sites as part of the functional analysis of new protein structures.

To analyze a new structure it is necessary first to locate potential functional sites and then consider what the structure of these sites can reveal about possible functions. It has been observed by Roman Laskowski and colleagues (1996) that enzyme active sites are located often in large cavities near the protein surface (more often than not the active site is the largest such cavity). The SURFNET software (*Table 2*) is able to locate such sites. To make a prediction about the actual function of any potential site on a protein surface is a problem currently at the forefront of bioinformatics research. In common with much of bioinformatics, most current approaches are knowledge based, and rely heavily on databases of information about active sites of known function. This is based on the idea that active sites that are in some sense similar are likely to have similar functions.

There are several databases, computer programs, and resources that implement similarity searches for functional sites in proteins. Such similarity searches are both geometrical and chemical: they search for sites where the relative geometrical positions and chemical type of atoms, residues, or chemical groups are similar. Catalytic Site Atlas is a manually curated database of three-dimensional atomic active site templates, which define particular enzymic activities. These have been extracted from the database of known structures, and a new protein structure can be scanned against the database to search for the presence of a known catalytic template. ASSAM and SPASM allow the user to define interesting residues in a protein structure and search similar sets of residues in the same relative spatial positions in other known structures, and the Leeds SITES database permits similarity searching of an active and ligand binding site database at both surface and residue levels. In a similar vein, Relibase is a database of protein–ligand complexes with sophisticated geometrical search facilities. More recent additions include PINTS, EF-SITE, SitesBase, and CavBase (part of the Relibase database). These programs offer variations of geometric search and comparisons of active sites and ligand binding sites. All these tools form a formidable armory for functional analysis of new protein structures.

**Structural
alignment**

It should be obvious from the material in Section J that the alignment of protein or nucleic acid sequences is a very important part of bioinformatics. An alignment defines a relationship between the elements of one sequence and those of another, and the degree of similarity of the sequences reflected in the alignment is the basic data on which we base judgments, by statistical or other means, of the likely biological meaning of the relationship. Proteins are likely to have similar structures and biological functions if their sequences can be aligned to show significant similarity.

Sequence alignment is relatively easy when the sequences are closely related, involving a high proportion of identical monomers, and few insertions and deletions. However, as we have commented earlier in this section, the sequences of evolutionarily and structurally related proteins can be very different. Divergent evolution can lead to sequences whose similarity is almost unrecognizable. In these cases naïve sequence alignment based on information only from within the sequences and standard substitution matrices (Section J3) can be very difficult. It can be almost impossible to find an alignment that is even approximately correct in biological or evolutionary terms.

Knowledge of protein structure provides a means by which highly divergent sequences can be aligned, and this process is known as **structural alignment**. This can be described conceptually as a process in which two similar structures are superimposed in three dimensions, so that peptide backbones of structurally equivalent residues lie close together in space. This superposition then is used to define a sequence alignment in which the aligned residues are structurally equivalent (see *Fig. 11*). Because structure is more strongly conserved during evolution than sequence, the structural alignment is much more likely to be correct in terms of biological function and evolution.

In *Fig. 12*, two alignments of a pair of distantly related globin sequences are shown, the first produced by sequence-only means and the second a structural alignment. The alignments are different, and although the sequence-only alignment shows a higher percentage of identical residues it is certainly not as good as the structural alignment. One very obvious advantage of the structural alignment is that the histidine residues that coordinate the heme iron in both globins are aligned (equivalenced), whereas they are not in the sequence-only alignment.

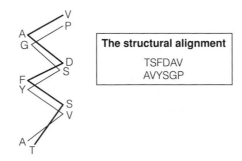

Fig. 11. A conceptual view of structural alignment. Two protein three-dimensional backbone structures are shown optimally superimposed. Structure 1 has the sequence TSFDAV and is represented by a thick line, and structure 2 has the sequence AVYSGP and is represented by a thin line. The implied structural alignment is shown also with structurally equivalent residues aligned.

(a)

```
SLSAAEADLAGKSWAPVFANKNANGLDFLVALFEKFPDSANFFADFK-GKSVADIKA-S
VLSPADKTNVKAAWGKVGAHAGEYGAEALERMFLSFPTTKTYFPHFDLSHGSAQVKGHG
                                    *
PKLRDVSSRIFTRLNEFVNNAANAGKMSAMLSQFAKEHVGFGVGSAQFENVRSMFPGFVA
KKVADALTNAVAHVDDMPNALSALSDLHAHKLRVDPVNFKLLSHCLLVTLAAHLPAEFTP
                  *
```

(b)

```
XSLSAAEADLAGKSW-APVFANKN-ANGLDFLVALFEKFPDSANFF-ADFKGKSVA--DIK
V-LSPADKTNVKAAWGK-VGAHA-GEYGAEALERMFLSFPTTKTYFPHF-------DLS-H
                                   *
ASPKLRDVSSRIFTRLNEFVNNAANAGKMSA-MLSQ-FAKEHV-GFGVGSAQFENVRSM-F
GSAQVKGHGKKVADALTNAVAHV-D--DMPNAL--SALSDLHAHKLRVDPVNFKLLS-HCL
                                   *
PGFVASVAA--PPAGADAAWTKLFGL-IIDA-LKAAGA-
LVTLAAHLPAEFTPAVHASLDKFLASVST-VLT-SKY-R
```

Fig. 12. Alignment of distantly related globin sequences (human hemoglobin, bottom; sea hare myoglobin, top), using (a) sequence-only (a local alignment of 87 residues showing 21% identical amino acids), and (b) structural alignment (an alignment of 139 amino acids showing 16% identical residues). The histidine residue that coordinates the heme iron is marked with an asterisk. Note these histidines are correctly aligned in (b) but not in (a).

Software

There is a large amount of software available to perform structurally based alignments. A good example is DALI WWW (http://www.ebi.ac.uk/dali). The detailed methods used by the various computer programs are different, and almost all of the methods do not conform exactly to the simple conceptual description of structural alignment given above. Some methods make use of sequence as well as structure information, and almost all make use of information about secondary structure. There is no real consensus about which methods are the best, but usually any alignment of distantly related protein sequences will be better if created using structural information.

Structural similarity

Most structural alignment methods produce measures that can be used to assess the level of similarity of the aligned structures. A variety of such measures are used, many of which are specific to particular structural alignment methods. Nevertheless most methods report a measure known as **root mean square deviation** (RMSD). When two structures have been aligned they can be optimally superimposed so the aligned residues lie as close as possible to each other in 3-D space (*Fig. 11*). With the structures superimposed, distances can be measured between the α-carbon atoms of the aligned residues, and the root mean square deviation is defined in terms of these distances by

$$RMSD = \sqrt{\frac{1}{N}\sum_i d_i^2}$$

where d_i is the distance between the ith pair of superimposed (aligned) α-carbon atoms, and N is the number of such superimposed atoms (i.e., the number of aligned residues). This expression represents the square root of the mean square distance between aligned residues. If the structures were identical then the RMSD would be zero, because all residue α-carbon atoms would lie directly on top of one another (zero distance apart), after optimal superposition. For less similar structures the RMSD becomes larger. Generally RMSD values between 0.0 and 1.5 Å represent very similar structures, with higher values indicating progressively increasing structural dissimilarity. Clearly a small RMSD computed over a large number of residues (N) is more significant than a small RMSD computed over a small number of residues.

Structural similarity searches

Just as we are interested often in searching sequence databases for sequences similar to a query sequence, it is sometimes desirable to search the structural database for structures that are similar to a query structure. A number of WWW sites provide this type of search, including DALI, SSAP, TOPS, VAST, CE and SSM, and the RSCB site. As with sequence searches, these search engines return a list of similar structures ranked according to some measure of similarity, such as RMSD.

L5 STRUCTURAL CLASSIFICATIONS

Key Notes

Why classify protein structures?
Classification groups together proteins with similar structures and common evolutionary origins.

Example classifications
CATH (http://www.cathdb.info/latest/index.html) and SCOP (http://scop.mrc-lmb.cam.ac.uk/scop)

Structural class
Proteins can be assigned to broad structural classes based on secondary structure content and other criteria. CATH has four such broad classes, but SCOP uses more, giving a more detailed description of structural class.

Fold or topology
All classifications gather together proteins with the same overall fold or topology. Proteins in the same fold or topology class contain more or less the same secondary structure elements, connected in the same way and in similar relative spatial positions.

Homologs and analogs
Homologs (homologous proteins) are related by divergent evolution from a common ancestor, and have the same fold. Analogs (analogous proteins) have the same fold, but other evidence for common ancestry is weak.

Super-folds
Super-folds are protein folds that seem likely to have arisen more than once in evolution. They are thought to have advantageous physico-chemical properties. They appear in SCOP and CATH as fold or topology levels containing several homologous super-families.

Related sections
Sequence analysis (J3)
Sequence families, alignment, and
 phylogeny (J4)
Domain families and databases (J5)
Structural bioinformatics (L4)

Structure prediction and modeling
 (L6)
Molecular dynamics and drug
 design (L7)

Why classify protein structures?

To classify protein structures means to group them so that each group contains similar structures. In Section J4 we learned that naturally occurring protein sequences can be grouped into evolutionary families, and in Section L4 we discovered that protein structure is much more strongly conserved by evolution than protein sequence. To classify proteins by structural criteria is therefore the most powerful way to assign them to families, and to reveal distant evolutionary relationships. Methods of protein structure classification rely heavily on the sequence comparison methods discussed in Section J3 and the structure comparison methods discussed in Section L4.

Example classifications

CATH (http://www.cathdb.info/latest/) and SCOP (http://scop.mrc-lmb.cam. ac.uk/scop) are example classifications. Both are hierarchical (tree-structured), and their structures are shown in *Figs 1* and *2*. Despite the apparent differences, the classifications agree to a large extent about which proteins should be grouped together.

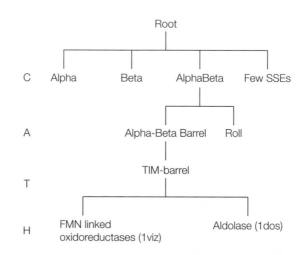

Fig. 1. The structure of the CATH classification illustrating the positions of the glycolate oxidase structure with PDB code 1viz and the class II aldolase structure 1dos, which share the TIM-barrel fold. 1viz and 1dos share the AlphaBeta broad structural class (C), Alpha-Beta barrel architecture (A), and TIM-barrel topology or fold (T). They belong to different homologous super-families (H).

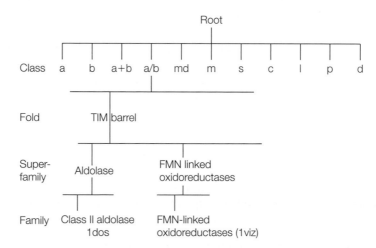

Fig. 2. The structure of the SCOP classification illustrating the positions of the FMN-linked oxidoreductase and aldolase super-families, and the classification of 1viz and 1dos to compare with Fig. 1. At the class level, SCOP has more divisions than CATH: a – all α, b – all β, a + b – α and β segregated, a / b – α and β alternating, md – multi-domain, m – membrane and cell surface, s – small proteins, c – coiled coil proteins, l – low resolution structures, p – peptides, d – designed proteins.

Structural class

At the top level, each classification places protein structures into broad structural classes. CATH uses four classes reflecting secondary structure content: all α, all β, α and β (αβ), and few secondary structures. SCOP on the other hand has more divisions, separating out more small groups, like membrane and coiled-coil proteins, and also splitting the α and β class into two, α + β and α/β, depending on whether the α-helices and beta-strands are segregated in the fold (α + β) or mixed up and tending to alternate (α/β). The next level in CATH further separates the structural classes into broad architectures reflecting the overall shape of the protein, and has no equivalent in SCOP.

Fold or topology

The CATH T (topology) and SCOP fold levels separate proteins into groups that have the same overall fold. This means that the proteins have the same core secondary structure elements, connected and arranged in space in more or less the same way. The example given in *Figs 1* and *2* are part of the TIM-barrel fold level. Within this level all proteins have the well-known TIM-barrel structure: a parallel eight-stranded β-barrel surrounded by α-helices. When proteins have the same fold or topology they usually can be structurally aligned to give large sections of superimposed backbone structure with low RMSD (Section L4) values.

Homologs and analogs

It is thought that many proteins with the same fold have emerged by divergent evolution from a common ancestor, but it is equally possible that they have no common ancestor and have the same fold simply because it is favorable from a physico-chemical point of view. In the former case the proteins are called **homologs** and are said to be **homologous**, in the latter they are called **analogs**, and are said to be **analogous**. It is almost impossible to know whether two weakly similar proteins have derived from a common ancestor or not. However, CATH and SCOP divide proteins with the same fold into homologous super-families. Within such a super-family, there is good evidence that the proteins are homologous. This evidence might have a striking similarity in sequence, structure, or function. When proteins with the same fold are in different super-families, it means that evidence for homology beyond the shared fold is weak, or non-existent.

Figs 1 and *2* show that although the two classifications agree on the fold classification of the proteins in question (TIM-barrel), they differ slightly at the homologous super-family level. Both assign the glycolate oxidase and aldolase examples to different homologous super-families, but CATH has a separate super-family for the class II aldolases, which are gathered into a general aldolase super-family in SCOP (containing class I aldolases in addition). This reflects a difference of expert opinion on whether there is sufficient evidence that the two classes of aldolase should be viewed as homologous. It should be noted also that the authors of these classifications do occasionally change their minds about the classification of particular groups of proteins at this level, and this is reflected often in changes made in newer versions of the database.

Super-folds

Only a relatively small number of the different folds in CATH or SCOP contain more than two homologous families. Such folds appear to have arisen more than once in evolution, and are called **super-folds**. Example super-folds are the TIM-barrel and the immunoglobulin fold. Characteristics of the super-folds are that they tend to exhibit approximate symmetries, and are characterized by repeated super-secondary structures. These suggest the possible physico-chemical favorability of these folds.

Families   Proteins within homologous super-families often are quite distantly related. Both classifications divide each super-family into a number of smaller families, within which the relationships between the proteins is much stronger, and often apparent at the level of sequence similarity.

L6 STRUCTURE PREDICTION AND MODELING

Key Notes

Why predict structure?

Structure prediction is a worthwhile exercise because experimental structure determination still is much slower than sequence determination. Structure predictions help us to understand function and mechanism and can be useful for rational drug design. The early work of Levinthal and Anfinsen made structure prediction a fascinating scientific problem.

What is structure prediction?

We will discuss methods in the categories of comparative modeling, fold recognition, secondary structure prediction, *ab initio* prediction, and trans-membrane segment prediction.

Theoretical basis

Sequences with more than 25% identity over an alignment of 80 residues or more adopt the same basic structure (Section L4). This is the basis of prediction by comparative modeling.

Ingredients

All that is needed is an alignment between a sequence of unknown structure (target) and one or more of known structure (template(s)) with the above property. Template structures can be found by standard sequence similarity search methods. Lack of suitable template structures is the main limitation of the method, but structural genomics projects are likely to change this in coming years. The accuracy of the alignment is crucial if a good prediction is to be obtained.

The process

Known structure(s) (templates) are used as the basis of the prediction. The process can then be viewed conceptually as comprising placement of conserved core residues, modeling of variable loops, side chain positioning and optimization, and model refinement. Conserved residues and some side chain positions can be obtained directly from structural information in the templates. Modeling of variable loops often makes use of the spare parts algorithm, and there are sophisticated algorithms for side chain placement to obtain an optimally packed hydrophobic core.

Accuracy

Accuracy is controlled almost entirely by the quality of the alignment. Good alignments yield good predictions with most of the main software packages. Of all prediction methods, comparative modeling produces the most accurate models.

Availability

Comparative modeling is available on the WWW and in several free and commercial software packages.

What is secondary structure prediction?

Secondary structure prediction predicts the conformational state of each residue in three categories, helical (H), extended or strand (E), and coil (C). Many methods are based on ideas related to secondary structure propensity, which is a number reflecting the preference of a residue for a particular secondary structure. Early methods had accuracies of around 60% (the percentage of residues predicted in the correct H/E/C state). Examples of early methods are the Chou-Fasman rule-based method, and the information theoretic GOR method.

Multiple sequence information

Using multiple alignments of related sequences can improve prediction accuracy enormously by revealing patterns of conservation indicative of certain secondary structures.

Accuracy of state-of-the-art methods

Current methods claim an average accuracy over trusted test sets of proteins equal to more than 70% of residues correctly predicted. This increase in accuracy can be attributed to the availability of more structural data, and the use of more sophisticated algorithms or methods.

Prediction of trans-membrane segments

Membrane spanning segments in integral membrane proteins can be predicted with reasonable accuracy. Most methods make use of a search for contiguous runs of hydrophobic residues that span a lipid membrane. Some methods also predict the orientation (in-out) or topology of the membrane spanning segments, but this is usually less accurate.

Availability of tools

Most of the secondary structure and trans-membrane segment prediction tools mentioned in this section are available from the ExPASy WWW site (http://www.expasy.ch).

Fold recognition

Fold recognition aims to detect very distant structural and evolutionary relationships. It aims to detect when a protein adopts a known fold even if it does not have significant sequence similarity to any protein of known structure. Methods generally try to find the most compatible fold in a library of known folds using both sequence and structural information. An alternative term for fold recognition is threading.

***Ab initio* prediction**

These methods rely on first principles calculation and are not sufficiently well developed yet to be of real use in practical structure prediction.

Strategies

After a thorough preliminary sequence analysis using the methods of Section J, the best strategy employs first comparative modeling, and if not successful, secondary structure prediction followed by fold recognition.

Related sections

Genomes and other sequences (J)	Structural classifications (L5)
Proteomics techniques (L1)	Molecular dynamics and drug
Structural bioinformatics (L4)	design (L7)

Why predict structure?

In the previous sections in this topic we have discussed protein three-dimensional structures and how these enable proteins to carry out their functions. An understanding of structure leads to an understanding of function and mechanism of action. Structural knowledge is vital therefore if we are to proceed to a complete understanding of life at the molecular level. Unfortunately structural knowledge is still rather limited, because the experimental process of structure determination is slow for most proteins and not possible for many. The database of known structures currently contains more than 15 000 protein structures, but the databases of sequences contain hundreds of thousands of sequences. This gap between sequence and structure knowledge is often termed **the sequence structure gap**; from a practical point of view, it is the main factor motivating the need for predictions of protein structure. To be added to this is the fact that many pharmaceutical drugs act by selective binding to target proteins, and knowledge of protein structures can aid the process of rational structure-based drug design (the design of drug molecules based on the structures of the proteins with which they are intended to interact).

As well as having enormous practical significance, structure prediction is a fascinating scientific problem that has interested scientists since before the first protein structures were determined. In fact, Pauling predicted the structure of the α-helix before it was observed experimentally. Much of the fascination of the problem stems from the work of two scientists, Anfinsen and Levinthal. By a set of elegant experiments on RNAse A, Anfinsen was able to prove that proteins can fold to their native structures spontaneously, without the intervention of any other agent. Therefore the protein fold is coded somehow in the amino acid sequence. Levinthal, on the other hand, pointed out that even relatively small proteins have an astronomically large number of possible structures, and that the process of finding the correct one cannot possibly proceed by a random search of the possibilities because this would simply take too long. To discover how the protein sequence codes for a three-dimensional structure, and to understand how this structure folds, therefore, has been a problem of much scientific interest for many years. Despite this, the problem is still far from solved, and even though there has been progress we still do not know in general terms how structure is encoded in sequence.

What is structure prediction?

In the most general form, structure prediction means to make a prediction of the relative position of every atom of the protein in three-dimensional space using only information from the protein sequence. However, not all prediction methods are so general, and some predict limited aspects of structure (e.g., secondary structure), without proceeding to predictions of actual atomic positions. In the sections that follow, we will discuss protein structure prediction methods in four categories, the use of comparative modeling, fold recognition, secondary structure prediction, and *ab initio* prediction. We will discuss also the prediction of membrane spanning segments and integral membrane protein topology.

It is useful to categorize prediction methods further according to their theoretical basis, either *ab initio* or knowledge-based. Like all physico-chemical systems, proteins are believed to fold to attain a state of minimum thermodynamic free energy. In our terminology, only methods that attempt to calculate and minimize this free energy, or a suitable approximation, are *ab initio* methods. They proceed from fundamental physical principles, using the accepted theories of quantum mechanics and statistical thermodynamics, to predict protein structures. This is very difficult. Proteins, when modeled with enough solvent molecules to be realistic, are enormous systems with many thousands of atoms. With current technology, detailed calculations of exact free energies are just not possible, and are not likely to be for many years to come. The task therefore is to find a suitable approximation to the free energy that captures the essentials of the folding problem. Such an approximation has not yet been found.

In contrast to *ab initio* methods, knowledge-based methods attempt to predict protein structure using information taken from the database of known structures. The simplest example of this is to predict that a sequence, similar to one of known structure will adopt that same basic structure. This was shown by Sander and Schneider in 1991 (Section L1). But, there are many other ways of using information from the database of known structures in predictions. The methods of comparative modeling and fold recognition are knowledge-based methods. We also view secondary structure prediction as a knowledge-based method. This is because most secondary structure prediction methods have been trained on data from known structures. However, it is sometimes considered as an *ab initio* method because it is applicable even to sequences that form tertiary structures that have not yet been observed in any structural relative, however distant. There is no doubt that knowledge-based methods are currently the most accurate and practically useful protein structure prediction methods.

In the past few years protein structure prediction methods have been subjected to some rigorous blind testing associated with the CASP (Critical Assessment of Structure Prediction) structure prediction competitions. The results of these competitions have added enormously to our understanding of the accuracy of the various methods. Details of the CASP competitions at http://predictioncenter. gc.ucdavis.edu/

Theoretical basis In Section L1 we discovered that two proteins share the same structure if their sequences can be aligned to show at least 25% identity in an alignment of 80 or more residues. This is the basis of the method of comparative modeling (also sometimes known as homology modeling). If a sequence of unknown structure (usually called **the target sequence**) can be aligned with one or more sequences of known structure, so as to satisfy this condition, then the known structures can be used to predict the structure adopted by the target sequence. The proteins of known structure used are referred to usually as **template structures**.

Ingredients All that is necessary for comparative modeling is an alignment between the target sequence and the sequences of the template structures. The first step is to locate possible template structures, and this can be done using the standard sequence similarity search methods discussed in Section J3. In this case it is necessary only to search the database of sequences whose structures are already known by experimental means. This is a relatively small database, and the lack of availability of suitable template structures is the main limitation to the use of comparative modeling at the present time. This situation may change in coming years. Large-scale projects aimed to create high-throughput structure determination, under the name of **structural genomics**, are now underway (see for example the New York Structural Genomics Consortium, http://www.nysgrc.org/). It is likely that suitable template structures will become available for most protein sequences within the next 10–15 years.

When template structures have been obtained it is necessary to align their sequences with the target sequence using a multiple alignment tool (Section J4). From a users point of view the alignment process is the most crucial step in comparative modeling. If the alignment is good then a good model will result; if it is poor then the model will be poor also. When the target and template sequences are closely related, with high percentage identities (for example 70% and above) then automated alignment methods usually produce good alignments. Nevertheless, even when percentage identity is very high the alignment should be inspected for any obvious alignment errors. For example, it is advantageous to check that conserved key structural and functional residues are aligned. If there are key structural cysteines, glycines, or prolines (Section L4) , then these should be aligned with residues of corresponding types in all the sequences.

When target and template sequences are more weakly related, with lower percentage identities, then alignment errors are more common and more manual checking is necessary. It may be very difficult to identify the correct alignment, and this can result in poor predictions. There are now many software packages that do completely automated comparative modeling, with no manual user input. These function well when target and templates are closely related, but should be treated with caution when relationships are more distant. Always check that the alignment used was acceptable from a structural and functional perspective.

The process

The various software packages carry out the comparative modeling process in slightly different ways, and what we describe here are generic features of the process that are common to all methods. A schematic illustration of the process is given in *Fig. 1*. The process begins by analysis of the template structures. If there is more than one such structure, an average structure is often calculated by superposing the structures in 3-D (Section L4) and calculating average atomic positions. The contributions of the various templates in this process often are weighted according to their degree of similarity to the target sequence, with

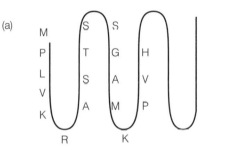

(b)

Known   MPLVKRASTSS_GAMKPVH
Unknown MPILKRGTSTSYGAMRPIY

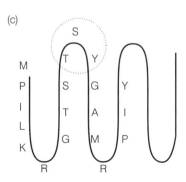

Fig. 1.   Comparative modeling. A schematic of a known (template) structure is shown in (a). Where a sequence of unknown structure shows a substantial pairwise similarity, as shown in (b), then one can substitute the aligned residues into the structure, reposition the side chains and extend or shorten loops to compensate for insertions or deletions – loop modeling denoted by the dashed circle – resulting in the model structure in (c).

more similar sequences more strongly weighted. In this process a framework of template atomic positions is calculated.

Next, using the alignment of the target sequence to the templates, the structure is divided into two distinct types of region, the structural core and the non-conserved loops. The core can be considered to comprise mainly the secondary structure elements (helices and strands) of the templates, within which each residue is aligned with a residue from the target sequence. The non-conserved loops are mainly surface loops for which, according to the alignment, target and template sequences have different numbers of residues (i.e., there are gaps in the alignment). Structure prediction for the core is easy, because the backbone atomic positions from the averaged template structure can be used directly as backbone atomic positions for the predicted (model) structure of the target sequence.

Within non-conserved loops structure prediction is more difficult, and more sophisticated structure prediction methods have to be used. The simplest method of loop structure prediction is to use the **spare parts** algorithm. This method makes use of a database of known loop structures from other proteins that are not similar in sequence to the target necessarily. For each loop needed, a spare part loop is found in this database which fits into the gap in the modeled structure, and this is taken as the predicted structure for the loop region. There are more sophisticated loop prediction algorithms, involving more complex calculation and minimization of estimated loop energies, but the simple spare parts algorithm seems to work well in practice. Prediction accuracy in the region of predicted loops generally is much lower than it is in the core, and while most loop prediction algorithms perform well on short loops (<6 residues) they become increasingly inaccurate as loop length increases.

When the steps above are complete there are predicted positions for all the backbone atoms of the target sequence, and the process turns to the question of side chain positions. Some side chain identities are conserved between target and template structures, and the structural information from the templates can be used to place these side chains in the predicted structure. However, not all side chains are conserved and the positions of the atoms in these need to be predicted. This is done usually by choosing from a library of allowed side chain structures (sometimes called a side chain rotamer library), so that the atoms of the side chains fill the available space in the protein interior without being too close (clashing) with any other protein atoms. The algorithms used to do this are often quite sophisticated, and aim to produce a protein hydrophobic core that is tightly packed, like those observed in experimentally determined structures.

At this stage, the preliminary structural model for the target sequence is complete and many predictions stop at this point. Sometimes however it is useful to refine the model further. Typically this involves the use of energy minimization software that can make small changes to atomic positions in order to produce a slightly lower energy model. This can be advantageous; for instance, to move any atoms that were left in too close proximity after the modeling process above, but the advantage in terms of the accuracy of the model is debatable.

Accuracy

We have commented already that the accuracy of comparative models is determined almost entirely by the accuracy of the alignment. When alignments are good, which they typically are when the sequences are very closely related, then very accurate models are possible. Typically accuracy is measured in terms of root mean square deviation (RMSD, see Section L4 for a definition) between the α-carbon positions in the predicted structure and the actual structure of the target sequence. RMSDs of less than 1.0Å represent very good predictions; this degree of difference between the two structures is similar to the degrees of difference between two separate experimental determinations of the same protein structure. When the percentage sequence identity between template structures and target sequence exceeds 70% it is reasonable to expect that the model should be accurate to an RMSD of less than 2–3Å even using completely automated methods. When the percentage identity drops below 40% then getting a good alignment, often with manual intervention, becomes crucial and automated methods can fail very badly.

It is, of course, not possible to know how accurate a model is without an independent experimental determination of the actual structure, but the above guidelines should be borne in mind. It is possible also to check some aspects of the predicted structure; for instance, software described in Section L4 for checking of stereochemical parameters of the structure will identify any unusual bond lengths or angles that might indicate potential problems in the prediction. Predicted structures should be viewed in general as quite low resolution models. For applications where a very accurate structure is required, for example detailed structure-based drug design, they are probably too low in accuracy, particularly in respect of non-conserved loops and side chain positions. Nevertheless, if it is possible to construct a structure prediction by comparative modeling then this prediction is likely to be more accurate than a prediction by any other method.

Availability

Comparative modeling is available on the WWW at the SWISS-MODEL site, and using the SWISS-PDBVIEWER software (http://swissmodel.expasy.org). It is also available in several commercial software packages. Other comparative modeling methods include 3-Djigsaw available from http://www.bmm.icnet.uk/~3djigsaw/ and Geno3-D (http://geno3d-pbil.ibcp.fr/cgi-bin/geno3d_automat.pl?page=/GENO3D/geno3d_home.html).

What is secondary structure prediction?

The previous section introduced comparative modeling as the method of choice for protein structure prediction. When suitably related template structures do not exist for a particular target sequence, secondary structure prediction is a viable alternative. Unlike comparative modeling it does not produce a full atom model of the tertiary structure, but rather just provides a prediction of the secondary structure state of each residue, either helical (H), strand or extended (E), or coil (C). The predictions are sometimes known as three-state predictions.

Most methods of secondary structure prediction have been trained on the database of known structures. Two early methods were those of Chou and Fasman, and GOR (Garnier-Osguthorpe-Robson, after its inventors). Although these methods work in different ways, the former being rule-based and the latter using information theoretic ideas, they both make use of the idea of **secondary structure propensity**. The amino acids seem to have preferences for certain secondary structure states, which are shown in *Table 1*. For instance, glutamic acid has a strong preference for the helical secondary structure, and valine a strong preference to be in strands. Glycine and proline have lower than average propensity for both types of regular secondary structure, reflecting a tendency to be found in loops, and some hydrophobic amino acids (e.g., phenylalanine) have strong preferences for both types of secondary structures reflecting their tendency to make up the structural core. However, none of the preferences are particularly strong, and all amino acids are found in each type of secondary structure quite often. This means that predictions of secondary structure cannot be based on single residues.

In the Chou-Fasman method, a helix is predicted if, in a run of six residues, four are helix favoring, and the average value of the helix propensity is greater than 1.0 and greater than the average strand propensity. Such a helix is extended along the sequence until a proline is encountered (these are known to break helices) or a run of four residues with helical propensity less than 1.0 is found. A strand is predicted if in a run of five residues three are strand favoring, and the average value of the strand propensity is greater than 1.04 and greater than the average helix propensity. Such a strand is extended along the sequence until a run of four residues with strand propensity less than 1.0 is found. This is a simple rule-based method dependent on finding runs of residues with preference for one type of secondary structure. The GOR method is different, considering the information carried by a residue about its own secondary structure, in combination with the information carried by other residues in a local window of eight residues on either side in the sequence of the residue concerned.

Table 1. Helical and strand propensities of the amino acids. A value of 1.0 indicates that the preference of that amino acid for the particular secondary structure is equal to that of the average amino acid; values greater than one indicate a higher propensity than the average, and values less than one a lower propensity than the average. The values are calculated by dividing the frequency with which the particular residue is observed in the relevant secondary structure by the frequency for all residues in that secondary structure.

Amino acid	Helical (α) propensity	Strand (β) propensity
GLU	1.59	0.52
ALA	1.41	0.72
LEU	1.34	1.22
MET	1.30	1.14
GLN	1.27	0.98
LYS	1.23	0.69
ARG	1.21	0.84
HIS	1.05	0.80
VAL	0.90	1.87
ILE	1.09	1.67
TYR	0.74	1.45
CYS	0.66	1.40
TRP	1.02	1.35
PHE	1.16	1.33
THR	0.76	1.17
GLY	0.43	0.58
ASN	0.76	0.48
PRO	0.34	0.31
SER	0.57	0.96
ASP	0.99	0.39

The accuracy of these early methods based on the local amino acid composition of single sequences was fairly low, with often less than 60% of residues being predicted in the correct secondary structure state. This should be viewed in light of the fact that for proteins containing roughly equal proportions of helix, strand, and coil the accuracy of random predictions would be 33% (i.e., the correct state would be predicted for one in every three residues).

Multiple sequence information

In the late 1980s it was realized that the information contained in multiple sequence alignments could improve secondary structure predictions significantly. Several factors contribute to this increase in accuracy, but they all relate to the patterns of conservation revealed in the multiple alignment. First, a residue with a high propensity for a particular secondary structure in one sequence may have occurred by chance, but if it is part of a conserved column in which all the residues have high propensity for that type of secondary structure then this provides much more predictive evidence. Second, multiple alignments can reveal more subtle patterns of conservation. For instance, a large proportion of α-helices in globular proteins is amphipathic. These helices (see Section L4) have hydrophobic and hydrophilic faces and are associated with periodic patterns of hydrophobic and hydrophilic residues in the sequence. The appearance of conserved patterns of this type is therefore highly predictive of α-helical structure. Finally, as we commented in Section L4, insertions and deletions are much more likely in surface loops (coil (C) secondary structure), so regions of the alignment where these are common are more likely to be the C secondary structure state. The use of evolutionary information from multiple alignments improved the accuracy of secondary structure prediction methods significantly, resulting in methods capable of an accuracy of around 66% at the time.

Accuracy of state-of-the-art methods

Since the late 1980s we have seen an explosion in the amount of data in the sequence and structure databases that has increased the amount of structural information available to secondary structure prediction algorithms. This, coupled with improvements to the algorithms themselves, has led to much increased accuracy. Examples of state-of-the-art prediction methods are PHD (Rost and Sander), DSC (King and Sternberg), PREDATOR (Frishman and Argos), NNSSP (Salamov and Solevyev), and PSI-PRED (Jones). These methods use a variety of techniques to make their predictions, many of which are based in the fields of artificial intelligence or machine learning. PHD and PSI-PRED make use of artificial neural networks (simple computational models of the neural networks in the brain), and DSC is based on a machine learning method known as linear discriminant analysis. NNSSP (Nearest Neighbor Secondary Structure Prediction) makes predictions based on locating the most similar sequence segments in the database of known structures. The accuracy of all these methods is above 70%, and, depending on the test set of predictions chosen, accuracies of above 75% have been achieved. An example secondary structure prediction from PSI-PRED in shown in *Fig. 2*. This prediction was carried out using a protein of very recently determined structure, and the actual secondary structure from that experiment is shown for comparison with the predicted structure. The overall accuracy in this case is higher than average (83%). It should be noted that many of the prediction errors are residues on the ends of helices or strands and that the secondary structure of these residues may not be well defined in the structure; it can be very difficult to decide whether a residue at the end of a helix is part of the helix or not. A new WWW server (JPred) now incorporates results from several of these prediction programs and uses a consensus approach to determine the best prediction.

One caveat that must be borne in mind always is that all secondary structure prediction methods have been trained using structures from the database of known protein structures. This set of proteins is biased for various reasons, principally because some very common proteins (e.g., integral membrane proteins) are under-represented. The prediction methods can be expected to work better for proteins that are in some sense similar to the proteins in this database, and less well for others. It makes no sense therefore to apply standard secondary structure prediction methods to integral membrane proteins.

```
Conf : 96882238889783789999999999998657887789999986159741478666 4032
Pred : CEEEECCCCCCCCHHHHHHHHHHHHHHHHHHHCCCCCCHHHHHHHHHCCCHHHCCHHHHHCC
Act  : NNNNNNNNNNNNNCHHHHHHHHHHHHHHHHHHHCCCCCCHHHHHHHHHCCHHHCCHHHHHHC
  AA : MHLYSSDFPLMMDEKELYEKWMRTVEMLKAEGIIRSKEVERAFLKYPRYLSVEDKYKKYA
             10        20        30        40        50        60

Conf : 237776523785312767999999972689889899976370689999996319879998
Pred : CCCCCCCCCCCCCEECCHHHHHHHHHHHHHCCCCCCEEEEEECCCHHHHHHHHHHCCCCEEEEE
Act  : CCCCCEECCCCCEECCHHHHHHHHHHHHHCCCCCCCEEEECCCCCHHHHHHHHHHHCCCEEEE
  AA : HIDEPLPIPAGQTVSAPHMVAIMLEIANLKPGMNILEVGTGSGWNAALISEIVKTDVYTI
             70        80        90       100       110       120

Conf : 229999999999988865998742676544345665789688998797231136789 8706
Pred : ECCHHHHHHHHHHHHHHCCCCCCEEEECCHHHCCCCCCCCCCEEEECCCHHHCCHHHHHHCC
Act  : ECCHHHHHHHHHHHHHHCCCCCEEEEEECCHHHCCHHHCCEEEEEEECCCCCCCCCHHHHHCEE
  AA : ERIPELVEFAKRNLERAGVKNVHVILGDGSKGFPPKAPYDVIIVTAGAPKIPEPLIEQLK
            130       140       150       160       170       180

Conf : 799999845797860689999626985568852555666152433 5413325229
Pred : CCCEEEEECCCCCCEEEEEEEEECCCEEEEEEECCEEEEECCCCCCCCCHHHHHCC
Act  : EEEEEEEECCCCCCEEEEEEEEECCEEEEEEEEEECCCCCCCCCCCCNNNNNNN
  AA : IGGKLIIPVGSYHLWQELLEVRKTKDGIKIKNHGGVAFVPLIGEYGWKEHHHHHH
            190       200       210       220       230
```

Fig. 2.   Secondary structure prediction. The PSI-PRED server was used to predict the secondary structure of a protein (L-isoaspartate O-methyltransferase) whose actual three-dimensional structure had very recently been determined. Shown above the sequence (AA) are the actual secondary structure (Act), the predicted secondary structure (Pred), and a measure of the confidence of prediction (0 (uncertain)–9 (very confident)). The percentage of residues predicted in the correct secondary structure is 83%. (C random coil, H helix (α- or 3_{10}), E extended (β-), N structure unknown).

Prediction of trans-membrane segments

In Section L4 we described the structure of integral membrane proteins as consisting of segments (usually helical) spanning the lipid membrane, connected by loops lying outside the membrane. The membrane spanning segments tend to contain a high proportion of hydrophobic residues and are often more than 20 residues in length, corresponding to 6–7 helical turns for trans-membrane helices. Such relatively long runs of predominantly hydrophobic residues are seldom found in water-soluble globular proteins. This means that often it is possible to predict, based on the runs of hydrophobic residues, whether or not a protein is an integral membrane protein, and if so, where membrane spanning segments are in the sequence.

Tools have been developed to predict trans-membrane segments, including TMPred, TMHMM, and TopPred. Many of these tools also predict the membrane spanning topology. This is a prediction of the orientation of the helices with respect to the membrane. For instance, and i-o (in to out) helices have their amino (N) termini inside whatever cell or organelle the membrane bounds, and their carboxy (C) termini outside. Clearly a consistent prediction of topology must have i-o helices followed by o-i helices and *vice versa*. Predictions of the membrane spanning topology are generally less reliable than predictions of just which segments span the membrane irrespective of orientation.

An example of a prediction of membrane spanning segments is shown in *Fig. 3*, where predictions have been made for the much-studied G-protein-coupled receptor rhodopsin. In this case the predictions are quite accurate, with all seven trans-membrane segments predicted in the correct sequence predictions. This is of course not always the case, and it is quite common for some real membrane spanning segments not to be predicted (false negatives), or for some non-membrane spanning segments to be predicted as membrane spanning (false positives). False negatives often occur in ion channel proteins, where some of the membrane spanning helices have hydrophilic sides which cluster together to make a hydrophilic path through the membrane along which charged ions can pass. The extra hydrophilic residues in these helices often means that they evade detection by methods searching for hydrophobic residues. False positives are often found in secreted proteins that contain an amino (N) terminal signal peptide. Such signal peptides tend to be hydrophobic, and this sometimes leads to their confusion with membrane spanning segments.

Helix	Predicted position	Actual position	Predicted orientation	Actual orientation
1	37–63	37–61	o–i	o–i
2	74–99	74–98	i–o	i–o
3	115–140	114–133	o–i	o–i
4	153–175	153–176	i–o	i–o
5	203–221	203–230	o–i	o–i
6	253–276	253–276	i–o	i–o
7	286–309	285–309	o–i	o–i

Fig. 3. Prediction of trans-membrane helical segments for human rhodopsin (a G-protein-coupled receptor). Predicted and actual predictions refer to residue numbers with respect to the amino acid sequence in SWISSPROT, where the actual position of helices is recorded. Predictions were made with TopPred.

Availability of tools

Most of the secondary structure and trans-membrane segment prediction tools mentioned in this section are available from the ExPASy WWW site (http://www.expasy.ch), under proteomics tools.

Fold recognition

In Section L4 we discussed the evolution of protein structures and established that similarity in sequence is sufficient to guarantee similarity in structure. The method of comparative modeling is based on this fact. However, we also established that during evolution, sequence is much more strongly conserved than structure. There are many cases of distantly related proteins with the same structure but whose level of sequence similarity is well below the 25% needed to **guarantee** structural similarity and the possibility of comparative modeling. Fold recognition methods are about detecting these distant relationships, and separating them from chance sequence similarities not associated with a shared fold. They operate by searching through a library of known protein structures (called a **fold library**) and finding the one most compatible with the query sequence whose structure is to be predicted. An alternative name for fold recognition is **threading**.

The product of fold recognition is usually an alignment between the query sequence and one or more distantly related sequences of known structure. Therefore fold recognition can be viewed as an extension of the comparative modeling method to very distant relationships. Once the alignment between the sequence and the distantly related known structures has been obtained, a full three-dimensional structure of the protein to be predicted can be obtained using the usual methods of comparative modeling. Because relatively few protein structures are known, most new sequences cannot have their structures predicted by ordinary comparative modeling. Fold recognition is valuable because it has the potential to extend significantly the number of protein sequences whose structures can be predicted.

In Section L5 we made the distinction between homologous proteins which share the same fold because they have evolved by divergent evolution from a common ancestor, and analogous proteins which share the same fold perhaps for physico-chemical reasons with little evidence for common ancestry. Fold recognition methods were intended originally to detect both analogs and homologs of the query sequence, but current evidence suggests that they can detect distant homology but not analogy.

Fold recognition methods employ a variety of approaches, most often mixing the similarity in sequence detected by amino acid substitution matrices with further structural information. For instance the **3-D-PSSM** (3-D-Position Specific Scoring Matrix) uses fold library structures described in terms of ordinary 1-D sequence profiles generated by PSI-BLAST (Sections J3 and J4), and also '3-D' profiles. The information within the 3-D profile includes secondary structure and solvation potential. The solvation potential takes account of the tendency of hydrophobic amino acids to occupy structural positions that are not accessible to solvent (buried in the hydrophobic core). The secondary structure component measures the degree of similarity between the predicted secondary structure of the query sequence and the secondary structure of the fold library member. There is very good evidence that the inclusion of structural information like secondary structure and solvation results in methods that are able to detect distant homology better than the most sensitive sequence only methods (e.g., PSI-BLAST), and therefore that these are useful in structure prediction. The URL for 3-D-PSSM is http://www.sbg.bio.ic.ac.uk/~3dpssm/; 3-D-PSSM is under development. The updated version is called PHYRE and is available from the same WWW site.

One problem with fold recognition methods is that while they are often able to recognize a distant homology to a known fold, they sometimes do not provide particularly accurate alignments to that fold. This situation is improving as the methods develop, but it must always be borne in mind that alignments may not be very accurate. Inaccurate alignments lead to poor structural predictions, using comparative modeling methods.

Ab initio prediction

Ab initio methods are methods that attempt to predict protein structures from first principles using theories from the physical sciences like statistical thermodynamics and quantum mechanics. The state-of-the-art software available at the time of writing is the Robetta server (http://robetta.bakerlab.org/). We will say little more about them here, except to set them in context of the methods we have already described. There is no doubt that first principles predictions are very attractive from an intellectual viewpoint. However, these methods still are not sufficiently well developed to be of real use in practical protein structure prediction. Proteins in their natural solvation environments are very large systems, and are currently beyond the scope of accurate calculation with accepted theories. This is not to say that research in *ab initio* methods is uninteresting, indeed to understand the process of protein folding in this way would be a huge scientific achievement. But, from the point of view of the practicing molecular biologist or biochemist, the methods are currently of little use by comparison with the methods of comparative modeling, secondary structure prediction, and fold recognition described previously.

Strategies

We close this section by outlining a strategy for protein structure prediction for a new query sequence of unknown structure. The first step perhaps should be to identify any features of the sequence that might affect the strategy to be adopted. The sequence should be examined for potential membrane spanning segments that might indicate the presence of an integral membrane domain. There are now some integral membrane proteins of known structure, but not many, and structure prediction for such domains remains difficult. Other sequence features that should be identified could be regions of low compositional complexity (Section J3), since few of these appear in the database of known structures and therefore structure prediction might be inaccurate in such regions. The presence of coiled coils could be tested (Section J3). Finally an analysis of the sequence by a tool like Interpro (Section J5) might reveal the presence of known domains and perhaps the overall domain structure of the sequence. This could be supplemented by a PSI-BLAST search (Sections J3 and J4) which might reveal other related sequences and the sub-sequences (domains) where they are related. If the protein is multi-domain, it would make sense to predict each domain separately, if the positions of domains in the sequence could be defined.

The most accurate and comprehensive structure prediction method is comparative modeling, so the first prediction step of the strategy is to see if this is possible. This requires the identification of sequences of known structure with sufficient similarity to the query sequence, and can be carried out at the SWISSMODEL WWW site. If this search is successful, then a comparative model can be built and there is no need for any further action. All other structure prediction methods are less accurate and less useful than comparative modeling. Even to make a secondary structure prediction is not useful, because the secondary structure of the comparative model itself is likely to be more accurate than any other secondary structure prediction method.

Comparative modeling is only possible for a minority of new protein sequences. When it is not possible, the next logical step is secondary structure prediction. This can be applied to any sequence, but of course is more accurate for globular protein domains and likely to be less accurate for integral membrane domains and low complexity regions. This will give a prediction of helix (H), strand (E), or coil (C) for each residue, which might be of use, for example in mutagenesis studies. Following secondary structure prediction, fold recognition methods might give an idea of how the secondary structures pack into the tertiary fold, but these methods should be applied with some caution. It must be remembered that fold recognition detects very distant relationships where even structural divergence can be significant, and that 3-D structure predictions from this method are generally much less accurate than those from standard comparative modeling.

L7 MOLECULAR DYNAMICS AND DRUG DESIGN

Key Notes

Molecular modeling
Molecular modeling is the name for a collection of techniques to create models of molecular systems in atomic detail, capable of reproducing and predicting physical properties of molecular systems.

Energy minimization
Energy minimization is the optimization of molecular geometry as computed by an energy function to produce a low energy molecular structure.

Molecular dynamics
Molecular dynamics is a term to describe a simulation of a molecular system, where the time-dependent interactions between atoms are modeled using classical Newtonian mechanics.

Structure-based drug design
Structure-based drug design is the process of designing a drug for a specific biological target based on the structural properties of the biological target. The design process identifies which structural features of the drug are suitable for use with the biological target.

Related sections
Essentials of physics (C)

Computation approaches to artificial intelligence and machine learning (I3)

Molecular modeling

The aim of molecular modeling is to create realistic models of chemical systems, in atomistic detail, that are capable of accurately reproducing the physical properties of the system. These physical properties can be calculated by a large number of techniques that differ in both resolution and complexity. The nature of the property to be calculated and the size of the system being studied are the main factors to consider when choosing an appropriate modeling method. Molecular modeling methods can be divided into three broad categories based on their description of the nature of atoms. The first category, quantum chemical methods, uses quantum mechanics to describe molecular systems. These methods explicitly represent the electrons within atoms, generally using approximations of the Schrödinger wave equation. The second category, semi-empirical methods, uses approaches similar to quantum chemical methods but introduce further approximations and empirical parameters into the equations used to describe the atoms. The third category, molecular mechanics methods, treats molecular systems as classical Newtonian objects ignoring an explicit description of electrons. Atoms are described generally as point charges with associated masses and the bonds connecting atoms are described using simple harmonic equations.

Quantum chemical methods
By treating atoms and molecules as quantum objects, quantum chemical methods provide the most detailed and accurate description of molecular structure by explicitly accounting for the behavior of electrons. Consequently these methods are the most computationally demanding and therefore owing to these demands are limited to describing smaller chemical systems. These methods use approximations of the Schrödinger wave equations to describe the behavior of the atoms and are described as *ab initio* methods (Latin 'from the beginning'). A number of approximations are made to the full Schrödinger equation to allow the practical application of these methods, such as the Born-Oppenheimer approximation, which separates electronic and nuclear wave functions owing to the large difference in timescales of the electronic and nuclear motion. This separation leads to the concept of a potential energy surface (PES) which describes the energy of a system as a function of only the nuclear coordinates, as the electrons are assumed to move instantaneously in response to nuclear motions. The concept of a PES for a molecule is a common theme used in molecular modeling methods because it is a useful simplification allowing the number of variables to be computed to be reduced.

In order to obtain the energy and other properties of the system, the approximated electronic wave function has to be solved for the electrons so that the energy and other properties can be calculated. Various quantum chemical methodologies have been developed, differing in the level and nature of the approximations used to describe the electronic wave function. The methodologies can be classified according to their inclusion or exclusion of the effects of correlated electron motions. The Hartree-Fock (HF) method does not include electron correlation whereas post-Hartree-Fock methods such as Configuration Interaction (CI), Coupled Cluster (CC), Møller-Plesset (MP), and Density Functional Theory (DFT) methods explicitly include electron correlation.

Semi-empirical methods
Semi-empirical methods are based on the equations of *ab initio* methods, but empirical parameters are introduced to replace some of the more computationally demanding equations. These simplifications greatly decrease the computational effort compared to rigorous quantum chemical methods and allow the application of the methods to larger systems than can be treated with purely *ab initio* methods. The parameters introduced are usually derived by fitting to either experimental or *ab initio* data.

Molecular mechanics methods
Molecular mechanics methods, also known as force field methods, use empirical energy functions to describe the interactions of atoms in molecular systems. Using the assumptions of the Born-Oppenheimer approximation, the total energy of a molecular system is calculated as a function of the nuclear coordinates, and the motions of electrons are ignored. The potential energy function, or force field, is expressed as a sum of the contributions from the covalent (bonded) and non-covalent (non-bonded) interactions of the atoms.

$$E_{potential} = E_{covalent} + E_{non-covalent}$$

The terms of the force field are empirically chosen to reproduce the behavior of particular aspects of the interactions. The covalent or bonding interactions are usually described as a summation of several terms describing the two-body

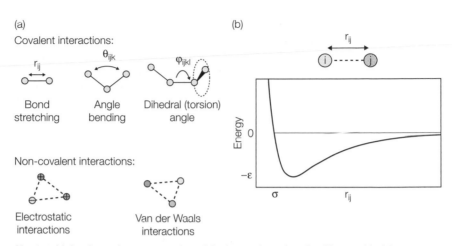

(a)

Covalent interactions:

| Bond stretching | Angle bending | Dihedral (torsion) angle |

Non-covalent interactions:

| Electrostatic interactions | Van der Waals interactions |

(b)

Fig. 1.   (a) A schematic representation of the interactions described by empirical force fields. Atoms are drawn as spheres with covalent (bonding) interactions represented using solid lines, and non-covalent interaction with dotted lines. (b) The Lennard-Jones interaction potential, showing the relationship of the parameters σ, the hard sphere radius and ε, the well depth, to the potential energy as a function of the separation of atoms i and j.

(bond stretching), the three-body (bond angle), and four-body (dihedral angle) interactions.

The two- and three-body (bond stretching and angle bending) interactions generally are described using harmonic potentials, where the equilibrium bond lengths, angles, and force constant for bond and angle stretching are parameterized from experimental or *ab initio* data. The four-body dihedral angle term usually is described by a Fourier series to represent the periodic nature of the energy barriers to free rotation about a single bond. The energy of a particular configuration is computed as a function of the dihedral angle formed by the four atoms.

The non-covalent interactions are partitioned into a sum of the energies arising from the charge (electrostatic) interactions and Van der Waals interactions.

$$E_{non\text{-}covalent} = E_{elec} + E_{VdW}$$

The electrostatic interactions commonly are described using Coulomb's law, where individual atoms i and j have charges Q_i and Q_j, respectively, and a separation distance of r_{ij}. The term ε is a constant and is related to the dielectric properties of the medium the system being simulated is in.

$$E_{elec} = \sum_{i<j}^{atoms} \frac{Q_i Q_j}{\varepsilon r_{ij}}$$

The Van der Waals interactions can be described using a number of different empirical energy functions and the particular function used is dependent on the force field. A popular empirical potential for the Van der Waals interactions is the Lennard-Jones potential where the interaction between atoms i and j is described as a function of their separation distance r_{ij} and two parameters ε_{ij} and σ_{ij} determined by the type of atoms interacting.

$$E_{VdW} = \sum_{i<j}^{atoms} 4\varepsilon_{ij} \left[\left(\frac{\sigma_{ij}}{r_{ij}} \right)^{12} - \left(\frac{\sigma_{ij}}{r_{ij}} \right)^{6} \right]$$

The parameters for a particular force field are generated by fitting the energy functions to experimental and/or *ab initio* data derived from a small test set of molecules. A key feature of force fields is transferability, which allows a force field (parameterized using a small test set of data) to be applied to a wider range of molecular systems. The transferability of a force field is inspired by the empirical observation that atoms, bonds, and chemical functional groups in chemically similar environments possess similar physical properties, such as atomic radii and equilibrium bond lengths. However, the parameters optimized for one specific force field generally cannot be transferred to another force field.

Despite the severity of the approximations made in the description of chemical systems, they have found widespread success at modeling the properties and behavior of systems ranging in size from small organic molecules to large biological systems such as proteins, DNA, RNA, and lipid membranes. Force fields commonly used for investigation biomolecular systems include AMBER, CHARMM, OPLS, and GROMOS. Programs for performing molecular mechanics calculations often share their name with the force field for which the program was originally developed. For links to these programs and force fields please see *Table 1*.

Energy minimization

As the potential energy of a chemical system can be calculated as a function of atomic coordinates using any of the previously outlined classes of methods, often it is desirable to optimize the atomic coordinates so that they correspond to an energy minimum on the potential energy surface. This is desirable as the structure of a given molecule, in thermal equilibrium with its surroundings, will tend to fluctuate about low energy structures; that is, local or global minima on the potential energy surface. A low energy structure can be achieved by minimizing the calculated energy of the system as a function of the atomic coordinates. The coordinates are usually iteratively optimized by either a defined number of successive applications, or steps, of the given algorithm to the coordinates, or until the change in the molecules calculated energy between successive applications has converged on a sufficiently small value. A number of different algorithms have been developed to optimize efficiently the atomic coordinates as a function of the potential energy, and the two most commonly encountered methods are steepest descent and conjugate gradient energy minimization.

The steepest descent algorithm finds the gradient of the energy function with respect to the current atomic coordinates and updates the positions of the atoms in directions calculated using the negative gradient of the energy function. A potential weakness of this algorithm is that it may take a large number of iterations to converge depending on the shape of the potential energy surface for a given molecular geometry.

The conjugate gradient algorithm is similar to the steepest descent algorithm, but the new directions for the movement of the atomic coordinates for each step are calculated using the energy gradient of the current coordinates and the direction moved in the last step. This is more computationally demanding than the steepest descent method but tends to produce geometries that converge to a minimum energy structure in situations where the steepest descent method performs poorly. Energy minimization methods are used in a variety of situations; for example, in conformational search procedures, and for preparing molecular structures for further types of calculation (e.g., molecular dynamics simulation).

Molecular dynamics

Molecular dynamics simulations compute the time-dependent interactions of the molecular systems by numerically integrating the classical Newtonian equations of motion for the system in a series of discrete steps.

$$F = ma$$

The forces acting on individual atoms at a given time-step are determined by negative derivative of the interaction potential function. Newton's second Law of Motion states that the force acting on a particle F is equal to the rate of change of the particle's momentum. As a particle's momentum is equal to its mass m multiplied by its acceleration a, the forces acting on the atoms in the simulation, combined with the current atomic coordinates, are used to update the atomic coordinates and velocities in the subsequent time-step. The potential function for atomic interactions is usually provided by an empirical force field, although purely quantum chemical or a mixture of quantum chemical and empirical energy functions can, in certain situations, be employed.

The nature of accurate interaction potentials requires the size of the time-steps to be very short, typically 1 to 10 fs (10^{-15} to 10^{-14} seconds). Consequently a considerable number of time-steps must be computed to allow the simulation time to become comparable to the process being studied. As a result a number of techniques are used to decrease the computational effort of simulation and increase the timescale of simulations. One of the most common techniques is to apply a cutoff distance, typically in the range of 10–15 nm, to the non-bonded interactions in the interaction potential function. Depending on the exact scheme employed, non-covalent interactions between atoms greater than the cutoff are either ignored or scaled, so that a vast majority of the possible interactions do not have to be computed. Despite the computational savings of these cutoff schemes, care has to be taken to ensure that the particular scheme used is compatible with the type of simulation performed. Another method for increasing the size of the time-step used is to remove high frequency bond stretching motions, such as those bonds involving hydrogen atom motions by constraining bond lengths, using algorithms such as SHAKE. An alternative to approximating the interactions within a given force field is to use a simplified force field, such as a united-atom force field, where some of the individual atoms are implicitly merged into larger groups; for example, hydrogen atoms connected to polar chemical groups are implicitly accounted for through the alteration of parent atoms' parameters.

An important consideration in the simulation of molecular systems, particularly for the simulation of biomolecules, is the treatment of solvent. Often the interaction of water and ions present in solution influence the biological process being studied. Simulations can be performed on a molecular system without any solvent molecules included, an *in vacuo* simulation; however, simulations of this type are known to have serious deficiencies in the quality of the structures produced when compared to solution structures. Solvent effects can be included either explicitly by the presence of solvent molecules in the simulation, or implicitly by adding additional terms to the interaction potential function. Implicit solvent models treat the molecular system in atomic detail but represent the effects of solvent–solute interactions by either using an empirical correction based on an atom's accessible surface area and its solvation parameter (ASA methods), or by applying equations describing the interaction of the molecule with a continuum solvent model such as the Generalized Born/Surface Area (GBSA) method.

Structure-based drug design

Structure-based drug design is the guided design of a drug for a specific biological target based on the known structural properties of the biological target, and/or a lead drug compound to modify the behavior of the biological target to inhibit or prevent a disease state. The aim of structure-based drug design is to predict which structural features of the drug make it interact efficiently with the biological target and, conversely, which features of the biological target are suitable for interaction with the drug. Lead drug compounds can be identified by techniques such as mass biological screening programs, previously identified successful drugs, and compound libraries. The biological targets for the drug design process can be identified in diseased states by analysis of the human genome, through systems biology approaches such as analysis of gene and protein interaction networks, or by more traditional high-throughput screening approaches. Once a biological target has been identified, the structure of the target is characterized by a technique like X-ray crystallography or NMR, as the knowledge of the target's structure guides the drug candidate selection process.

The prediction of the activity, specificity, toxicity, and drug likeness of the potential drug candidates are crucial to guiding the process of candidate selection. The physico-chemical properties of a potential drug compound are investigated through the identification of the pharmacophores, which are identifiable chemical groups responsible for the drug's biological activity. By studying the relationship of the physico-chemical properties and the pharmacophores present in several candidate drugs to their pharmacological properties and biological activity, a quantitative structure–activity relationship (QSAR) can be identified. One example of a QSAR, Lipinski`s rule of five, is used to predict the drug likeness of a candidate molecule. Poor oral absorption and/or distribution are likely to be observed in a candidate molecule when it has the following properties.

- MW (molecular weight) > 500.
- LogP (partition coefficient) > 5.
- More than 5 hydrogen-bond donors (the number of OH and NH groups).
- More than 10 hydrogen-bond acceptors (the number of N and O atoms).

Potential drug candidates can be screened quickly and efficiently for desired properties using QSAR. The search for drug candidates can proceed through several routes, by the virtual screening guided by lead compound and by the *de novo* design of drug molecules. The virtual screening of the 2-D structure of drug candidates can by performed by encoding the pharmacophores of a previously identified candidate drug into a query feature vector. A feature vector is a binary representation of all the pharmacophores present or absent in a molecule, where the presence or absence of one particular pharmacophore is encoded by a one or a zero, respectively, at a specific position in the vector. Libraries of candidate molecules are also encoded into feature vectors according to their own pharmacophores. Candidate molecules are selected from the library on the basis of their similarity to the query feature vector, where the similarity of two feature vectors is computed by a scoring function. An example of a frequently used scoring function is the Tanimoto coefficient, T_c, which for two feature vectors A and B is calculated by the sum of the number of bits common to both A and B, divided by the sum of the number of bits in both A or B. A simplified example calculation of the Tanimoto coefficient between two important benzodiazepine drugs, diazepam and flunitrazepam, is shown in *Fig. 2*.

$$T_c = \frac{c}{a + b + c} = \frac{A \cap B}{A \cup B}$$

a = count of bits in feature vector A **only**
b = count of bits in feature vector B **only**
c = count of bits **both** in feature vectors A and B

(a) **Compound A:**
  Diazepam (Valium)

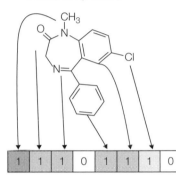

(b) **Compound B:**
  Flunitrazepam (Rohypnol)

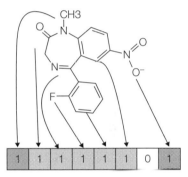

(b)

									Sum (bits)
A	1	1	0	1	1	1	0		6
B	1	1	1	1	1	1	0	1	7
Only A				1					1
Only B			1				1		2
A and B	1	1	1		1	1			5
A or B	1	1	1	1	1	1	1	1	8

$$T_c = \frac{5}{(1 + 2 + 5)} = \frac{5}{8} = 0.65$$

Fig. 2. (a) The 2-D chemical structure of diazepam (Valium), with various chemical functional groups highlighted, showing one possible way of encoding the functional groups into a feature vector. (b) The 2-D chemical structure of flunitrazepam (Rohypnol), with various chemical functional groups highlighted, showing one possible way of encoding the functional groups into a feature vector. The differences in the presence or absence of various functional groups relative to diazepam are encoded by differences in the feature vector. (c) An example comparison of the features vectors for diazepam and flunitrazepam, illustrating the calculation of the Tanimoto coefficient, T_c, from the feature vectors. For these feature vectors, the Tanimoto coefficient is 0.65.

3-D structure matching techniques are similar to 2-D techniques but they use not only the nature of the pharmacophores but also their spatial arrangement to filter libraries of compounds.

The *de novo* design of drug ligands involves the construction of a model of the binding site of the biological target and drug candidates created using this model as a guide, usually by the docking of drug ligands into the model of the biological target. In fragment-based construction methods, chemical fragments are docked with the biological target and scored using some function.

High-scoring fragments identified from the docking can be both chemically synthesized and tested, or virtually linked, and further rounds of docking and scoring undertaken. Techniques such as NMR can be used to identify fragments that will definitely interact with the biological target, so as to guide the docking process. In incremental construction techniques, a potential drug candidate is split into pharmacophore fragments. These fragments are sequentially added and built onto previously docked fragments so the candidate drug is rebuilt on the biological target. Spatial constraints, such as a specific arrangement of pharmacophores, can be used to filter and score the docked ligands. Genetic algorithms (Section I) start with set of docked target/ligand complexes and create new child complexes based on the parents and random factors, equivalent to mutation in evolutionary processes. The fitness of child solutions is tested and poorly scoring solutions are eliminated. The process is repeated a large number of times, so that over successive rounds of scoring the population of possible solutions converges towards an optimal solution.

Table 1.   A summary of popular software for performing quantum chemical and semi-empirical calculations and molecular dynamics/biomolecular simulations

Type of calculation	Name of software	Web site
Quantum chemical	GAMESS-UK	http://www.cfs.dl.ac.uk/
	GAUSSIAN	http://www.gaussian.com/
	Q-Chem	http://www.q-chem.com/
Semi-empirical	GAMESS-UK	http://www.cfs.dl.ac.uk/
	GAMESS-US	http://www.msg.chem.iastate.edu/gamess/
	MOPAC	http://www.openmopac.net/
Molecular dynamics/ biomolecular simulation	AMBER	http://www.ambermd.org/
	CHARMM	http://www.charmm.org/
	GROMACS	http://www.gromacs.org/
	GROMOS	http://www.igc.ethz.ch/GROMOS/index
	TINKER	http://dasher.wustl.edu/tinker/

As this is a very large subject area, some highly recommended textbooks are given in the Further Reading section.

M METABOLOMICS

Key Notes

Introduction
Metabolomics is the investigation of the entire range of metabolites found in biological samples. Metabonomics is very similar but focuses more on changes in metabolite profiles in response to toxins, disease, and diet.

Instrumentation
Chromatographic devices separate the compounds in the sample before they enter the detector where they generate a signal. In spectral analysis the intact sample can be placed in the instrument, where it is scanned to form the signal.

Data formats
There are two main data types: spectral and chromatographic. The first has values continuously along a wavelength axis; the second generates values where peaks occur along the chromatogram.

Chromatographic data
Can be 2-D with signal intensity vs. time. Some detectors produce 3-D data by repeatedly scanning the sample as it leaves the separating column. Peak area data are extracted from the chromatograms by integration.

Spectral data
Spectra can be processed to give peak areas for signals occurring at specific wavelengths, or the whole profile can be used, with data lined up by wavelength rather than peaks at a specific wavelength.

Data standards
For metabolomics, this has lagged behind developments in other -omics areas, but the Metabolomics Standards Initiative is now effectively developing the necessary standards.

Related sections
Data standards and experimental design (K5)   Proteomics techniques (L1)

Introduction

This section looks at the final class of biological molecule, namely the metabolites. In keeping with the terminology, the **metabolome** is the entire set of metabolites found in cells or other biological samples. As is the case with proteomics, the metabolome in practice refers to the entire set of metabolites **that can be detected and characterized with current instruments**. In some areas of the literature, the word **metabonome** appears. This refers to the changes in metabolite profiles in response to exposure to toxins, disease, and diet. Historically it sprang more from medical research than molecular biology and sometimes has greater emphasis on the use of nuclear magnetic resonance (NMR).

Instrumentation

Chromatographic equipment relies on a column to perform the separation of compounds before they enter the detector and result in a signal. In gas chromatography compounds are carried by a stream of gas into the detector as the oven surrounding the column heats up; separation is mainly dependent on the boiling point of a compound. With liquid chromatography it is the affinity of the compound for the solvent relative to column packing that affects the time required for it to exit the column and reach the detector. Separation in this case relies on the compound's polarity (hydrophobicity).

In NMR, the sample is placed in a magnetic field and a pulse of radio frequency is applied to excite the appropriate nuclei (e.g., carbon, hydrogen, etc.) of the compounds. This response is transformed then to give the final signal that forms the spectrum. Different combinations of radio pulses and intensities can lead to some very sophisticated structural analyses, and the more powerful the magnet the better the resolution of equipment.

Data formats   There are two main types of metabolomics data. Chromatographic data, where a series of peaks occur with time as molecules leave a chromatographic device and are detected (*Fig. 1*), or spectral data. Spectra are generated when a sample is scanned to determine its response at specific wavelengths; for example, in NMR analysis (*Fig. 2*).

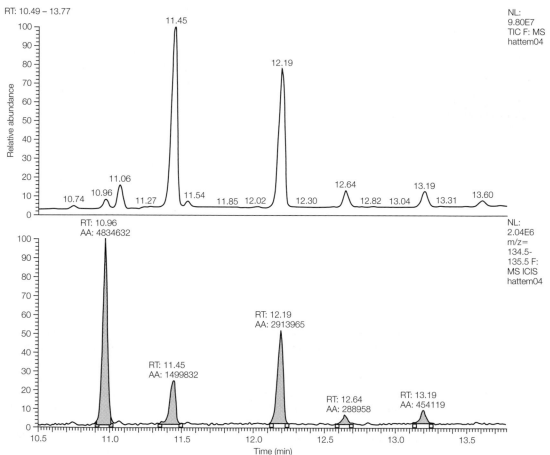

Fig. 1. Part of a chromatographic trace showing fatty acid esters eluting from a gas chromatography column and detected using a mass spectrometer. The upper trace shows the signal generated by summing the signal for all ions formed in the mass spectrometer (TIC), while the lower trace shows the signal for the ions at one mass, m/z 135, after integration to determine peak areas (AA:).

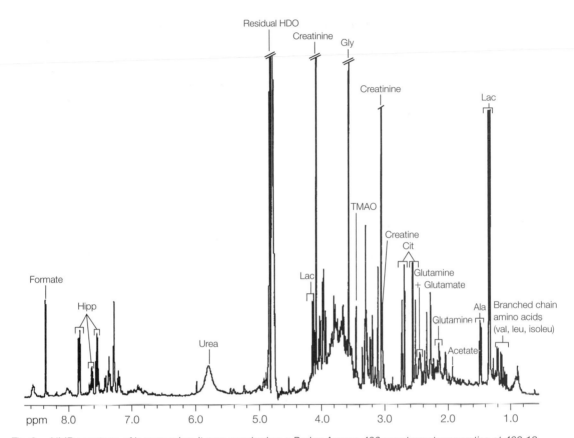

Fig. 2. NMR spectrum of human urine. It was acquired on a Bruker Avance 400 spectrometer operating at 400.13 MHz 1-H observation frequency, in c. 10 min using 500 μl of urine mixed with 100 μl of deuterium oxide, which acts as a field frequency lock. Certain peaks have been labeled, and abbreviations mean the following. Ala – alanine, Cit – citrate, Gly – glycine, HDO – hydrogen deuterium oxide, Hipp – hippurate, Lac – lactate, TMAO – trimethylamino N-oxide. It is not unusual for a single metabolite to generate more than one peak as they correspond to the environments in which different nuclei are found.

In NMR the wavelengths occur and are measured at specific frequencies, and the signal at each frequency can be compared across the samples, whereas the elution of peaks (retention time) from liquid or gas chromatography equipment will depend on local conditions such as the state of the column. Peak areas for a specific compound need to be lined up with each other for further comparison. This process should take into account the fact that there will be some variation in retention time from run to run.

Chromatographic data

Data from simple chromatography systems such as a high performance liquid chromatography system equipped with an ultraviolet (UV) light absorption detector will produce two-dimensional data. Here the signal for UV absorption increases or decreases with time to produce a series of peaks very much like the upper trace in *Fig. 1*. The data extracted from each chromatogram are the retention time for each chromatographic peak and its peak area. The differences in peak areas occurring at each retention time then can be evaluated.

More complex detectors such as mass spectrometers generate three-dimensional data. In addition to the time and intensity axes, there is the further dimension of the mass of the ions generated by the molecules entering the detector. This is dependent on a compound's atomic composition and molecular structure. Data values for chromatographic peaks can be generated by the integration of the summed signal for all of the ions formed in the detector known as the total ion chromatogram (TIC) (see upper trace in *Fig. 1*). This effectively uses the mass spectrometer as a complex device for creating a 2-D signal. Greater specificity and enhanced sensitivity can be obtained by focusing the analysis on specific ions characteristic of compounds at specific retention times. This can be seen in the example in *Fig. 1*. It would be possible to determine the peak area of the compound eluting at 10.96 min using the TIC trace. However, it is only 0.1 min away from another peak, and the two could be confused if peak alignment were based solely on retention time, which can vary from run to run. The lower trace shows the signal for only one of the ions monitored (m/z 135). This results in a much stronger signal for the peak at 10.96, making it more likely that the correct value will be recorded for the compound.

The final outcome of this process is a table of peak area values for compounds eluting from the chromatographic system at different times, which then can be processed further.

Spectral data

Simple spectral data are 2-D in format, consisting of signal intensities at a range of wavelengths collected by UV-visible, infrared, or NMR analysis (*Fig. 2*). Data can be processed in a similar way to the 2-D chromatographic data, to produce peak areas for signals occurring at specific wavelengths, which then can be tabulated. Alternatively, the entire spectrum can be tabulated for further processing with each data column corresponding to the signal intensities at a given wavelength. This approach is ideal for screening experiments where the target compounds and their wavelengths are unknown, while the processing of spectra to generate peak area data can focus on specific features of the spectral profile.

Data standards

As for transcriptomics and proteomics, the raw output from these experiments is a matrix of numbers. Their real value resides in the meta-data that are associated with them, such as which metabolites the different peaks correspond to. However, the conventions for recording and laying out these meta-data have lagged behind those for transcriptomics (MIAME and MAGE, Section K5) and proteomics (MIAPE, Section L1). At the time of writing, there are several initiatives working in parallel: MIAMET, ArMeT, SMRS, and the Metabolomics Standards Initiative. It is likely that the latter will integrate and develop the other activities further (http://msi-workgroups.sourceforge.net/)

N1 SUPRAMOLECULAR ORGANIZATION

Key Notes

Biological structures at different physical scales

Cells are the basic units of life. They contain a range of multi-macromolecular complexes and organelles, including the nucleus, mitochondria, and Golgi bodies. In multicellular organisms, cells have specialized functions that form tissues and organs.

Anatomical and medical classification schemes

The different physical scales at which these biological structures are found lend themselves to ontological classification schemes. MeSH terms, SNOMED, and UMLS are the main medical ontologies used, and others are being developed continually for the main organisms studied in biology.

Related sections

Data and databases (D)

Biological structures at different physical scales

The earlier Sections J–M focused on different types of biological molecules. This and the following two sections look firstly at how these interact to form biological structures (this section), then biochemistry (Section O), and finally physiology (Section P).

The basic unit of life is a cell. It consists of a surface bilayer of phospholipid molecules (**plasma membrane**) forming a bag in which many proteins, nucleic acids, and metabolites are concentrated. This vastly reduces the time needed for moving molecules around, speeding up the processes of life. This membrane has protein complexes embedded in it to regulate what may enter and leave a cell, and may have a strong outer coat (usually of polysaccharides and proteins) to stop the cell exploding as a result of osmotic shock. Within the cell, there are complexes of proteins and nucleic acids to carry out cellular processes: energy production, biochemical conversion, translation of RNA, transcription of DNA, and DNA replication. These combine together to produce the so-called characteristics of life. Eukaryotes have internal phospholipid bilayers analogous in composition to the plasma membrane (known as **vesicles**) that partition cellular functions into **compartments,** such as the **nucleus**, where genetic material is held, copied, and transcribed; **mitochondria** for energy production, and **Golgi bodies** for directing vesicles to relevant parts of the cell. **Organelles** are subcellular components seen through microscopes, which include the above compartments and large complexes such as **ribosomes** – the biological machine for translating mRNA into proteins.

Moving up the physical scale, eukaryotic organisms may be multi-cellular, in which case the cells usually have different biochemistries and physical characteristics to enable them to fulfill some specialist role; for example, **neurons** transmit electrical signals, **myocytes** (muscle cells) contract, and **hepatocytes** (liver cells) carry out a broad range of biochemistry. Furthermore, cells may group together into **tissues** that have a specific range of physiological functions. Tissues might consist of a single cell type but usually contain a variety of cells, which have a three-dimensional organization that is consistent across and often between biological species. A higher organism has various organs in a specific anatomical arrangement, and these organs may contain one or more tissues.

Anatomical and medical classification schemes

The above description of biological structures clearly lends itself to the formation of ontologies (see Section D), as they have a hierarchical organization based on physical scale – molecules **are a component of** organelles, which **are a component of** cells, which **are a component of** tissues, which **are a component of** organs, **which are a component of** organisms. For a variety of reasons it is useful to organize our knowledge of anatomy and histology. Now that this knowledge is being computer encoded, it can be used in the most sophisticated and bioinformatics analyses.

Organizing this knowledge was originally done to help classify and provide keywords for research work, particularly in medicine. One of the earliest attempts to classify biological entities was the Medical Subject Headings (MeSH), associated with **Medical Abstracts**, subsequently Medline, and now PubMed (http://www.ncbi.nlm.nih.gov/pubmed/). In biomedicine, there are two main ontologies, covering anatomical, physiological, and disease terms: SNOMED (now managed by the International Healthcare Terminology Standards Development Organization http://www.ihtsdo.org/snomed-ct/) and its derivatives, and **UMLS** (Unified Medical Language System), which is an amalgamation of 130 medical vocabularies. Previous rivalries between devotees of these two systems now have been largely resolved. Indeed, the US National Library of Medicine has a perpetual license to distribute SNOMED within the UMLS metathesaurus.

For organisms other than humans, there has been less effort and hence far less duplication. However, there has tended to be a partition between ontologies for general anatomy and embryonic development for these other species. This is especially the case for *Caenorhabditis elegans*, *Arabidopsis thaliana*, *Drosophila*, and cereal plants. For other organisms, these two categories have been integrated; for example, in mouse, Medaka-fish, and zebra-fish. The more recent plant ontology (PO: http://plantontology.org/) consortium is an attempt to unify some of the earlier initiatives in the plant biology area. The level of detail and completeness varies.

Further development of these ontologies is an active area of work, so the Open Biomedical Ontologies (http://www.obofoundry.org/) is the recommended place to find out about latest developments. The field has been concisely reviewed also by Bodenreider & Stevens (2006).

N2 TISSUE- AND ORGANISM-SCALE ORGANIZATION

Key Notes

The 3-D mouse atlas   The 3-D mouse atlas combines a 3-D atlas of development with a gene expression database.

The rat brain project   The rat brain project combines anatomical atlases with magnetic resonance imaging (MRI) data.

Models of human bodies   A number of projects include the Visible Human, Virtual Human, and Virtual Soldier model sections of human bodies.

Related sections   Computation (F)   Physiology (P)
                        Modeling and mathematical
                          techniques (H)

3-D mouse atlas   It is important in the understanding of complex systems, such as our bodies, that we consider the context of particular processes in the light of those others that influence them. Perhaps the grandest aim of all, modeling the structure of a complete human body, may seem like an ambitious goal. However, there are projects in existence today working towards this, as we shall see below. Increases in computing power and storage have accelerated the development of such complete organism projects in recent years. Often they are a combination of advanced imaging techniques, producing an accurate 3-D visualization, with physiological information and models of relevant processes.

Three-dimensional 'atlases' containing a multitude of information categories have arisen recently. These atlases typically represent geometric models of the organism, and can be populated with further data and functions. For example, a 3-D mouse atlas has been under development, as a collaboration between the UK MRC Human Genetics Unit and the University of Edinburgh (http://genex.hgu.mrc.ac.uk/intro.html), since the early 1990s. Managing the information on the genetic changes that are produced during the development of a mammalian embryo is crucial for understanding the genetic interactions that are producing those changes. Time and space information is key to understanding such interactions, and so the 3-D mouse atlas also contains anatomical information of what the observed structures are, so that it can be closely linked to a gene-expression database, the GXD. One of the key goals and challenges of this work is the easy transfer of data, first into the atlas, and then to communicate information with the GXD. The ability to view and dissect 3-D models of mouse embryos is one output the project has produced. This allows interactive sectioning of the model, allowing a user to view the anatomical components. An internet accessible database of standard anatomical nomenclature has been produced also; this was seen as the first step in providing the ability to study the genetic basis of mouse development, where the accurate anatomical context of gene expression is essential.

The rat brain project

Combining imaging and anatomical data can aid with the analysis and visualization of MRI studies. There has been recent interest in MRI studies of rat brains, driven largely by the lure of translating knowledge to human brain studies. Anatomical atlases have been created by cryosectioning frozen samples, organizing the sections into a volume, and registering or aligning the sections. MRI data can be co-localized with existing volumetric anatomical atlases, therefore permitting the use of information already normalized to the atlases with MRI data. The combination of such techniques has potential advantages such as improved segmentation of areas of interest, which are less subjective than manually defined areas.

Models of human bodies

It is not just small mammals that are being modeled as complete organisms. Virtual human models aim to increase understanding of how the body as a system functions, and to simulate responses to stimuli such as drugs or injury. Such a large-scale simulation involves the tight integration of many smaller-scale models, such as models of individual organs, and even cells and neurons. One such project is the Virtual Human Project at Oak Ridge National Laboratory in Tennessee (www.ornl.gov/sci/virtualhuman). Again, the aim is to integrate physiological models with image data, specifically to integrate image model data from the National Library of Medicine's Visible Human Project (www.nlm.nih.gov/research/visible/visible_human.html) with physical models describing tissues and models of gene function.

Integrating the various models gave rise to computational issues, as there simply was not enough processing power on personal computers. Therefore, a client-server system was used, allowing simulations run on client desktop computers to perform calculations on high performance servers, and send the results back to the client. An example of just such a system as used by the Virtual Human project is the NetSolve software (http://icl.cs.utk.edu/netsolve). Moving the computations to a remote computer allows users to use a lightweight front end, such as a web browser, to control the simulations and view results. The system could therefore be available wherever network access is available and a suitable web browser is installed.

Another, practically grounded virtual human project is the Defense Advanced Research Projects Agency (DARPA) Virtual Soldier Project (http://www.virtualsoldier.us). The aim of this project is to diagnose battlefield injuries by modeling affected areas of the human body. The simulator will range from modeling cell level events such as cell death, to organ scale reactions such as lung collapse. It aims to integrate anatomical torso models with models of heart function, and models of thorax anatomy and physiology. These will exhibit respiratory and cardiac activity. The geometry will be able to model the impact and exit effects that result from being hit by a projectile, such as a bullet. The visualization of the model is planned to be free-warping, with the hope of being able to overlay the predicted model over actual video of a wounded person.

These examples illustrate that organism scale models, and the techniques to produce such models like combining image data and physiological models of different scales, indeed are being developed, and have clear practical applications, particularly in the field of medical research.

01 NETWORK STUDIES OF METABOLISM

Key Notes

Introduction
Biochemistry and metabolism are the fundamental processes of life. However, making the study amenable to bioinformatics has been plagued by issues over the non-standard terminology of metabolites and enzyme reactions.

Holistic networks
From genome annotations and other data, it is possible to reconstruct networks representing the whole of metabolism of an organism. Analysis of such networks can reveal much information about alternative reactions and especially gaps in our knowledge. Their main use is as a reference.

Subgraphs
From biomolecular profile data, it is possible to generate subsets from the holistic network of an organism to show the relevant parts of metabolism pertaining to the data.

Graph comparison
This is an approach to comparing metabolism in different states. This could be healthy versus disease, or a pathogen versus its host. Apart from helping to understand disease processes, such analyses can help to identify drug targets.

Elementary modes analysis
Any given path between two metabolites is an elementary mode. Analysis of alternative paths can help to understand diseases and identify potential drug targets.

Stoichiometric matrices
These represent both the topology and stoichiometry of metabolic networks. They have applications in helping to understand cellular metabolism better and showing which reactions to modify through biochemical engineering for over-production or bioremediation.

Related sections
Essentials of physics (C)
Graph theory and its applications (H2)
Ordinary differential equations and algebra (H3)

Analyzing differential gene expression (K3)
Use of calculus and algebra (O2)

Introduction
Biochemistry and metabolism (i.e., the combination of multiple biochemical reactions) are the smallest physical-scale processes in biological systems. Their origins lie in work done during the late nineteenth and early twentieth centuries but they did not begin to blossom until the 1950s and 1960s. The Enzyme Commission was established during this period to classify enzyme reactions as they were discovered. They developed a simple, hierarchical, four-segment code

(known as the EC number) to label the reactions they defined. Had the original Commission members realized the impact that computers would have in future decades they might have been more careful in the design of their scheme and the use of biochemical terminology. Despite its now obvious shortcomings, the EC classification scheme is still widely used.

Table 1 lists the main databases of enzyme reactions and kinetics. Apart from non-standard terminology for metabolites and inhibitors, they provide a useful resource for constructing biochemical pathways and networks. The terminology issues fall into three types.

1. Synonyms – for example **2-aminopropionate** and **alanine**;
2. Close homonyms – such as **phenylacetate** (a carboxylic acid) as opposed to **phenyl acetate** (the aromatic ester of phenol and acetic acid);
3. Generic terms – such as **an alcohol** or **a fatty acid**.

The third of these can be especially challenging because it is not always obvious when terms are generic. For example, **alanine** is well known to molecular biologists, but it is actually a generic term that might refer to the L-stereoisomer (found in many proteins), the D-stereoisomer (found in some bacterial cell walls), and the D-/L-racemic mixture (sold by some biochemical manufacturers). Enzymes generally react with one specific isomeric form of a biochemical, so it may be important to know which it is.

Table 1. Enzyme databases

Name	URL	Comments
BRENDA	http://www.brenda-enzymes.info/	This is a large database that uses EC codes as the primary key. Apart from the standard reaction equation, extra details include alternative substrates and inhibitors, kinetic constants, enzyme purification and reaction buffers, and much more. One snag is that relying on EC codes, novel reactions are parked in the unassigned table until a code has been defined after which the details have to be moved within the database
EMP	http://www.empproject.com/	This database contains many of the same details as BRENDA. However, it was not tied to EC codes. For many years it was the database of choice, and was the primary reference for the first metabolic reconstructions from genome annotation data. At the time of writing, its development has stopped
ENZYME	http://ca.expasy.org/enzyme/	This is the online version of the Enzyme Commission database. For each EC code, there is a reaction equation, free-text notes, and Uniprot codes of proteins thought to carry out that reaction
KEGG	http://www.genome.jp/kegg/	This links genes with EC codes and reactions, but contains more alternative substrates and products which are derived from the generic reactions maps
SABIO-RK	http://sabio.villa-bosch.de/	This is a web-accessible database containing similar information to BRENDA, but aims to make life easier for those producing dynamic models of biochemical systems
UMBBD	http://umbbd.msi.umn.edu/	This database focuses on pertaining to xenobiotic degradation and bioremediation

Attempts to harmonize these terminologies are underway. The KEGG database aims to be internally consistent through its use of **KGML** (KEGG Markup Language), though mapping of generic to specific terms is still incomplete. The other main activity is **CheBI** (Chemical Entities of Biological Interest).

The next scale of complexity involves the linking of these metabolites into chains of biochemical reactions. From the early studies of the fate of radio-labeled metabolites (i.e., the other metabolites in which the radio-label was found over a period of time), the concept of **biochemical pathways** emerged, along which the atoms comprising biochemicals flowed. This idea became so entrenched that many biologists have viewed these pathways to be biological units in their own right. Indeed some groups have constructed databases of metabolic pathways (see *Table 2*), but these are fraught with difficulties. Given the highly interconnected nature of metabolism, there might be disagreements over where a given pathway starts and ends, which pathways connect together and how, and aberrant or non-standard pathways arising in disease or rare physiological states might be overlooked entirely. Furthermore, in otherwise well established pathways, some organisms (especially microbes) have variant metabolic intermediates. This explains why the metabolic pathways database (MPW) had over a dozen pathway records for glycolysis!

The alternative approach initially adopted by KEGG was to have a set of generic pathway-charts, which were then color-coded on the basis of whether or not a given enzyme was thought to be present. This approach has some anomalies. For example, the chart for glycolysis extends down to ethanol, which may be true of yeast but usually is not considered to be the case for mammals. The charts are also weak on defining connections between metabolites on one chart and another. However, KGML circumvents some of these issues.

Table 2. Pathway databases and holistic networks

Name	URL	Comments
BigG	Duarte *et al.*, (2007)	A manually annotated network of human metabolism from Palsson's group, not easily compatible with anything else
BIOCYC	http://biocyc.org/	Links genes, enzymes and metabolites. Has many species-specific subsets, which can be downloaded in standard formats. Useful
GeneGO	http://www.genego.com/	A proprietary database primarily of human metabolism
Ingenuity Pathways Analyst (IPA)	http://www.ingenuity.com/	A proprietary, manually curated database encompassing metabolic, signal transduction, and gene regulation of a variety of species, but using humans as the principal reference. Good but expensive
KEGG	http://www.genome.ad.jp/kegg/	Best known for its color-code generic pathways, but complete reaction sets to construct holistic networks can be downloaded
Pathway studio	http://www.ariadnegenomics.com/ products/pathway-studio/	A proprietary database similar to IPA, with data collated more by text mining
Reactome	http://www.reactome.org/	A highly curated database of specific biological processes with excellent links to all other relevant databases

Holistic networks   The above issues have led some groups to attempt to capture all the potential reactions in an organism into one very large graph. Since they attempt to represent the whole of metabolism, sometimes they are referred to as **holistic networks**. The most holistic network would include all gene regulation and signaling networks as well. Progress towards these networks can be achieved by using genome annotation data as the initial guideline for constructing initial large networks. These can later be refined by looking for and filling in gaps, caused by incomplete genome annotations, some reactions taking place spontaneously or mediated by mineral ions, peroxides or free radicals.

The literature contains many examples of these networks shown in their entirety, though often this is not very useful as there are so many nodes and edges that node labels are illegible and it is almost impossible to trace any given path through the network. The main use of these networks is as a reference, which can be interrogated in a variety of ways. The four most common are as follows.

1. For a given enzyme/metabolite, which others are some specified number of reaction steps away?
2. What are the metabolites that connect only to one enzyme? These could genuinely be essential substrates or fermentation products, but are more often an indication that reactions are missing.
3. Which parts of the network are disconnected from the main network? These are sometimes isolated activities, such as toxic compound or metal transporters, but may also indicate problems with the network arising from overlooked enzymes or errors in metabolite terminology (see above).
4. Do members of a subset of metabolites or enzymes (from some biomolecular profile experiment such as in Section K3) lie close together on the network? If so, what does that sub-network look like?

The most widely used metabolic networks of this type are listed in *Table 2*.

Subgraphs and comparison   The above networks are valuable as a reference but, just like databases, it does not usually help to look at the entire dataset in one go, rather subsets/subgraphs can be extracted which correspond to particular aspects of the overall metabolism of the organism. This approach can highlight alternative pathways specific to particular physiological or disease states. Discoveries can also be made by graph comparison. For example, comparison of the human network with that of a pathogen can help to identify new drug targets. This approach can show also how waves of changes in the network take place in cells in response to stimuli. A prime example is leukocyte stimulation in the inflammatory response. Even an analysis of general network properties (hubs, between-ness nodes, cliques/modules, etc., see Section H2) can yield useful information.

Elementary modes analysis   A more subtle investigation of metabolic networks involves the search for alternative pathways through a holistic network between defined pairs of metabolite nodes. In this context, a given path is known as an **elementary mode** (of metabolic flux). Different modes may predominate depending on fluxes elsewhere in the network, or changes in gene expression leading to substantial changes in either the levels or kinetic activity in one mode versus another. Perhaps the most common example is the pentose phosphate shunt (see *Fig. 1*).

Elementary modes analysis can also be used to identify potential drug targets against infectious agents. Where the latter lack alternative elementary modes that are present in human metabolism, they are dependent upon a specific pathway,

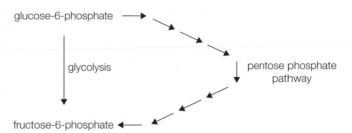

Fig. 1.   Elementary modes. Through a series of sugar inter-conversions, molecules of glucose-6-phosphate can connect up to fructose-6-phosphate by an alternative route (elementary mode), through the pentose phosphate pathway, from that of glycolysis.

whereas humans are not. So, developing drugs against enzymes in that pathway is likely to interfere with the metabolism of the pathogen leaving human cells unaffected. Software to carry out such analysis includes METATOOL and YANA.

Stoichiometric matrices

The first step towards a quantitative understanding of metabolism is through the construction of so-called **stoichiometric matrices**, in which one dimension of the matrix (usually the rows) corresponds to the reactions, and the other (usually columns) to the metabolites. Section H2 outlined how a hypothetical metabolic cycle (depicted in Section H2, *Fig. 2*) can be converted into a stoichiometric matrix, duplicated in *Table 3* for convenience. As a reminder, the first row corresponds to the forward reaction of enzyme E1 (labeled E1f), in which 1 molecule of M4 and 2 molecules of M5 are consumed to produce 1 molecule of M1, denoted by -1, -2, and 1 in the respective matrix elements. It is also clear from this table that most elements are empty. Hence, it would be wasteful of computer memory to store this **sparse matrix** as a conventional 2-D array. In practice more complex data structures are employed that use combinations of arrays and hashes.

Table 3.   Stoichiometric matrix

	M1	M2	M3	M4	M5
E1f	1	0	0	-1	-2
E1r	-1	0	0	1	2
E2f	-1	1	0	0	0
E2r	1	-1	0	0	0
E3	0	-1	1	1	0
Net	0	-1	1	1	0

One can draw various conclusions from analysis of these matrices. First, summing the values of the columns results in the net reaction for the system. At first glance, the result might seem counter-intuitive, but this net reaction actually corresponds to the sum of the irreversible reactions. In this example, the net reaction is M2 converts to M3 plus M4, the only irreversible reaction present. To comply with the Law of Conservation of Mass (see Section C), the mass (and charge if it has been recorded in the matrix) should be equal on both sides of the equation. If it is not, then either the source data and/or the matrix construction are at fault and need to be corrected.

Having satisfied oneself that the matrix is correct, it is straightforward to discover both the essential substrates (needed for the cell to survive, such as aromatic amino acids in humans) and fermentation products. This is done by looking at the columns to see which have only negative and positive numbers respectively. In *Table 3*, M3 fits the description of a fermentation product. However, one should not draw conclusions from the network naively. The early matrices derived from annotations of the *E. coli* genome appeared to indicate that X-ray-damaged DNA is an essential substrate. Of course, this is not the case, because the bacterium grows better in the absence of X rays. Rather, it shows that when *E. coli* DNA is damaged somehow in this way, it is repaired by an irreversible reaction. These same matrices also declared certain amino-acyl tRNAs to be fermentation products because the equations for protein synthesis had not been specified adequately.

Stoichiometric matrices also give the first indications of comparative reaction rates. If there are many more reactions consuming rather than producing some particular metabolite (e.g., ATP), then to ensure homeostasis the enzymes generating the metabolite must be working faster. However, the main value of these matrices lies in their use for biochemical engineering.

Let us suppose that we wish to over-produce some metabolite, at commercially viable quantities. Simply elevating levels of the enzyme responsible for its synthesis is no guarantee of higher levels, because other reactions in the cell may either continue to limit supply of precursors or consume extra metabolites as fast as it is produced. If an enzyme in the synthesis pathway is known to be slow then elevating its levels might result in higher metabolite levels. However, a stoichiometric matrix immediately shows us which reactions are consuming the metabolite and should be knocked out or retarded by genetic constructs. In the case of reversible reactions, alternative enzymes that mediate an irreversible reaction can be sought. Of course, the inverse might be sought also; that is, developing strains of organisms that consume some given metabolite to very low levels. This approach has potential applications in bioremediation of contaminated land.

Biochemical engineers might also carry out other analyses. With regard to over-production, one can look at other reactions on the pathway(s) between central metabolism and the metabolite of interest. There may be side reactions consuming an intermediate that can be blocked, thereby diverting more material to the desired metabolite. For example, a patent published by Joe *et al.* (2002) describes a way to elevate glutamate levels in tomatoes by blocking a side reaction, as shown in *Fig. 2*.

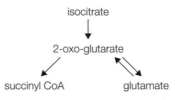

Fig. 2.   A side reaction on the pathway to glutamate synthesis. From central metabolism, isocitrate converts to 2-oxo-glutarate and then, by a reversible reaction, to glutamate. However, 2-oxo-glutarate can also convert to succinyl CoA through reactions that are part of the TCA cycle. Blocking the latter reactions stops this drain on 2-oxo-glutarate, while still allowing succinyl CoA to be produced by the other half of the TCA cycle. Thus, glutamate levels increase because a precursor is not consumed by other reactions.

Cellular metabolism is exquisitely designed to keep metabolite levels low, and one of the mechanisms to do this involves **feedback inhibition**. This occurs when a metabolite near the end of a pathway can bind to an enzyme near the start of the pathway to (partially) inhibit its activity. If the concentration of the metabolite is raised for some reason, then flux through the pathway slows by feedback inhibition to stop it increasing any further and allow the metabolite level to decrease through other reactions. One might think that making the enzyme insensitive to such inhibition will increase metabolite levels, but this is often not the case. Such inhibition acts to maintain levels at whatever they should be, rather than determine what they should be. Absolute levels are determined by the dynamics of the cell and the controlling reactions can be discerned using metabolic control analysis (see Section O2).

However, blocking feedback inhibition could be effective if the feedback inhibitor can be kept low by diverting its excess to some other cellular compartment. A clinical example of this relates to the genetic predisposition of some people to gout, a disease caused by precipitation of uric acid in the joints (especially of feet,

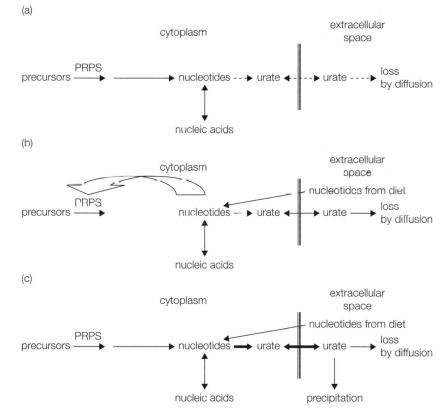

Fig. 3. Misregulation leading to gout. (a) For people on a low-nucleotide diet, these nucleotides are synthesized from a pathway that involves the enzyme PRPS. Under these circumstances only a trace of urate is produced (depicted by dotted arrows), which is lost by diffusion and excretion through the kidneys; (b) when a person's diet is rich in nucleotides, these come into cells and block nucleotide synthesis by allosteric regulation, while urate synthesis increases to remove excess nucleotides; (c) in the case of people lacking the allosteric regulation of PRPS, intracellular nucleotide synthesis continues, even more urate is produced and excreted from cells (depicted by thick arrows), which stops metabolic control taking effect, but this excess extracellular urate is so high that it precipitates out as well as diffuses away. It is the urate precipitate that causes the symptoms of gout.

see OMIM:300661). *Fig. 3* shows what is happening. In most of the population, the enzyme phosphoribosyl pyrophosphate synthetase (PRPS), at the start of nucleotide synthesis, is subject to allosteric feedback so that the nucleotides and uric acid levels are kept low. Intracellular metabolic control only detects low levels. Some people have lost this feedback control mechanism. Hence, metabolic control does not detect any problem because the flow of uric acid out of the cells and its subsequent precipitation lead to a large flux through the pathway while maintaining low intracellular nucleotide levels. The only option for such people is to maintain a low nucleotide diet.

Finally, where efficient production of biomass is the objective, then the matrix can be used to identify the energy-consuming reactions: those involving nucleotide triphosphates and reducing hydrogen carriers such as NADH and FADH. It is sometimes possible to substitute these reactions for less energetically demanding alternatives. This approach has been successfully applied in a commercial setting.

O2 USE OF CALCULUS AND ALGEBRA

Key Notes

Flux balance analysis	This approach can be applied to metabolic systems in a steady or otherwise linear state. Owing to this simplification, it can be employed to represent the entire metabolic network of cells.
Differential equations	Dynamic behavior is usually modeled as systems of ordinary or sometimes partial differential equations. They have a variety of uses for gaining a quantitative understanding of changes in metabolite levels, and also can be used to represent regulatory mechanisms, but require kinetic parameters that might be difficult to measure. Other modeling approaches also can represent changes in enzyme levels.
Complex behavior	Where feedback regulation is found, then the metabolic system may behave in a complex manner, finding itself in a range of different stable states. Sensitivity, phase-plane, and bifurcational analysis can all contribute to an understanding of this complex behavior.
Metabolic control theory	This is a laboratory technique capable of uncovering which enzymes in a pathway are controlling flux and absolute metabolite levels. It is analogous to sensitivity analysis.

Related sections	Ordinary differential equations and algebra (H3)	Metabolomics (M)
	algebra (H3)	Network studies of metabolism
	Advanced modeling techniques (H4)	(O1)

Flux balance analysis

Cellular systems in a steady-state have constant enzyme reaction rates. Because of this simplicity, it is possible to model all the reactions in a cell simultaneously. This technique is known as **flux balance analysis** (FBA), and falls into the more general category of **constraint-based methods**. It entails converting a cell's stoichiometric matrix (see Section O1) into a form where reaction rates can be either calculated from metabolomic profile data (see section M) or applied to quantify metabolite levels. Various software tools have been developed for this, notably CellNetAnalyser.

In view of these steady-state conditions, these constraint-based models can be remarkably predictive. Shortly after the first bacterial genome was sequenced, Palsson's group constructed FBA models of this organism from genome annotation data and rudimentary data on their biochemical composition. These models predicted steady-state growth rates within the error margins of the laboratory measurements for a broad range of culture media and growth conditions. They only failed when catabolite repression (of gene expression) took place. This is because the models took no account of gene regulation. These approaches have subsequently been applied to other bacteria, yeast, mitochondria, and erythrocytes.

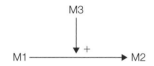

Fig. 1.   Allosteric reaction. An enzyme, which converts metabolite M1 into metabolite M2, is allosterically activated by metabolite M3.

Differential equations

The dynamic behavior of individual reactions can be represented algebraically in various ways, depending on what kinetic parameters are known or can be calculated. With simple rate constants, these can be converted into systems of ordinary differential equations (ODEs) as shown in Section H3. They can be refined in various ways; for example, reaction rates can be modified to take into account allosteric regulation (feedback inhibition or activation) using a so-called **Hill function**. Suppose we have the reaction shown in *Fig. 1*. If it were not subject to regulation, its rate equation would take the form

$$\frac{d[M2]}{dt} = K_c \times [M1]$$

where K_c is the rate constant. A Hill function replaces the right-hand side with the following

$$\frac{d[M2]}{dt} = K_{max} \times \frac{[M3]^n}{HM^n + [M3]^n} \times [M1]$$

where K_{max}, HM, and n are, respectively, the maximum rate of the enzyme, concentration of M3 causing half-maximum kinetics, and Hill coefficient. The latter is usually an integer corresponding to the number of M3 molecules that bind to the enzyme. The higher the Hill coefficient, the steeper the curve, as shown in *Fig. 2*. The Hill function for allosteric inhibition has the modified form

$$\frac{d[M2]}{dt} = K_{max} \times \frac{1}{1 + ([M3] / HM)^n} \times [M1]$$

For many years it has been common biochemical practice to represent enzyme behavior in terms of **Michaelis-Menten** kinetics. When enzyme (E) and substrate (S) form a complex that converts to enzyme and product, this takes the general form

$$v = \frac{Vmax \times S}{Km + S}$$

where v is the rate of formation of product, **Vmax** is the maximum reaction rate of the enzyme, only theoretically possible at infinite substrate concentration, and **Km** (the Michaelis-Menten constant) is the substrate concentration at which the initial enzyme reaction rate is half that of Vmax. *Fig. 3* outlines how these constants might be determined. This converts the earlier rate equation to

$$\frac{d[M2]}{dt} = \frac{Vmax \times [M1]}{Km + [M1]}$$

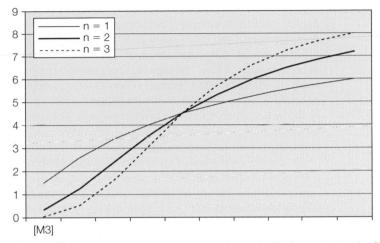

Fig. 2. *Hill function characteristics. This plot shows the kinetic constant value for different Hill coefficients (n), where Kmax = 9, and half max (HM) concentration of M3 = 0 5. The higher the value of n, the more pronounced the sigmoidal shape of the curve.*

Graphical representation

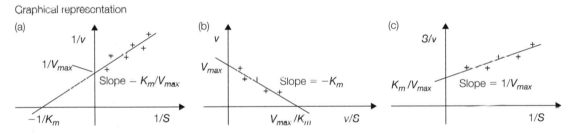

Fig. 3. *Determination of Michaelis–Menten kinetics. For a series of different substrate concentrations (S), the initial rate of formation of product (v) by the purified enzyme is measured. Simply plotting v against S would lead to a hyperbolic curve from which Vmax and Km cannot be determined accurately. However, if the numbers are manipulated in different ways, then far more reliable values can be obtained. The two most common approaches can be attributed to (a) Lineweaver & Burk, and (b) Eadie & Hofstee. In the former, the reciprocals of the numbers are taken, whereas for the latter, v is plotted against v/S. The axis intercepts lead to the Michaelis-Menten parameters as shown. (c) depicts a Hanes and Woolf plot.*

There are modifications of these equations where inhibitors are involved, the specific equation depending on the form of inhibition, which may be competitive, uncompetitive, non-competitive, mixed, or partial.

Dynamic models of pathways can be constructed as a system of ODEs, as outlined above. Where diffusion is a significant aspect of the study; for example, when the biochemistry is embedded in some environment such as a fermenter, then reaction-diffusion equations can be used (see Section H4). Different mathematical approaches might be needed when enzyme levels and changes of gene expression need to be modeled. This is because enzyme concentrations are often several orders of magnitude lower than their substrate and product concentrations, and enzyme turnover is several orders of magnitude slower than the reactions they catalyze. In such situations **S-systems** are a common option (Torres and Voit, 2002).

Complex behavior   Any systems that contain cycles or feed-back (or indeed feed-forward) loops may behave in a complex, non-linear way. This is also true of metabolism. A mathematical model of some part of metabolism can be used to explore its behavior under a broader range of circumstances than could be determined experimentally. Thus, biochemical engineers might carry out **sensitivity analysis** – the systematic perturbation of every parameter in turn – on a model to reveal which parameters have the biggest effect upon some specific property of the system, such as rate of consumption or over-production of some defined metabolite. This same technique is used, of course, to determine which parameters need more careful laboratory determination, to make the model more realistic.

On a more sophisticated level, there are various things that can be discovered from **phase-plane analysis**. This involves plotting the value(s) of one variable against defined values of another, and is most often achieved by running a simulation of the model and simply noting what the different values are at the different time points. The first provides indications of the stability or otherwise of the system. For a simple pathway with no regulation, the loss of substrate would match the rise in final product. The pathway would reach equilibrium when all the reactions have reached their thermodynamic equilibrium, and the absolute concentration of product would depend on the initial substrate concentration. However, in complex systems the plots on phase planes can be a variety of shapes including S-bends, ellipses, and spirals. Where there is negative feed-back, then the final product concentration is likely to be lower, but there may be oscillations along the way if there is some form of delay in the feed-back.

A more formal branch of mathematics known as **bifurcational analysis**, whose output is represented in phase planes, can reveal both the potential occurrence of such behavior, and the likelihood of switching from one metabolic state to another. This type of analysis can reveal unexpected results, such as one parameter value might occur in two or more steady-states, say the rate of ATP consumption in a pathway at different metabolic flux rates. Such work leads into studies of **system stability**; that is, whether or not a perturbed system will return to its original or some other fixed state. However, this falls outside the scope of this work.

Metabolic control   As already mentioned, enzymes subject to feed-back inhibition do not control
theory   pathway fluxes or final metabolite levels. **Metabolic control analysis** is a technique developed during the 1970s and 1980s to identify the controlling points of metabolic pathways. Its origins were controversial, as it flew in the face of the prevailing hypothesis that metabolic flux was controlled by the enzyme whose reaction was always **rate-limiting**. In metabolic control analysis, each enzyme (or part of a pathway) is perturbed in turn and the change in metabolic flux at the end of the pathway measured. From this, a measure of each enzyme's control on the pathway can be evaluated. Thus,

$$\text{Flux control coefficient for enzyme n } (J_n) = \frac{\text{quantitative change in flux caused by enzyme n}}{\text{sum of changes by all enzymes}}$$

By definition, the sum of flux control coefficients for all enzymes in a pathway equals 1.

Several striking features arise from such studies. The first is that all enzymes exert a little control ($J = 0.05 – 0.2$). A single enzyme can be considered to control a pathway only when J is above ~0.6, and a given pathway might not have any specific controlling enzyme. Control enzymes tend to be at the end of pathways, unlike allosteric regulatory enzymes which are usually at the start of pathways, showing that metabolic flux is driven by demand for the end products of the pathway, not the supply of precursors. Finally, the enzyme controlling a given pathway may change depending on the physiological state of the cell(s) being investigated.

Metabolic control analysis is the laboratory equivalent of carrying out a sensitivity analysis of a mathematical model. However, laboratory work can be expensive and time consuming, but may still be quicker than the laboratory work needed to determine relevant parameters to input into a mathematical model of the pathway. For those seeking further information on this, a lucid and well-written book by David Fell is recommended (Fell, 1997).

P1 PHYSIOLOGY

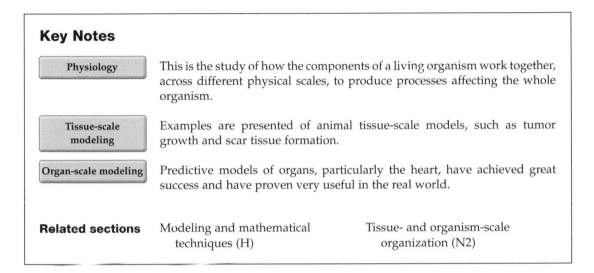

Key Notes

Physiology

This is the study of how the components of a living organism work together, across different physical scales, to produce processes affecting the whole organism.

Tissue-scale modeling

Examples are presented of animal tissue-scale models, such as tumor growth and scar tissue formation.

Organ-scale modeling

Predictive models of organs, particularly the heart, have achieved great success and have proven very useful in the real world.

Related sections Modeling and mathematical   Tissue- and organism-scale
techniques (H)   organization (N2)

Physiology

Physiology refers to the study of the components of a living organism that work together, across different physical scales, to produce life processes in the organism. Physiological modeling therefore refers to the modeling of these multi-scale components, such as organs, tissue, cells, biochemical processes, etc., and ultimately the modeling of the complete system composed of these components.

The approach of **integrative systems biology** lends itself to the development of such models to describe mammalian function from sub-cellular to tissue and whole-organ scales. Models of these systems have clear benefits in their ability to make predictions about how the tissue or organ will behave under particular circumstances. For example, a model of a heart can be used to identify how it reacts under normal and pathological conditions, and a model of tumor growth can help predict the effect of drugs. Recent increases in computational power and the amount and quality of experimental data available lend themselves to the development of mathematical models. These models are allowing quantitative descriptions and predictions of biological systems to be made in a variety of domains, some examples of which are illustrated below. Recent work has also seen the beginning of an amalgamation of these component models into whole system models, with the long-term aim of producing a complete model of a human being.

Tumor growth

Tissue-scale modeling

Tumor growth is an inherently multi-scale progress. In the spatial scale it ranges from sub-cellular processes to whole-tissue behaviors. Timescales range from sub-second for signaling pathways, to several months for tissue-scale growth processes. Modern models use approaches that cover more than one scale to make predictions about the effects of drugs on growing tumors. Models can be developed from experimental data, or based on theory alone in an effort to drive experimental studies, or via a collaborative effort between the two. The collaborative model-experiment cycle is a hallmark of integrative systems biology, as this inherently places biologists and mathematicians in an ideal position to work together on such models. The experimental data from the biologists can be used to conceive and refine models, which in turn can be used to make further predictions that can be tested by experiment (see *Fig. 1*). Such a symbiotic relationship is proving a powerful tool in the advancement of scientific knowledge.

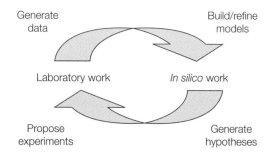

Fig. 1.   *Illustration of the 'wet-dry' cycle of collaborative, integrative research. Building 'dry' models generates hypotheses, which can be tested by 'wet' experiments in the laboratory. Results of these experiments can be fed back into existing or new models.*

Scar formation

Another tissue-scale modeling example is the formation of scar tissue. Models have been developed to predict the fiber alignment in scar tissue. Skin wound-healing consists of a number of processes such as blood clot formation and fibroblast migration culminating in the laying down of remodeled tissue, known as a scar. This scar tissue is typically of a different structure to, and is less functional than, the original tissue. The most significant difference appears to be the orientation of the fibrous matrix. The developed model consists of a multi-scale representation of the fibrous matrix (collagen and fibrin) and cells (fibroblasts), allowing a representation of the density of the fibrous protein at each location. The direction of movement and orientation of the cells in the model is determined in part by their history and by the properties of the matrix. The resulting multi-scale model predicts collagen alignment and scar formation, and provides suggestions for novel anti-scarring treatments.

Organ-scale modeling

The modeling of complete organs is a complex task because the component tissues, blood vessels, electrical networks, etc. do not act in isolation but by definition form integral parts of the complete functioning organ. Successful modeling of an organ needs to address the function and interactions of its component parts. Activities at the organ scale can influence molecular-scale activities (such as the biochemistry and gene expression), and, importantly, vice versa. In other words, these influences affect both higher and lower scales. An understanding of the tissue mechanics, signaling mechanisms, blood flow, etc., and how they interact across different scales, might be required. A grand challenge in this kind of integrative work is to model the scales in the correct amount of detail. Clearly it is not possible to model everything, as modeling all the parameters of millions of cells is well beyond the computing capabilities of current state-of-the-art technology. Therefore, the implemented model must capture sufficient information to be useful across other physical scales. Combining input from biologists, mathematical modelers, computer scientists, and engineers with modern powerful computational resources, integrative approaches allow this problem to become tractable. Models of the physical shape of the organ also become important at the organ scale. We will look at two well-studied examples of organs modeled using these approaches: the heart and the lung.

Heart

The heart is the classic example that exhibits all the complexities inherent in organ modeling. It is a very dynamic system, pumping blood around the body enabling the delivery of nutrients to the various organs. It is composed of four chambers. The two atria act as blood reservoirs for the two ventricles, which are the primary pumping mechanisms in the heart. Muscle cells within the walls of the heart contract upon initialization by an electrical charge.

The heart is a good example of the importance of multi-scale approaches to modeling. A heart beats in a coordinated fashion, illustrating that there must be some form of coordination from individual cells, which produces a large-scale effect at the organ level, while at the same time the structure of the whole organ dictates which cells contract when.

Sub-cellular models exist to model, for example, calcium-induced calcium release in the cardiac dyad. These spatio-temporal models range from the molecular scale up. Cellular models of the heart have been in development for four decades, and can represent, even now, the expression levels of various genes. Building biochemical information into cellular models has led to new and counter-intuitive predictions about function that were not obvious from existing models. The combination of well-developed cellular and sub-cellular models, modern imaging techniques, and the advent of greater computing power has allowed these models to be combined to create organ-scale models. Whole organ models can be used to model electrical activity in the heart. The spread of an electrical wave front throughout an anatomically detailed heart model has been simulated. Such electrical models can allow the prediction of electrocardiogram (ECG) traces, and are therefore of great interest to those developing new drugs. These models can influence the release of new drugs, an example of which was the drug Posicor. Model predictions were able to explain effects occurring in the ECGs of some participants in clinical trials, which had a bearing on the adoption of the new drug. These models have also been used to look at the causes of arrhythmias, energy conservation in the cardiac cycle, calcium balance, etc. Often the successes and failures of these models drive further research and discovery in modeling other systems.

Dynamic physical models of the heart have been built also. The mechanics of the pumping mechanism have been modeled using **finite element** methods. By modeling the anatomy and tissue structure, the finite element method can be used to predict how the organ will deform, predicting pressure and volume at certain regions of the heart.

Such virtual models of the heart can be used to simulate and develop real-world techniques, such as defibrillation. The equations and parameters of these models, once developed, also can be used to drive model development in other tissues. An example of this is the modeling of gastrointestinal waves in the small intestine, which, as a foundation, uses the parameters and models that were developed in the more advanced field of heart cell modeling.

Lung

High-quality medical imaging techniques have recently permitted the building of models of the complex airway, arterial, venous, and capillary networks of which the lungs are composed. The key to modeling such a system is to integrate models of the different functional sub-systems.

For organs with such complex shapes as the lungs, accurately modeling the geometry of the system is important. Geometry may affect the other sub-systems, by constraining air flow or blood supply, for example. Understanding this geometry traditionally has been achieved via anatomical studies of excised lungs. With modern imaging techniques, however, it is now possible to image the lungs on a per subject basis with the organ in a variety of functional states, as opposed to having only a few rigid casts of non-functional lungs with which to infer geometry. Clearly this is a major advantage in the study of the geometry of the lungs and in the acquisition of parameters for the resulting models.

Finally, there has been work integrating heart and lung models *in situ* in a virtual torso. This kind of approach has been widened with the aim to produce a virtual physiological human in the near future. More details can be found in Section N2.

P2 INTEGRATIVE BIOLOGY AND PLANT MODELING

Key Notes

Plant modeling	The modeling of plant growth and development lends itself to integrative systems approaches. Separate models of the function of various component systems of plants are under constant development. Some examples are presented below.
Example models	Models of petal growth have been developed, with promising models arising from a surprisingly small number of parameters. Root growth and development is of great interest as clearly roots have a fundamental bearing on the successful growth of the rest of the plant. Other models include those of leaf development, seed germination, and fractal-based systems simulating plant growth.
Getting data into these models	There has been great interest in how these various models can share data to produce a unified model of plant growth. Integrative plant biology modeling presents a clear example of why this work is key to future models of plant growth.
Related sections	Modeling and mathematical techniques (H) Image analysis (Q)

Plant modeling

We saw in the previous section how well tissue-, organ-, and body-scale physiological modeling is suited to an integrative biology approach. However, integrative modeling projects are by no means restricted to building models of mammalian physiological function. An increasing number of integrative biology projects are researching plants, how they grow, and how they respond to environmental cues. The attraction of taking an integrative approach should be clear here as well: bringing together computer scientists, mathematicians, biologists, and bioinformaticians allows us to bring types of research together that might otherwise have been worked on in isolation. The motivation behind plant modeling can be to understand and develop certain traits of crops, producing increased yields, or understanding the spread of disease in plants, for example. Models, once mature, allow scenarios and predictions to be developed and tested. Again, the wet-dry cycle of mathematical predictions driving experiments driving more models is applicable here, as it is to any integrative research.

Clearly, modeling the whole plant involves combining several different kinds of models and approaches, such as molecular-scale regulatory models and spatial models of cell organization. This is demonstrated in the Computable Plant project, the purpose of which is to provide a quantitative description of plant growth. This project combines imaging technology with bioinformatics, mathematics, and biological experimentation.

As with the physiological modeling in Section P1, modeling a whole plant consists of modeling a number of smaller-scale processes and structures that act together at the organism scale. These modeling approaches are often driven and validated by biological experimentation and techniques such as confocal imaging. These modeling frameworks often involve multiple diverse approaches, and so some way is needed of gluing them together, enabling different models to talk to each other. A number of such examples exist and a more detailed discussion of data standards to make such tasks easier is presented in Section E. Building these holistic connections into the system, which allows communication between the component parts, is one of the more challenging aspects of integrative modeling approaches.

Example models   Models of petal growth have been developed. This was considered a first step on the way to developing a complete model of plant growth. The researchers used a technique called clonal analysis to track the development of cells over time. From these observations, parameters that described the growth could be inferred. In fact, only three parameters were required: the growth rate, the principal direction of growth, and the degree to which growth occurs in this preferential direction. Computer simulations confirmed that these parameters were sufficient to produce realistic models of petal growth.

Approaches have also been developed to model the appearance of plants. Very realistic models have been produced using modeling techniques such as L-systems, which are fractal-based methods that apply rules iteratively to define how a plant appears across multiple scales. The models work by simulating the emergence and growth of individual plant components, which together produce realistic-looking growth and architecture models of plants. These methods are very effective at producing branching structures and repeating features, such as tree branches and flowers.

L-system models have been extended to allow for environmental sensitivity, and also to allow components to be influenced by near or far elements of the same structure. This allows for mechanisms such as plant signaling to be modeled. Awareness of context for the plant model also allows for the modeling of factors such as plant orientation and the competition for light.

Major research is being undertaken also to try to understand and model the functions of plant roots in the model plant *Arabidopsis thaliana*. A multi-scale approach has been adopted, with functional processes being studied from the sub-cellular scale right through to complete organ-level behavior. This will result in multi-scale models being produced, drawing on data across a range of physical scales.

Many other areas of plant growth have received attention also. Seed germination is one example. Understanding this process allows more efficient crop production and the prediction of crop emergence rates within particular environmental constraints. Understanding how leaves grow is another research area. Image analysis techniques can be used to study the relative growth rates across the leaf surface.

Getting data into   One of the challenges within such projects is in analyzing the large quantities of
these models   data generated. Much of the data is in the form of images, and so many of these projects often invest a lot of resources in developing image analysis techniques to

try to automate and objectify such analysis. Section Q of this book looks in more detail at the field of image analysis.

Developing ways for these diverse projects to communicate and share data with other projects is also a hot research area (see earlier sections in this book for more details). Once able to share data, the ultimate goal is for the separate models, such as those of root development and of stem development, to be combined via the sharing of data between models. This is a very challenging goal, but one that would truly allow the detailed modeling of a complete virtual plant. In turn, these virtual plants can be used to predict crop genotypes for new strains that can mitigate such effects as climate change and increases in the human population.

P3 INTEGRATIVE BIOLOGY – CONCLUSIONS

Key Notes

Integrative physiological models	Integrative approaches have allowed the development of models of complex mammal- and plant-based systems.
Imaging techniques	Advances in imaging techniques have allowed the generation of high-quality datasets from living samples. The data can often be gathered at frequent intervals, producing large, high-quality datasets, which are of great value in model development and testing.
Computing power	Advances in available computing power and data storage capabilities have allowed other advances such as imaging techniques and complex modeling to become available to a larger number of scientists. This has been necessary for the generation of the highly complex models evolving today.
Integrative biology researchers	Technological advances clearly offer great promise for producing sophisticated models of physiological systems. However, this work remains very challenging because it involves researchers from different disciplines working together for long periods of time, and who may use widely different technical languages and have different ways of working.

Related sections   Computation (F)   Image analysis (Q)
                         Modeling and mathematical
                            techniques (H)

Integrative physiological models

We have seen in Sections P1 and P2 some of the varied models of parts of physiological systems in development (e.g., those for the lungs, heart, leaves, roots, etc.), and we introduced the idea that to produce full physiological system models (e.g., a whole human or a whole plant), it is necessary to combine smaller-scale models. Each of these models requires an understanding, to varying degrees, of mathematics, relevant biology, physics, image analysis techniques (to quantify the large amounts of data collected), and bioinformatics (to organize, search, and condense the various large datasets into meaningful results). The success of these models depends on the integration of researchers from the various disciplines to work together, making hypotheses, testing models, and refining parameters until a satisfactory model is produced. One of the key formats for this is the wet-dry cycle where models provide predictions for experiments, which can then be evaluated, thereby providing further data for the modelers. This is repeated in an iterative process of model development.

Once models exist for the separate processes, one of the main grand challenges in this work is combining the models to form a complete system, such as a virtual torso, or a complete growing seedling model. One of the biggest challenges in doing this is how to make the models talk to each other in a meaningful way. The key to this is developing data standards for information exchange. At the time of writing, standards are emerging such as systems biology, cell, and field mark-up languages, and BioPAX. However, these require further development if they are to capture the entire spectrum of modeling approaches that integrative biologists use.

Imaging techniques

Modern imaging techniques have been crucial in the development of these models. *In situ* live imaging has allowed accurate models to be developed of living samples, allowing accurate geometric models of the shape of the organ or tissue under investigation to be built in a variety of real-world situations. This was illustrated with the example of lung models. Lung geometry is very complex, and such live imaging techniques allow much greater insight into the function of this geometry.

Computing power

Such imaging techniques tend to generate large amounts of data and have a high processing cost. Therefore, such advances have been possible only with an accompanying advance in the availability of high-performance computing. This has occurred both at the level of high-performance desktop PCs as well as the increased accessibility to very high-performance shared computing resources. Such shared resources often have multiple processors and large memory stores, making it possible to do very complex computations in feasible timescales. The quality of the image data is improving all the time, with new techniques and equipment being developed, and an increasing sensitivity of the imaging chips themselves. Image analysis techniques are being developed which allow the analysis and quantification of the large and complicated datasets these systems produce (more can be read about this in the following section).

Integrative biology researchers

These technological advances clearly offer great promise for producing sophisticated models of physiological systems. However, this work remains very challenging because it involves researchers from different disciplines working together for long periods of time. These disciplines often have different technical languages, mind sets, and modes of working. This can lead to a significant management overhead and training in others' disciplines, and the resolution of 'cultural' differences. However, since bioinformatics is implicitly a multi-disciplinary subject, bioinformaticians are well placed to act as mediators and ambassadors in this context. We can be the 'glue' that holds such teams together.

Q1 WHAT IS IMAGE ANALYSIS?

Key Notes

What is image analysis?	Image analysis is a field that aims to extract numerical information from an image or sequence of images.
Digital images	Digital images are stored as arrays of numbers representing the intensity of the light in the scene, either in one channel (grayscale) or three channels (color). The resolution in pixels, and bit depth of the intensity levels stored affect the amount and fidelity of the image data that are represented.
Color	Images can be color or grayscale. A color camera can have either one light sensor chip or three (one for each wavelength of light). Three-chip cameras are more expensive but higher quality. Single-chip cameras often use a Bayer mosaic arrangement of color-detecting elements on a chip, from which measurements are interpolated to give color intensity estimates.
Storing image files	There are many formats for image files (e.g., TIFF, BMP, JPEG, RAW). Some are compressed (the file size is less but the data quality may be compromised) and some are uncompressed (large file size but no reduction in data quality).
Related sections	Data and databases (D) Tissue- and organism-scale Data categories (E) organization (N2)

What is image analysis?

Image analysis is a field that uses techniques to extract information from a digital image, or sequence of digital images. The use of image analysis is common in a wide variety of domains; for example, biosciences, engineering, industry, medicine. Wherever pictures can yield useful information, there is a desire to recover that information in an efficient, accurate, and objective way. Today, pictures are taken normally with a digital camera, or they might be generated by some other imaging mechanism, such as a scanning confocal microscope or MRI scan. Once captured, the images are stored on a computer and then processed, either automatically or with some degree of interaction from a user, to produce the desired output data. The kind of processing that can be done might be an automated form of a traditional manual method; for example, measuring something in the image, or it may be something that is only possible with computerization. An example of the latter is computing a 3-D representation of a physical object in CT (computed tomography) applications, which is calculated from a series of 2-D X-ray images.

The ultimate result of image analysis is some form of quantification of the data in the image. The precursor to performing image analysis is usually some form of **image processing**. Image processing is different from image analysis in that the output is another image, such as one where image noise has been removed.

Image analysis computations are often complex, and it was only recently that reasonable processing speeds could be achieved on desktop PCs; a decade ago specialist hardware would have been required to do some of the heavier processing. With modern computers performing image analysis tasks on standard office or home computers is certainly possible. Scientific researchers may not have much to do with the 'behind the scenes' development of image analysis software, but the chances are that you will come across software packages that make use of image analysis techniques. It is useful therefore to have a general understanding of the underlying technology and techniques, as these will enable a greater comprehension of what is happening to a set of data that is being processed.

Digital images

Many people today are familiar with digital cameras and the images they produce, but it will be useful to also know how that image is produced and how it is stored. In this section we shall take a brief look at what makes a digital image, as understanding that should help us understand the processing and analysis techniques themselves. It is useful also to understand the basic formats of some images that will be encountered when using imaging systems and software.

Traditional film cameras work by capturing light with a lens and directing it onto a photosensitive emulsion on the film. This emulsion contains components that react to the light falling on the film in proportion to the intensity of the light, thus producing an image of the scene on the film. In a digital camera, the film is replaced by an electronic sensor that is also sensitive to light. This sensor takes the form of a chip; for example, a CCD or charge-coupled device, onto which light is focused. This chip is pixelated, which means it is divided into a grid of very small regions, each of which is sensitive to the brightness of the light. When light falls on each square in the grid, a variation in electrical charge is generated and recorded as an intensity value in the final image. Digital images themselves are divided also into corresponding grids of many small squares. A typical image might be in the order of 1024×1024 squares in size – this is called the **resolution**, and each individual square is called a **pixel** which stands for 'picture element'. Resolution might also be defined in terms of the total number of pixels in the image; for example, a 1024 x 1024 image contains 1 048 576 pixels, so might be referred to as being one megapixel (a megapixel being a million pixels). The maximum resolution of a digital camera image is largely determined by the resolution of the light sensor chip itself.

So, you can think of the light-sensitive grid squares on the imaging chip as being sensors, which set the value of the corresponding image pixels in the final digital image. The measured intensity at the chip is converted to an integer so that it can be represented in a digital image. The range of values available in the digital representation is referred to as the **bit depth** of the image, and in real terms relates to the number of levels of brightness available in the image. A truly black-and-white image has only two levels, black or white. This could be represented with one bit per pixel, where a bit is a **binary digit** (see Section D), which can only have the value of one or zero. Black and white often does not provide us with enough information about the world, so we normally use bit depths greater than 1-bit. For example, an 8-bit image uses eight bits to store the number. It is not useful here to explain binary numbers in detail, but suffice to say that an 8-bit number has a range of 256 levels, considerably more than the two levels of our 1-bit, black-and-white image. So, for an 8-bit gray-level image, the intensity as stored at each pixel in the image can range from 0 to 255. In other words there are 256 **shades of gray**, which can be used to compose an image. *Fig. 1* below illustrates the effect bit depth has on the available number of gray levels with which to represent a scene.

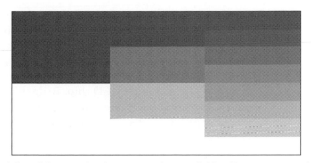

Fig. 1.   From left to right: gray-level representations available with one, two, and three bits. Note how the number of available shades of gray doubles with each extra bit. With an 8-bit image there are 256 gray levels available.

This is useful to know because it can allow you to fully exploit the system you are using to produce the best quality images you need. Some image capture systems offer to store images at 12-bits per pixel – this means you get 4096 gray levels from which to compose your image. This gives you a potentially larger range of values to differentiate between the darkest black part of your image and the brightest white part, or greater resolution in your recording of these values.

All imaging technologies (film and digital) suffer from what is called **over-saturation**. This is an effect which happens when the physical brightness of a scene is too bright for the equipment recording it. So, as we mentioned earlier an 8-bit image has 256 levels of brightness. The maximum brightness that can be stored is a value of 255. The imaging system is tuned normally so that the brightest value you expect to appear in the scene falls under this 255 level. If, however, the scene gets brighter than expected (e.g., you have been working on an overcast day and the sun comes out), then it is likely that the image will over-saturate – that is, the maximum recordable level of 255 will be reached and exceeded. In this case, all physical values that get converted to digital values above 255 will get clipped to 255. Therefore, data will be lost in this case. If however, we use a 12-bit system with 4096 levels, we can have much more leeway in the range of values available to us. The technical term for this is increasing the **dynamic range** of the imaging system. As an example of the kind of ranges of intensity (luminosity) found in the real world, the sun has about one million times the luminosity of the moon. Therefore to accurately record a scene with both the sun and the moon in it, without any over- or under-saturation, we would need about 1 000 000 gray levels (or 20 bits). The development of high dynamic range cameras and display methods is an active research area.

Color

In the human eye, we have three classes of cones – cells in the retina of the eye that are selectively responsive to either short-, medium-, or long-wavelength light. Wavelengths of light correspond, broadly speaking, to the colors we perceive; for example, short-wavelength visible light is what we term 'blue,' medium-wavelength is 'green,' and long-wavelength is 'red.' A certain wavelength of light will stimulate the three cones to varying levels, and from these signals we can infer a color.

For a digital color image, normally the intensity of light falling within red, green, and blue bands is measured by the sensor and stored in one of three separate **channels** in the image. This is analogous to the three cones in the human eye or three layers in color film, which also record the three component colors separately. There are two common methods by which the CCD array detects the three color wavelengths. The most expensive is to use three sensors to record each of the color wavelengths separately. These are called 3-CCD or 3-chip cameras. The light is passed through a beam splitter that directs red, green, and blue light to the respective CCD sensor. The less expensive option is to record all three wavelengths on a single CCD chip. To do this, elements of the CCD sensor are set normally to detect only one of the wavelengths of light, and arranged in a pattern on the chip, the results of which are then interpolated. A common pattern for this arrangement is called the Bayer filter mosaic, and is arranged as follows

Fig. 2.   The Bayer filter mosaic layout of color-detecting CCD cells. White represents green light detection, gray represents red, and black represents blue. Note that there are twice as many green-detecting cells as red or blue.

It can be seen in *Fig. 2* that there are more green-responsive cells than red or blue. The eye is more sensitive to green wavelengths so it is sensible to record green light with greater fidelity. Note, however, that this arrangement may not be appropriate where high accuracy color images are required, for which a 3-CCD camera may be a good investment.

Once the light has been recorded by the sensor, it is converted into numbers that represent the scene. With our gray-scale image we might have had 256 gray levels to represent the image in an 8-bit system. In a color image we would have three numbers to represent at each pixel, one for each color band, not just the one representing the overall brightness in gray-level images. Therefore we might end up with a commonly seen image format, which is called **24-bit color**. This refers to the fact that there are three numbers per pixel, representing the intensity of red, green, and blue light at that point, using 8-bits (256 levels) to represent each of these color channels. This therefore produces an $8 \times 3 = 24$-bit image. A lot of common image formats such as BitMaps (.BMPs), Tagged Image File Format (.TIFFs), etc. can save the image in 24-bit format. This is a common format for image analysis data.

Storing image files

It is worth mentioning here a little about file types and compression. Some imaging equipment offers to save images in RAW format. These files often contain raw, unprocessed data from the image sensor. There is no one format for these files as they tend to vary with manufacturer. Data stored in this format need further processing before they can be viewed or analyzed as an image, usually using some software the manufacturer supplies. RAW files have the advantage that the data are as close as possible to what the image sensor chip records; you could consider these 'primary data'. Disadvantages are that the file can be larger than other file formats, and it requires further processing to get to common exchangeable file formats.

Broadly speaking, there a two kinds of common image formats: those which are **compressed** and those which are **uncompressed**. Uncompressed formats (e.g., BMP, RAW, uncompressed TIFF) can represent the data in as high a quality as the capture equipment allows, at the expense of relatively large file sizes. Compressed formats (e.g., JPEG) algorithmically 'compress' the data (i.e., reduce the amount of space required for storage) by removing components of the data that are least visible in the final image. For example, JPEG compression uses a number of techniques, including reducing the amount of data used to represent the color component of the image data as opposed to the brightness information, as the human eye is much more sensitive to brightness information than color. JPEG and similar compression techniques can produce images that are almost identical to the original to the human eye – however, the underlying data may have been altered significantly and so, for image analysis, uncompressed formats should be used. If compression must be used to save space, lossless compression is an alternative form where the underlying data are **not** changed. Examples of this are files compressed in the ZIP format, and some forms of PNG and TIFF files. The reduction in file size is often not as great for lossless compression compared to full, lossy compression (lossy meaning that the original data cannot be restored completely).

We have now seen what goes into making a digital image file, and introduced a little technical information that will help you when you need to work with image analysis; now we can take a look at what systems produce these files, and the kind of analytical results they can extract from them, in the domain of bioscience research.

Q2 HOW IS IMAGE ANALYSIS USED IN BIOSCIENCE RESEARCH?

Key Notes

Constrained images	Constrained images are images where the data they contain are organized in a predictable way. This organization is enforced usually by the way the image was captured. Examples include microarray images.
Semi-constrained images	These images are less formally structured than the constrained images, but they are still generated by a mechanism that, to some extent, can predict how the underlying data are arranged. Examples include 2-D gel electrophoresis images.
Unconstrained images	These images contain any kind of data arranged in a non-structured way. Examples include digital camera images.
Real-world applications	Image analysis has a wide variety of applications in bioscience, as the examples in this section show. Part of its wide domain of application is because of its non-invasive nature and flexibility.

Related sections	Transcriptomics (K)	Physiology (P)
	Tissue- and organism-scale organization (N2)	

By now you should have a background understanding of how an image is formed, so it will be useful to see what kinds of images can be captured and analyzed within bioscience research. Remember, image analysis typically allows us to extract meaningful numbers from, or quantify, images – bear this in mind as we look at the examples below. Then in further sections we shall look at how the processing and analysis actually happen.

Images can be grouped into one of three general categories that reflect how they are formed. A constrained image is formed in such a way that the data it represents are organized in a rigid and predictable pattern. Examples of this type include microarrays, and 1-D gel electrophoresis. The second category contains the semi-constrained images. These images are formed in a less structured way than the constrained images, but there is still an underlying method that can describe how the data might be laid out. 2-D gel electrophoresis produces semi-constrained images. The final category represents the complex, unconstrained images. These images are formed in a way that means you cannot predict where the information they contain is going to be located. Examples include microscope images, photographs, and medical scans. The more unconstrained an image, generally speaking, the harder it is to analyze automatically. We will take a look at each of these areas in turn, using examples to illustrate each case.

Constrained images

Microarray analysis

For a detailed description of microarray experiments refer to Section K1 and Stekel (2003).

There are two types of microarrays: spotted arrays and Affymetrix-type arrays. With spotted microarrays, also known as two-channel microarrays, two different samples are spotted onto the arrays. One of these two samples will be a control sample usually, and the other from an experimental condition. The samples are labeled with two different fluorophores. This means the relative expression levels for both the samples can be compared. Absolute quantitative measures are not usually possible with these array types. Single channel arrays, such as Affymetrix arrays, can produce absolute quantitative expression levels and only operate on a single sample at a time.

Both types make use of image analysis to probe their outputs, and both have their own particular problematic features. Affymetrix arrays suffer from light refraction onto surrounding spots as a result of the nature of their production. This can cause a leakage of signal between spots. The proprietary software for analysis of the spots compensates for this, however. The proprietary nature of this software often means it can be considered a black box. We shall look here at two-sample spotted-arrays image analysis in more detail, as there is a wider variety of software available for these array types.

The output of a spotted microarray is an array of many colored dots, which represent the expression levels of potentially thousands of genes simultaneously. We are interested in automatically determining the intensity of each dot in both the red and green color channels, so that we can compare the two. To achieve this, there are three basic stages of processing. First, the underlying grid structure must be extracted so that the dots can be reliably located and indexed. Second, the dots must be segmented – that is, the data at each dot must be reliably isolated from the noisy data of the background. The last step estimates the relative expression levels – that is, the intensity of each spot in each channel must be measured, giving a quantitative indication of the expression level of the underlying gene for both samples. We shall take a look at each step more closely.

The first step, determining the grid structure, can be done either interactively, where a user indicates where the blocks of dots begin and end and gives the separation between the rows and columns, or automatically by an algorithm that identifies and then groups spots together, correcting for alignment as well. The motivation here is to exploit the rigid grid structure inherent in microarrays so that we can automatically locate the center of the dots to a sufficient accuracy. The software may have to correct for warping of the grids, caused by the misalignment of pins on the spotting head, for example. This 'gridding' step feeds into the segmentation step, where the data within each dot need to be separated from background noise. The estimated centers of the dots from the previous step are used to locate them approximately in the image, and from there segmentation algorithms separate the dot itself from the background. A number of possible algorithms are common in microarray dot segmentation, including circle-based methods that try and fit a circular geometric model to the dot, and non-spatial methods such as histogram segmentation, which model the distribution of intensities across a dot and its local background and then divide this histogram into background- and foreground-originated values.

Once the dots have been segmented, and we have access to the image data recorded for each spot, a quantitative measure is required for each spot. Normally, a mean or median intensity in each color channel is calculated for each spot, representing the average fluorescence at that spot for the two different fluorescent dyes. Final calculations are based usually on ratios of these values, so conversion to absolute units is not required. It is desirable also to have a measure of the quality of these final quantifications, and this can be achieved by providing statistics such as the standard deviation of the values within each spot, and the variation in spot size or shape.

1-D gel electrophoresis

Gel electrophoresis is the process by which the molecules in a sample can be separated according to their size. The sample is placed in a well in a layer of gel, and a high voltage set across the gel. The resulting current forces the molecules through the gel matrix, and smaller molecules can move faster. So, over time, smaller molecules are able to travel further along the gel than the larger molecules. The resulting positions of the molecules can be visualized by staining and imaging, as illustrated in *Fig. 1*.

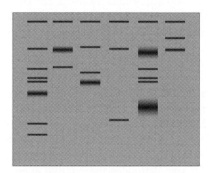

Fig. 1.   Illustration of a typical outcome of 1-D electrophoresis after staining. The molecules have been pushed through the gel from the top where the samples are placed. Each vertical 'lane' represents a different sample; the left-hand lane holds a control sample called a ladder, with which the other lanes can be compared.

The way the samples are organized into lanes allows image analysis software to automatically locate these regions and then extract the information from the bands within each lane. An **intensity profile** can be constructed by measuring the intensity of the pixels down the middle of one of the lanes. This gives an indication of the staining level as distance increases from the original sample well. It is possible then to compare the profiles generated in the different lanes. The profiles can be calibrated also against the known molecular weights in the control ladder.

Software to carry out such analyses often is available under license; for example, TL120 (Non-linear dynamics 2007), Lab image 1D (Kapelan Bio-Imaging 2007), and Un-scan-it Gel (Silk Scientific 2007).

Semi-constrained images

2-D gel electrophoresis

2-D electrophoresis commonly is used to separate complex protein mixtures. Separation in the first dimension is accomplished by pushing the molecules along a pH gradient using an electric charge (called isoelectric focusing). At the molecule's isoelectric point, there is no longer a net charge on the molecule so its migration ceases. The molecules then are separated along the second dimension according to their molecular weight.

The images this technique generates spread data across the two dimensions of the image plane. Therefore, the data do not lie in constrained lanes as in 1-D electrophoresis. However, we do know the method that causes the molecules to spread across the gel, so we can make some predictions about how the data will be located. The images often contain some inherent variability and noise, caused by variations in sample preparation, streaking during the separation process, gel warping during staining, etc.

It is possible to label multiple samples differentially and run them all on the same gel. This technique is called **differential in gel electrophoresis** (DIGE). By labeling proteins with different fluorescent dyes, they can all migrate together on the same gel, but the results can be visualized separately. This eliminates some of the variability present when different gels must be run individually, and allows a control sample to be run on each of the gels.

When complete, the gels are scanned, and image files produced. Usually multi-channel images are used to store the differing fluorescent spectra results of DIGE experiments, allowing the results to be overlaid in a single image. Multiple gels can be compared by warping images to match a reference image. This helps counter some of the inter-gel differences. It is possible to select proteins in the control sample across multiple gels, and software can be used to calculate complex transformations to map these points onto a reference set of points. From these transformations the images can be warped and aligned so that the other samples on the gels can be compared across the set of gels. Image analysis can aid this process by automatically detecting reference spots in the gel images. A variety of resources including software packages can be found online (e.g., Non-linear dynamics 2007; Swiss Institute of Bioinformatics 2007).

Unconstrained images

Microscopy

Image analysis can be used often to enhance images and help quantify microscopic images. The digital images may be produced by the microscopy system itself (e.g., in confocal microscopy), or may be produced by attaching a digital camera to more traditional light microscopes. Either way the result is a digital version of a highly magnified image of the sample. The resolution of the features in the image can be limited by the resolution of the camera, such as the number of megapixels on the sensor, or the resolution of the optics of the microscopy system itself. Some examples are presented in *Fig. 2*.

Scientists might be interested in measuring distances in the resulting images, looking for and quantifying stained areas, or counting instances of something; for example, counting the number of cells of a certain type. A **time-series** of a sample may be taken, where an image is stored at fixed time intervals. From these, a time-lapse movie can be generated, or motion within the image can be quantified over the time window.

There are a number of processing steps that can be applied to a typical microscopy image to increase the quality of the image. One such example is **deconvolution**. This refers to removing the problem of out-of-focus light contributing to the final

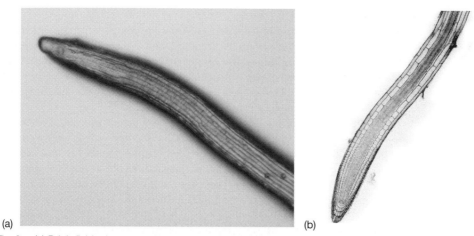

(a)   (b)

Fig. 2.   (a) Brightfield microscope image of a growing Arabidopsis root *, and (b) confocal images of fluorescing*
Arabidopsis root *(courtesy of Susana Ubeda-Tomas).*

image; the intensity of a pixel in the final image may not only be produced from
light originating in the focus plane, but also from sources above and below the
focus plane. This manifests itself in a blurring of the resultant image. While we do
not go into detail here about how to deal with this problem, the general solution
is to model the contribution of out-of-focus light to a point on the final image,
and eliminate this from the image. Deconvolution is sometimes used as a way
of approximating the images produced by a confocal microscope using a non-
confocal microscope. Both techniques aim to eliminate light from outside the plane
of focus, but they go about it in different ways. Confocal techniques eliminate out-
of-focus light inherently through the technical operation of the microscope, while
deconvolution techniques offer a post-capture software solution.

You might think that the relatively sharp and clearly labeled fluorescent images
produced by a confocal microscope lend themselves to image analysis techniques
and quantification. However, it is prudent to exercise care when quantifying any
image, as you must consider all the factors that may affect the data during image
capture. For example, typically we might want to use the fluorescence intensity
in the confocal image to estimate the level of protein activity in the sample; but
beware! Assuming the two are directly related might cause errors in our analysis.
Always think about how images are formed. A fluorescent molecule is excited by
a laser, which then emits light of a certain wavelength. Then this light is captured
by the optics of the microscope and stored as pixel values in an image. It is not
the case that the only factor affecting the intensity of the pixels is the number of
fluorescing molecules in the sample. Consider changes in the intensity of the laser,
the depth of sample that the exciting and emitted light has to travel through, the
consistency of illumination across the image field of view, the optical properties
of the microscope, etc. All these factors can contribute to the final recorded pixel
intensity in the final image. When using any image capture equipment, care must
be taken to ensure, as much as is practical, that the image data are reliable and
represent what you intend them to represent.

Medical image analysis
The images produced by X-rays, CT (computed tomography) scans, MRI
(magnetic resonance imaging) scans, etc. are clearly of great importance to

modern medicine. Medical image analysis differs from other domains in that, typically, it is important to use data in three dimensions, and software is seen often as a tool to aid expert specialists rather than to produce final numerical results. One area where computers excel in this domain, however, is the measuring of areas and volumes in these images. To do this, an image analysis technique called **segmentation** is widely used. This is a way of (semi-)automatically defining regions of interest in images, such as regions in the brain, whole organs, blood vessel structures, etc. Some segmentation techniques are examined in more detail in Section Q5, but a thorough examination of segmentation techniques for image analysis can be found in textbooks.

Quantitative photograph analysis
In some areas of biology, for example measuring phenotypes or an organism's response to a stimulus, photographic images have been used often to illustrate the results. Traditionally, the analysis of the images has been largely qualitative ('X is more blue than Y') or simple quantitative judgments may have been made ('X is 6 mm long'). Image analysis can provide the tools to carry out much more objective and consistent analysis of the images.

However, with the use of these techniques extra care must be taken that the images themselves are reliable. By this it is meant that the data in the images should be taken in a controlled fashion. In the same way as the images produced by a microscope have to be carefully controlled, so do those taken with a digital camera. Both settings in the camera and changes in the environment can affect the pixel values recorded by the camera. For example, white balance settings can change the way certain hues are recorded. White balance is designed to compensate for certain colored lighting situations and correct the hues in the image back to a more accurate representation, as if taken in neutral or white colored light. You may have noticed how fluorescent lighting tubes produce a different color of light to incandescent bulbs – in digital cameras this effect is magnified as human eyes are very good at compensating for such changes. Not understanding the white balance settings of a camera may lead to unintentional misrepresentation of colors in the images, especially if the white balance changes during the image capture session, either because of manual adjustment or changes in the automatic selection made by the camera. Likewise, physical changes in the environmental lighting affect the perceived and recorded color of objects. Morning light appears 'cooler' than the 'warm,' orange light at sunset. This is because the evening light is shifted more towards the red end of the spectrum, a phenomenon owing to the optical effects in the atmosphere. So, generally an object will appear 'more red' in the evening than the morning in natural light. It is important to be aware of such effects if they alter the features of your sample that you are trying to measure.

Likewise, distortions introduced by the lenses of cameras or perspective might misrepresent the true geometry of the scene being photographed. This might affect the quality of any geometric measurements made from the image, such as measuring lengths.

Being aware of these things will allow researchers to take the best possible images before any analysis takes place. Once images have been captured, it can be very expensive in time, or even impossible, to go and recapture the images again. So, before any images are taken it pays to understand fully the capture process and limitations and what bearing these have on future analysis.

Real-world applications   Image analysis is used often in some very practical applications in bioscience work on larger physical scales. The fact that it is a non-invasive technique and is very flexible means that the applications have been wide and varied. Image processing has been used to identify the outlines of feeding pigs in real time in order to bring an ultrasound sensor into contact with them to measure back fat thickness. Three-dimensional techniques also have been applied to live animal conformation analysis. Mechanical control of inter-row hoeing of crops using machine vision techniques to guide the hoes has proved viable. Software that uses image analysis algorithms exists for estimating the growing rates of plant roots. Hopefully these examples highlight the enormous variety of applied image analysis work in bioscience.

Q3 IMAGE ENHANCEMENT

Key Notes

Image noise	Digital images can be corrupted by different kinds of noise, which affect pixel values during the capture process. Two common types are 'salt and pepper' and 'Gaussian.' Noise can be filtered from images using mean, Gaussian, or median filtering, dependent on the noise type.
Image transformations	Images may represent a distorted perspective on the world, and images may need to be rotated, skewed, scaled, etc. to provide an accurate representation of the real-world data.
Improving contrast	There are two common approaches to improving the contrast of an image so they are easier for the eye to interpret: histogram normalization and histogram equalization. Both manipulate the frequencies of intensities in the image to produce a more pleasing image for people to view. Care must be taken if quantitative measures are required, however.

Related sections	What is image analysis? (Q1)	How is image analysis used in bioscience research? (Q2)

When images are produced, they often contain not only the data of interest but also other artifacts. These unwanted additions can include noise, deformations, clutter, etc., and are inherent to varying degrees in all digital image capture processes. Image enhancement techniques look at ways of handling these effects so that the images become a more accurate representation of the underlying data. This is referred to sometimes as image processing – the process of tidying an image prior to further analysis or inspection.

Image noise

All digital images are affected by noise. Noise manifests itself as variations on the true pixel values that would be recorded by a perfect sensor, and is normally caused by the capture equipment itself. Imperfect sensors lead to a less-than-perfect representation of the original data. One form of image noise is caused by faulty cells in the array of sensors on the imaging chip. When a cell gives either a false low response (producing a black pixel) or a false high response (producing a white pixel), this is known as **salt-and-pepper noise**. Examining *Fig. 1* below should reveal why – the image contains erroneous black and white dots.

Another form of image noise is caused not by the cells on the recording chip being completely broken, but by them recording slight errors in addition to the true values. These errors tend to be small, and clustered around the true values. They essentially consist of the sum of small random numbers, and for this reason the Central Limit Theorem can be applied. The Central Limit Theorem is an important concept in probability, which states that a sum of random

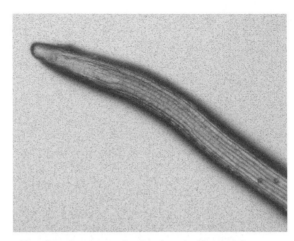

Fig. 1.   Image with salt-and-pepper noise (black and white spots).

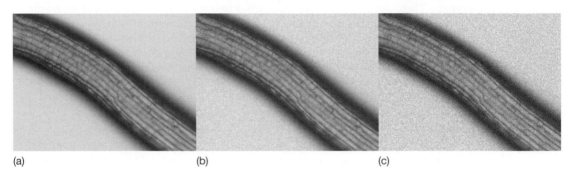

(a)   (b)   (c)

Fig. 2.   Image of root cells with added Gaussian noise with a standard deviation of (a) 1, (b) 10, and (c) 20 gray levels.

numbers tends towards a normal distribution. Therefore, the errors in the sensor measurements can be thought of as being drawn from a normal (Gaussian) distribution and added to the true value. Therefore, this is known as **Gaussian noise**, one of the most common types of noise encountered in image analysis.

Gaussian noise is described in terms of the mean μ and variance σ^2 of the normal distribution from which the noise is drawn. To model Gaussian image noise, the mean of the noise distribution is taken to be zero (called 'zero-mean'), and this distribution is assumed to be centered on the true value. This means that on average the recorded values approximate the true value. The higher the standard deviation of the normal distribution (a measure of the spread of the errors), the more extreme the noise is, as can be seen in *Fig. 2* above.

Removing image noise

To remove these types of noise, we can **filter** it from the image. The two types of noise discussed so far require two different types of filter to remove them most effectively.

As we have seen, for Gaussian noise the values on average produce the correct measurements. Therefore, one way of estimating the true value of a pixel is to assume the values are distributed around the pixel according to a Gaussian distribution, and apply a Gaussian smoothing filter to the image, which calculates a Gaussian weighted average value for each pixel. This can be approximated by **convolving** a Gaussian **kernel** across the image. This simply means that at each pixel a value is calculated by centering a grid of numbers (the kernel) on the pixel of interest, multiplying all pixels by their respective value in the grid, summing these and multiplying by a weighting term. This value then replaces the current pixel of interest value in the final image. In essence, this value is a weighted mean of the surrounding pixels. *Fig. 3* below shows two example kernels, one for mean filtering and one for a Gaussian filtering. Both have the effect of averaging pixel values over space.

Mean filter kernel   Gaussian filter kernel

Fig. 3.   Example kernels for mean and Gaussian filtering. The kernel is centered on the pixel of interest. This and the surrounding pixels are multiplied by the respective numbers from the relevant kernel above. These results are summed, and then multiplied by the relevant fraction shown above. The final number is the new value for the pixel under inspection, and is stored in a new image.

Both filters are suitable for removing sensor noise, and both have the drawback that they effectively blur the image. However, the Gaussian filter has an advantage in that it is capable of blurring edges to a lesser extent than the mean filter.

Another method of reducing Gaussian noise in an image is to record several images one after another, and average the sensor values over time. This assumes that the conditions in the scene do not change; the only varying factor in the data is the Gaussian sensor noise which, due to its zero-mean random nature, can be removed by averaging over time. This technique is called **image integration** and is an effective way of increasing the quality of the image, though clearly it must be planned for at the time of capture; it cannot be computed later if you only have one image of your scene captured.

To remove salt-and-pepper noise, another approach must be adopted. Calculating an average value to 'smooth out' the extreme but rare error values present with this kind of noise is not appropriate as the noise is not normally distributed. A much better way to remove this kind of noise is using a **median filter**. This still examines the pixels surrounding the pixel of interest, but this time instead of the mean, a **median** value from the surrounding pixels is selected. This is very effective at removing extreme values, and has the benefit of not destroying edge information as mean filtering methods can do.

Image transformations

It is possible that as well as noise, the image is corrupted by a transformation caused by the imaging optics. This might result in the image appearing 'warped,' for example, producing an effect like **barrel-distortion**, where the center of the image appears to bulge out. However, this is less of a problem in expensive scientific imaging systems, and most high-end processing packages can remove this distortion anyway, so it will not be examined in detail here – suffice to say there are methods of removing it if it is present. Other types of transformation include affine transforms, necessary where the plane of interest in the real world is not orthogonal to the image plane; for example, the camera might be at an angle to the bench or rotated relative to the subject of interest. It is possible to correct for these transformations by knowing the **internal and external parameters** of the capture device, and the relationship between the camera and the scene, which can be determined by imaging special calibration targets, though the details are beyond the scope of this introduction.

Improving contrast

Another common type of image enhancement is to balance the gray levels in the image. An example is **histogram normalization**. A histogram of an image is simply a frequency distribution of the number of pixels at each intensity level. Histogram normalization (sometimes called contrast stretching) is able to take an image of low dynamic range (i.e., recorded over a small range of graylevels) and stretch it so that it is more appealing to the human observer. **Histogram equalization** is a similar technique that tries to force each intensity level in the image to be represented by an equal number of pixels. This produces a more uniformly distributed range of frequencies in the resulting histogram. This can be used to enhance detail and to correct for non-linear effects in the digitization process. However, it can reveal **too much** detail, producing a grainy-looking image, owing to noise. Both techniques should be used with care when quantification of the images is required, as the pixel values no longer directly represent the underlying data; they should be used only to improve images for human inspection.

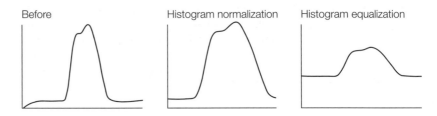

Fig. 4. Example of histogram normalization and histogram equalization.

Q4 FEATURE DETECTION

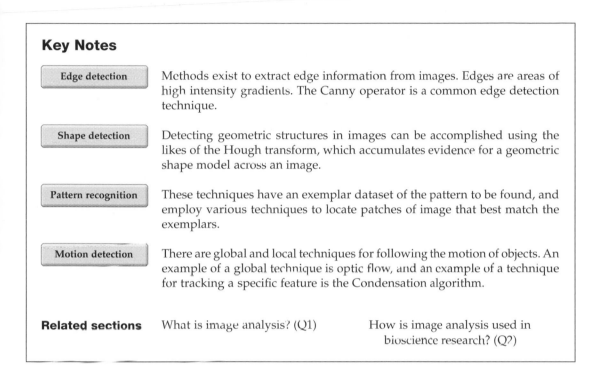

Key Notes

Edge detection	Methods exist to extract edge information from images. Edges are areas of high intensity gradients. The Canny operator is a common edge detection technique.
Shape detection	Detecting geometric structures in images can be accomplished using the likes of the Hough transform, which accumulates evidence for a geometric shape model across an image.
Pattern recognition	These techniques have an exemplar dataset of the pattern to be found, and employ various techniques to locate patches of image that best match the exemplars.
Motion detection	There are global and local techniques for following the motion of objects. An example of a global technique is optic flow, and an example of a technique for tracking a specific feature is the Condensation algorithm.
Related sections	What is image analysis? (Q1) How is image analysis used in bioscience research? (Q?)

Once the images you are working with have had any noise suppressed and have been corrected for any distortions as necessary, it is possible to process the images to reveal information about features. Image features are typically the simple and fundamental components of a scene, such as the edges between objects or areas of texture or lighting, and geometric shapes such as lines and circles. Some techniques for identifying fundamental features such as edges have close relations to how we think the eye may process such information. Our visual system is very good at detecting such low level information; so good, in fact, that it is often all too easy for us to assume such tasks are trivial. When programming a computer to find features, often the process is quite complicated, a testament to how well developed our visual system is.

The very simplest form of processing that can be done is called **thresholding**. This simply produces a binary image that represents all pixel intensities above a threshold in the original image with a value of one and all other pixels with a value of zero. It can be used, for example, to differentiate between a foreground object and the background, but only if the two produce substantially different intensities in the image. Here we will have a look at some more sophisticated feature detectors.

Edge detection **Edge detection** is useful as edges often mark physical boundaries between objects or features within a scene. An edge in an image is where a high gradient exists in pixel intensities between pixels in a local neighborhood.

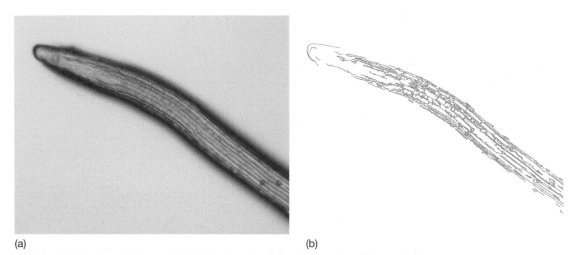

(a)   (b)

Fig. 1.   Gray-level image of an Arabidopsis *root (a), and the result of a Canny edge detector, isolating some of the edges (b).*

An example of a classic edge detection filter is the **Sobel** operator. It works by essentially looking at intensity gradients over very small patches of the image, and from this calculates a strength and direction of an edge. The quality of the edge detection is affected by the strength of the edges to begin with – a blurry image has very indistinct edges, and operators such as Sobel are likely to perform poorly at detecting them. Probably the most famous classical edge detector is called the **Canny filter**. The image in *Fig. 1b* shows an output of the operator. This filter really is a sequence of techniques to produce optimal edge detection. First, the image is smoothed with a Gaussian filter (as described previously). This is followed by processing with the Sobel operator to detect the edges. It then traces the edges using a technique called **thresholding with hysteresis**, which attempts to remove falsely identified lines. This works by making use of the assumption that lines are continuous. Edges whose strengths fall below a low threshold are disregarded, but there is a higher threshold also, below which edges are disregarded if they are unconnected but kept if they are connected. The result of the filter is a binary image of lines.

Shape detection   Detection of geometric shapes is more complex than just edges, and typically requires a model of the shape in question. An example technique is the **Hough transform**. This is used often for finding the parameters of straight lines in images, and works by accumulating evidence for lines in a parameter space. A **parameter space** is simply a way of representing all the different values of each parameter in a model; the dimensionality of the parameter space is equal to the number of parameters. The image is examined pixel by pixel, and if a line segment is detected at a region, the parameters that might have created that line are stored. If there truly exists a line across the image, the same parameters will be identified at each point of the line, accumulating evidence that a line does indeed exist as defined by the respective parameters in the straight line model. Using similar techniques, parameters for other models, such as circles, can be used. The advantage of this technique is that it is truly global, considering evidence across all of the images. This means that a heavily occluded line still can be identified. The disadvantage is that the time required to process the entire image can be large, as each location must be examined individually.

Pattern recognition

Another major technique for detecting features is **template pattern recognition**. A template, or exemplar image of the feature of interest is passed across the image areas of interest. A template matching algorithm rates each area of the image for how closely the template model and the actual image patch match. Thus locations with high similarity to the template can be identified. This method requires a method of handling the possible transformations of the image if flexibility is required when matching the model to a target; for example, if the orientation of the target is not important. Multiple, simple templates can be combined in sequence in such a way as to detect complex shapes such as faces efficiently; however, the technique can be very slow to train on example images, despite working swiftly on test images once trained. Training only needs to be carried out once, and its purpose is to build a representative model of what the pattern is. Problems of over-specificity arise where the training set does not contain enough variation, and the system is unable to recognize targets that differ from the over-specified examples. Problems of over-generalization occur where the training set contains too much variation; the system then has too vague a model of the pattern to be able to differentiate a specific instance from clutter. *Fig. 2* illustrates the result of a pattern recognition algorithm that has been trained to recognize the appearance of a certain collection of cells in images of an *Arabidopsis* root.

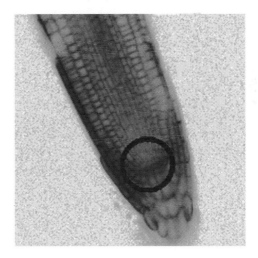

Fig. 2. Pattern recognition of the quiescent center, a set of cells in Arabidopsis roots that are always located in the same place relative to the tip. A small training set of images of the quiescent center was used with a Viola-Jones pattern-recognition algorithm. The circle here shows the algorithm is able to detect this location, once trained.

Motion detection

One approach to motion detection is to use algorithms that calculate the amount of global motion across an image. These techniques typically look at small intensity patches and, assuming the intensity remains constant to the next frame, they locate the new position of the patch and calculate the velocity vector of the patch. Certain assumptions are imposed; for example, it is assumed the flow field varies smoothly in the local neighborhood. An example output of an optic flow algorithm can be seen in *Fig. 3*.

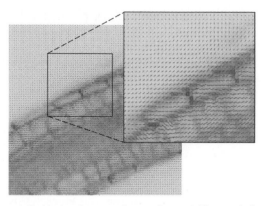

Fig. 3.   Optic flow applied to two images of a growing root. The regularly spaced lines indicate the flow vector, the speed, and direction of optic flow at those locations.

As well as global approaches that measure the motion across the whole image, there are also techniques that **track** the motion of individual features or objects within the image. These algorithms have some model of what an object looks like, and a model of the object's dynamics. The tracking usually works by having an estimate of the target's position at time *t*, and uses this plus the motion model to predict a position for the object at time *t+1*. Normally the exact location of the target is found by searching locally around the estimated position, looking for something that matches the model of the object's appearance. This process is repeated throughout the sequence, and at each time-step the location and motion parameters of the object are stored.

There are a number of common tracking algorithms. A classic example of a single target tracking method is the **Kalman filter**. Recently there has been much interest in methods that use probabilistic descriptions of the problem, as these can be more robust to distracting clutter in images.

By tracking an object or feature, the algorithm is able to measure motion information, such as position and velocity across the image plane, for each time-step in the sequence. Care is required that the targets are correctly tracked throughout the sequence, otherwise clearly there will be errors present in the recorded motion information.

Q5 DATA EXTRACTION

Key Notes

Segmentation
Segmentation is the name of a technique that divides an image into a number of homologous regions, such as foreground objects and the background. The Watershed algorithm is a segmentation method that divides up an image by treating the intensity gradients within the image as heights in a 3-D terrain, which is then flooded producing stranded ridges that define the segmentation.

Measuring color
Measuring the color of objects is not as simple as reading off the pixel channel values. The environment and technical settings of the capture equipment must be considered.

Measuring size
Software exists to aid the measurement of objects within images. Some form of calibration is needed to convert the measurements from image-based pixel measures to real-world measurements.

Shape analysis
Statistical methods are available to help examine the shape of images.

Related sections
What is image analysis? (Q1)

How is image analysis used in bioscience research? (Q2)

So far we have illustrated a number of common techniques to clean up an image and extract some basic features, and we will now turn to the problem of how to actually extract high-level data from these images. Increasingly biological research is requiring quantitative evidence from images to support conclusions. This means numbers must be extracted from the representations of the world we have in our images.

The kind of quantification required depends on what the researcher needs. There are several general categories of quantification that might be required; these include determining color measurements, and size, location, and shape analysis. To do this, the object of interest needs to be identified, and this is usually done using a combination of techniques discussed in the previous section, and segmentation methods.

Segmentation
Segmentation refers to dividing up an image into regions within which the pixels share some similarity. For example, you might want to segment the foreground objects in a photograph from the background. There are a number of possible ways to segment an image in such a way. Grouping pixels according to their color is one method. Clustering algorithms such as k-means clustering can be employed for this purpose. This algorithm splits the data (in this case, pixel channel intensities) into k-groups. Initial estimates for the means of each of

the k-clusters are assigned, and each data point is assigned to whichever cluster has the nearest mean. Once all the data have been assigned, new means are calculated for the groups and the procedure is repeated until the data settle into k-stable clusters. The expectation-maximization (EM) algorithm is an alternative clustering algorithm that attempts to estimate the parameters of a model that can generate the color distributions in the regions that are to be segmented. For example, it tries to describe two models, one that generates the foreground pixel colors, and one the background pixel colors. Segmentation algorithms such as these can segment the colors well; however, there is nothing that forces the regions to be continuous areas – a single foreground-color pixel hanging over a background region will be labeled as foreground, even though it is likely to be noise.

Normalized cuts is an approach that segments an image based on boundary information. The image is treated as a graph, with adjacent pixels being connected by an edge, the weight of which is smaller the more different in appearance the two pixels are. Therefore, the image can be segmented by finding ways of 'cutting' low-weight edges, and separating two pixels that were highly different in appearance. A full cut is a route through the graph cutting adjacent edges in just such a way. This approach explicitly looks for region edges in the image, and so the resulting segmentations have clearly defined appearance boundaries.

The last method of segmentation that is very popular uses the intensity gradient levels in an image as though they were height measures on a map. These gradients are calculated using the Sobel operator, described in Section Q4. Therefore, strong edges (high gradients) can be thought of as the peaks of the terrain, and the lack of an edge (low gradient) produces a valley. The name watershed comes from the idea of 'flooding' this terrain with increasing levels of water (see *Fig. 1*). The areas of the image with no edges (the valley floors) become flooded first. By the end, only the peaks produced by the very strongest edges remain above the flood level. The separate 'lagoons' that are formed between the ridges become the segmented areas.

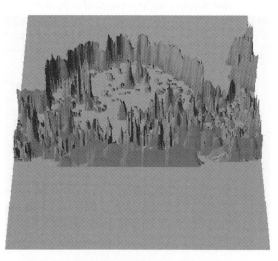

Fig. 1.   *3-D 'terrain map' representing the intensity of edge gradients in the image. Peaks represent strong edges. The image is partially flooded.*

Segmentation is often important in medical images, where the size of various imaged structures is important. 3-D volumes can be estimated by segmenting a structure in a collection of 2-D slice images through an object. The segmented regions can be combined then to give an estimate of the surface of the 3-D structure, and therefore volume can be estimated.

Measuring color

The accurate determination of color in images is more complex than just simply reading pixel values. The recorded color of an object may be determined by the light with which it is illuminated. The extreme example of this is that a white object takes on the color of the illuminating light – a white object illuminated by blue light appears blue, and in an image it may be impossible to determine whether it is the object or the light which is blue. Colored objects are likewise affected. Therefore if measuring color is critical, then the nature of the illuminating light needs to be accounted for or controlled. A light meter can be used to measure the wavelength (and hence true color) of the illuminating light. Some cameras have an **auto white balance** setting which attempts to correct the appearance of an object as if it were being illuminated by a white light source. So for example, an incandescent white balance setting will try to remove the yellowish hue imposed by incandescent lights. It is important to understand that these corrections might be taking place or else errors may occur. For example, if one set of objects are photographed outside and the other in the laboratory, the kind of light illuminating them is likely to be very different, and so care must be taken if their colors are to be determined. It is worth noting that the human eye is particularly good at masking out differences in appearance caused by lighting changes – generally speaking, white paper looks the same outdoors as it does indoors, but to the trained eye or a camera the corresponding images can be very different.

Likewise, how bright an object is in an image depends not only on the amount of light reflected from/transmitted through/emitted by the object in question. Altering the aperture (size of the window through which light passes in the lens) or exposure time (how long the sensor or film is in contact with the light) on the camera affects the brightness of the image. Long exposures and large apertures produce a brighter image than short exposures and smaller apertures. Again, our eyes are very good at doing this without conscious effort, so we do not realize how large an implication these technical changes can have on the resulting image.

Measuring size

As a scientific researcher we may be interested in measuring the size of an object across multiple images; we may want to measure a root as it grows (e.g., *Fig. 2*), or see if a leaf from one experimental group has a larger surface area than a leaf from another group. Of course, this means knowing how the image units relate to real-world units. A calibration target (sometimes a simple ruler) is imaged along with the object of interest, so that the number of image pixels per millimeter can be measured, for example. However, it is important to note that the unit relationship may vary across the image plane according to transformations of the image, such as warping because of the lens or on account of the imaging device not being orthogonal to the plane of interest. This needs to be corrected if it will have a bearing on the measurement.

Shape analysis

As well as detecting shapes, sometimes it is important to analyze their form. Statistical shape analysis uses the statistics of landmark points on the objects in

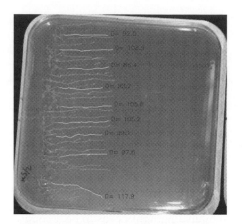

Fig. 2.   Screenshot of semi-automatic root length measuring software.

question to provide a quantification of their shape. Landmark points can either be physical points on the object, mathematical points of interest (such as points of high curvature), or interpolation points spaced around a shape. Statistical shape analysis allows comparisons between the morphology of objects. Typically, these landmark points are either manually or automatically marked on each object in a set. Statistics can then be calculated describing the variation in shape between the objects that were landmarked. Principal component analysis (PCA), for example, provides a means of identifying the main modes of variation in shape.

R TEXTUAL ANALYSIS

Key Notes

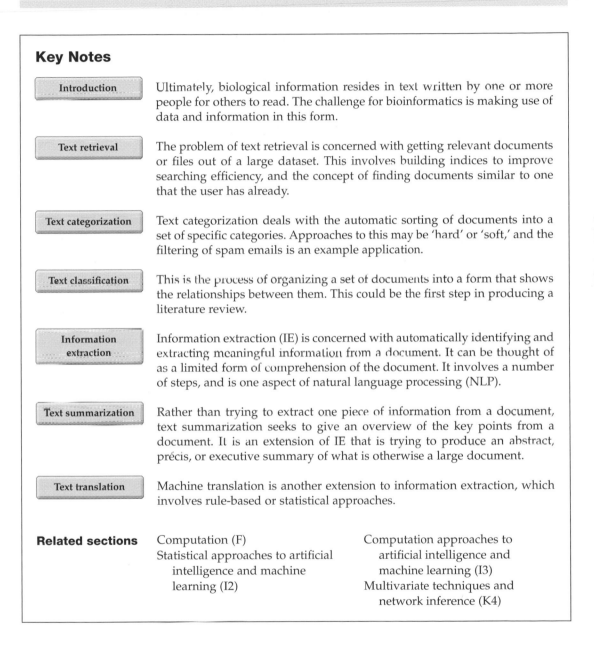

Introduction

Ultimately, biological information resides in text written by one or more people for others to read. The challenge for bioinformatics is making use of data and information in this form.

Text retrieval

The problem of text retrieval is concerned with getting relevant documents or files out of a large dataset. This involves building indices to improve searching efficiency, and the concept of finding documents similar to one that the user has already.

Text categorization

Text categorization deals with the automatic sorting of documents into a set of specific categories. Approaches to this may be 'hard' or 'soft,' and the filtering of spam emails is an example application.

Text classification

This is the process of organizing a set of documents into a form that shows the relationships between them. This could be the first step in producing a literature review.

Information extraction

Information extraction (IE) is concerned with automatically identifying and extracting meaningful information from a document. It can be thought of as a limited form of comprehension of the document. It involves a number of steps, and is one aspect of natural language processing (NLP).

Text summarization

Rather than trying to extract one piece of information from a document, text summarization seeks to give an overview of the key points from a document. It is an extension of IE that is trying to produce an abstract, précis, or executive summary of what is otherwise a large document.

Text translation

Machine translation is another extension to information extraction, which involves rule-based or statistical approaches.

Related sections

Computation (F)
Statistical approaches to artificial
   intelligence and machine
   learning (I2)

Computation approaches to
   artificial intelligence and
   machine learning (I3)
Multivariate techniques and
   network inference (K4)

Introduction

Textual analysis refers to the process of finding and deriving information from large amounts of text or documents. Its major aspects include finding a particular text in a large database (text retrieval), creating subsets of documents on the basis of various criteria (text categorization), identifying the relationships between a set of documents (text classification), finding specific details from one or more documents (information extraction), abstracting key information (text summarization), and translating to another language (machine translation). Feldman and Sanger (2007) is recommended for further reading on this subject.

Text retrieval

The problem of text retrieval is concerned with getting relevant documents or files out of a large dataset. Examples of common problems include searching for appropriate web pages on WWW, or searching for literature in an abstract database, such as PubMed (http://www.ncbi.nlm.nih.gov/pubmed/).

Often, an index is built from the source texts and maintained, to speed up the searches. Then this index, rather than the source documents, is searched. Companies such as Google have developed complex algorithms to provide the indices for the search, which not only maintain lists of keywords but also provide other measures of how suitable a page is, given a particular search. These include the number of links pointing to the page, the frequency it has been visited in recent times, the number of subscribers to news feeds on the page, etc. The key concept is that an index is maintained, which facilitates a very fast search; accessing and searching every single web page when someone searches for a keyword is clearly impractical. Instead, an efficiently designed index allows a very fast response to searches.

It is important to bear in mind when designing such retrieval indices that some keywords seriously reduce the accuracy of searches, and should be excluded from the list. These are known as **stop words**, and include *a, an, the, and, but, so, is, are, have, very,* etc. They are words so common that they are liable to appear in most documents or even literature abstracts. Apart from returning links to almost all the documents in the database and hence having very little value, they slow down the process of finding the documents that are of real interest.

Text-retrieval search-engines often allow a combination of keywords to be entered in conjunction with Boolean operators, such as AND, OR, and less commonly NOT. More detailed searches with regular expressions may be possible as well (see Section F), most commonly using the wild-card '*' to denote any number of extra printable characters.

It may be that rather than searching for specific text, a user may wish to find documents similar to the one he/she already has. In this case, approaches that use word frequency profiles may be appropriate. A profile of the frequency of the words in the user's document (excluding stop-words of course) is created, and this is compared with the profiles of all the documents in the database. The similarity of the word frequency profiles can be used as a measure of similarity between the documents. The search can be made more powerful by considering not just word frequencies but other measures of similarity as well. These might include, for example, replacing individual words with semantic categories: the words **hot, cold, Celsius, thermometer,** and **degrees** might be replaced with the semantic tag **temperature**, or replacing **alanine, glycine, cysteine, tryptophan,** etc. with the **amino acid** tag.

Text categorization

Simply retrieving documents is the first task; the user will then wish to do something with them. **Text categorization** deals with the automatic sorting of documents into specific categories. This might take the form of tagging a document with keywords; for instance assigning a set of keywords to a collection of academic papers. Most text classification adopts a machine learning approach, where a classifier is learned from a number of manually classified training examples (see Sections I2 and I3), and again stop words are excluded.

Software for this may perform **hard** or **soft categorization**. For the former, the system assigns a definitive label to a document; that is, a decision is made as to whether the document is in a category or not. Conversely for the latter, a firm decision is not required. Instead, for a document, a confidence score is assigned to each possible category, followed by the user choosing the categories of interest from an ordered shortlist supplied by the system. Hard categorization can subsequently be used on the basis of a **confidence threshold** set automatically or by the user.

The detection of spam emails is a form of text categorization, in this case the categories being 'spam' or 'not-spam.' A system can learn the profile of a spam email, and then assign incoming mail as either probably being spam or not. The people who send spam emails often try to confuse such detection systems by including paragraphs of legitimate text within the email. Detection systems often need to learn the labeling process by asking the user to categorize their first emails manually, but in addition they may have been 'pre-taught' from large datasets of already-classified normal and spam emails. Statistical methods are used to learn the classifications; for example, a naïve Bayes' classifier (see Section I2) can learn to identify spam emails by considering the probabilities of the individual elements (such as the words) that make up the email, and thus calculating the probability that the email as a whole is spam. Again semantic tagging can play a role.

There are two common measures of error for text categorization systems: recall errors that arise when a category fails to contain a document it should, and precision errors when a category contains a document it should not.

Text classification

The obvious extension from categorization is **text classification**, in which the relationships between all the documents are shown, usually in some form of tree structure. This is analogous to sorting gene expression profiles (see Section K4), and similar techniques are used: hierarchical clustering and self-organizing maps. This approach can be very useful for gaining a broader understanding of the body of documents retrieved, and could provide the basis for producing a literature review.

Information extraction

Doubtless some students (and high-pressure business executives) dream of the day when a computer reads a document for them and supplies the key pieces of information they are seeking. This is the aim of **information extraction** (IE). It is concerned with automatically identifying and extracting meaningful information from a document. It can be thought of as a limited form of comprehension of the document. Having a complete understanding of a document (working out what it means, and what knowledge and relationships are contained within it) is something still only reliably achievable by humans. However, it is possible to obtain a limited understanding, in the form of identifying certain semantic information within a document, under certain circumstances.

In order to perform sophisticated information extraction, often a degree of **natural language processing (NLP)** is necessary. NLP is concerned with interpreting text written 'naturally'; that is, making human-written text understandable by a computer. The process of information extraction consists of a number of steps, as follows.

1. Break the text down into words, sentences, and paragraphs.
2. Perform a lexical analysis on the words, and tag accordingly. For example, identify and tag obvious nouns, verbs, prepositions, conjunctions, etc., and also tag special semantic meanings.

3. Perform syntactic analysis on the organization of the words.
4. Parse sentences.
5. Search for pre-defined scenarios.
6. Extract relevant information.

The initial processing of a text breaks the document down into subunits which might comprise words, sentences, and paragraphs. Where possible, words are tagged then according to their role in the text (noun, verb, etc.), and additionally words may be tagged using particular dictionaries and custom rules. A dictionary of chemical compounds would be able to tag compound names encountered in the text as being chemicals. An example of a rule might be:

Words with more than five letters AND ending with **ase**, but NOT **increase, decrease**, or **database** should be tagged as **enzyme**.

Proper nouns can be identified using a combination of context and regular expression matching.

Groups of words are identified next. Word combinations are tagged into grammatical units; a process referred to as **chunking**. For example, the components of an address may be tagged as a location. These groups are created based on manually constructed templates, which help identify groups of nouns and verbs, etc. This process is referred to as **shallow parsing**. Once complete, relations are built between the units of the document. One such task is resolving the co-referencing of pronouns; in other words, determining what words such as **he, they**, and **it** refer to. Consider the sentence

James wore a red coat. He always did.

He refers to James in the real world, so the process of resolving the co-reference would equate **He** and **James** as the same physical thing. Other types of co-referencing include equating variations of proper names, such as identifying 'Mr S. Smith' and 'Sam Smith' as the same person.

Scenarios are logical clauses that contain tagged words of particular interest. Let us suppose that the task is to extract information about enzyme substrates. A scenario might take the form

$(\{enzyme\ tag\}) + ((metabolize^*)\ or\ (convert^*)\ or\ (catalyze^*\ the\ conversion\ of)) + (\{compound\ tag\})$

which is looking for an enzyme and compound separated by possible verbs denoting that the compound is a substrate. This scenario provides the clue for inferring the enzyme-substrate relationship implicit in the text. This approach can be useful for filling in missing or incomplete information, and has been used to attempt to generate enzyme databases automatically.

Poorly written text can introduce ambiguity that the rule-based systems cannot resolve properly. From the text:

The cat sat on the mat. It was made of cotton.

What is the carpet is made of? Splitting the words and tagging reveal that **mat** is a synonym of **carpet**. However in this case, **It** would co-reference **cat** and not **mat**. A person would be able to guess that **it** should refer to the **mat**, as prior knowledge suggests cats are seldom made of cotton, whereas mats might be. Without the background knowledge and inference capabilities of a human being, the system would struggle, even on text as simple as this. Out of interest, the correctly formed sentence should have been:

The cat sat on the mat, which was made of cotton.

There are a number of software packages to help with NLP, a popular example being the General Architecture for Text Engineering (GATE) environment. This software can be used as a visual development environment for developing tools to perform NLP tasks, and comes complete with a toolkit for doing various tasks within NLP, such as information extraction (http://gate.ac.uk). This area of development is very sophisticated, so most people usually decide it is quicker simply to read the document.

Text summarization

One obvious extension of the above is to try to extract all the key points out of a document. This **text summarization** is an area of NLP in its own right. It aims to provide précis, abstracts, or executive summaries of larger documents. Used recursively, it distils down to a single sentence, which would probably be an irrelevant sound-bite.

Text translation

Building on information extraction systems, there are now grammar checkers available in most word processing packages to highlight linguistic errors. Even automated mechanisms for translating text from one language to another have been developed. However, translation is a more complex problem than simply replacing a word in one language with the equivalent in another, as often the structure of the sentence itself needs to be altered. For example, in some languages such as German, verbs can appear at the end of the sentence, so when translating from languages such as English, the order of words in the sentence will change. So to successfully translate text, software must have an 'understanding' of how sentences are formed in each language. Decoding the sentence to be translated is an information extraction problem, where as much meaning as possible must be extracted from the sentence to enable successful translation. This knowledge is used to reform the sentence in the other language.

Two common approaches to the translation problem are rule-based translation and statistical approaches. In the former, rules are entered manually into the system by a human linguistics expert. This method requires a lot of manual effort to define the rules, and as with any rule-based approach, it is hard to cover every possible scenario with a suitable rule. Statistical approaches rely on a large, structured dataset of linguistic examples (called a **corpus**), from which the aim is to learn probabilistic rules about how parts translate from one language to another. While its advantages lie in processing partial or poorly formed sentences, statistical approaches can require much computing power and struggle with relationships that are spatially distant in the text.

Whichever approach is used, there are still problems, usually to do with semantic meaning, which are very hard to resolve without the real-world knowledge we, as humans, have. We can draw on a lifetime of knowledge to help us understand documents. As yet this is still beyond the realms of feasible computing capability, and fully automatic machine-translation remains a challenging problem.

Most of these textual analysis systems usually rely on large sets of rules, especially rules to catch 'exceptions to the rule' – things which break from the norm. One of the hardest problems is the variation of meaning of words in a natural language. For example, in most usage, the word **bad** refers to something that is not good. However, in some situations, perhaps especially in parts of the US, referring to something as **bad** actually means it is **extremely** good. We leave it up to the reader to decide whether or not this book is bad.

FURTHER READING

Section A   Benigni, R. and Giuliani, A. (2003) Putting the predictive toxicology challenge into perspective. *Bioinformatics* **19**, 1194–1200.

Cann, R.L., StoneKing, M. and Wilson, A.C. (1987) Mitochondrial DNA and human evolution. *Nature* **325**, 31–36.

Jobling, M.A. and Gill, P. (2004) Encoded evidence: DNA in forensic analysis. *Nat. Rev. Genet.* **5**, 739–51.

Leach, A. (2001) *Molecular modelling: Principles and applications.* Prentice Hall, Upper Saddle River, NJ.

Paabo, S., Poinar, H., Serre, D., *et al.* (2004) Genetic analyses from ancient DNA. *Annu. Rev. Genet.* **38**, 645–679.

Schena, M., Shalon, D., Davis, R.W., *et al.* (1995) Quantitative monitoring of gene expression with a complementary DNA microarray. *Science* **270**, 467–470.

Southern, E. (1975) Detection of specific DNA sequences among DNA fragments separated by gel electrophoresis. *J. Mol. Biol.* **98**, 503–517.

Section B   Hodgman, T.C. (1998) Eating the seed corn – inhibiting the growth of Bioinformatics. *Bioinformatics* **14**, 829.

Hodgman, T.C. (2000) A historical perspective on gene/protein function prediction. *Bioinformatics* **16**, 10–15.

Letovsky, S. (1999) *Bioinformatics: Databases and systems.* (ed. S. Letovsky) Kluwer Academic Publishers, Boston, p.5.

Marchant, J. (2000) Careers in bioinformatics. *New Scientist* **167** (2252), 54–55.

Rybak, B. (1968) *Psyché, Soma Germen.* Gallimard (France).

Rybak, B. (1978) A program for teaching bio-informatics. *Biosci. Communications* **4**, 158–159.

Section E   Bourne, P.E., Berman, H.M., McMahon, B., *et al.* (1997) The macromolecular crystallographic information file. *Methods Enzymol.* **277**, 571–590.

Brune, R.M., Bard, J.B.L., Dubreuil, C., *et al.* (1999) A three-dimensional model of the mouse at embryonic day 9. *Dev. Biol.* **216**, 457–468.

Daylight, I. SMILES Tutorial. http://www.daylight.com/dayhtml_tutorials/languages/smiles/index.html (2007) 30–10.

Dolan, M.E., Holden, C.C., Beard, M.K., *et al.* (2006) Genomes as geography: using GIS technology to build interactive genome feature maps. *Bioinformatics* **7**, 416.

Schierz, A.C., Soldatova, L.N. and King, R.D. (2007) Overhauling the PDB. *Nat. Biotechnol.* **25** (4), 437–442.

Section F   Jeffries, R., Anderson, A. and Hendrickson, C. (2000) *Extreme programming installed.* Addison-Wesley Professional, Salt Lake City, UT.

Weston, P. (2004) *Bioinformatics software engineering: delivering effective applications.* John Wiley & Sons, New York, NY.

Section G   Ennos. R. (2007) *Statistical and data handling skills in biology.* 2nd edn. Prentice Hall, Upper Saddle River, NJ.

Section H   Albert, R., Jeong, H. and Barabási, A.-L. (1999) Diameter of the world-wide web. *Nature* **401**, 130.

Becker, A.A. (2003) *Introductory guide to finite element analysis*. ASME Press, New York, NY.

Burrage, K., Hancock, J., Leier, A., *et al.* (2007) Modelling and simulation techniques for membrane biology. *Brief. Bioinform.* **8**, 234–244.

Calvano, S.E., Xiao, W., Richards, D.R., *et al.* (2005) A network-based analysis of systemic information in humans. *Nature* **437**, 1032–1037.

Elowitz, M., Levine, A.J., Siggia, E.D., *et al.* (2002) Stochastic gene expression in a single cell. *Science* **297**, 1183–1186.

Gibbons, A. (1985) *Algorithmic graph theory*. Cambridge University Press, Cambridge.

Gross, J.L. and Yellen, J. (2004) *Handbook of graph theory*. CRC Press, Boca Raton, FL.

Hiroi, N. and Funahashi, A. (2007) Kinetics of dimension-restricted conditions. In: *Introduction to systems biology*. (ed. S. Choi). Humana Press, Totawa, NJ, pp 261–281.

Jeong, H., Tombor, B., Albert, R., *et al.* (2000) The large-scale organization of metabolic networks. *Nature* **407**, 651–654.

Karp, P., Ouzounis, C.A., Moore-Kochlacs, C., *et al.* (2005) Expansion of the Biocyc collection of pathway/genome databases to 160 genomes. *Nucleic Acids Res.* **33**, 6083–6089.

Klipp, E., Herwig, R., Kowald, A., *et al.* (2005) *Systems biology in practice*. Wiley-VCH, Weinheim.

Klipp, E., Liebermeister, W., Helbig, A., *et al.* (2007) Systems biology standards – the community speaks. *Nat. Biotechnol.* **25**, 390–391.

Noble, D. (2006) *The Music of life*. OUP, Oxford.

Paun, Gh. (2000) Computing with membranes. *J. Computer Syst. Sci.* **61**, 108–143.

Petri, C.A. (1962) Kommunikation mit Automaten. PhD Thesis, Univ. Bonn, Bonn.

Wilkinson, D. (2006) *Stochastic modelling for systems biology*. Chapman Hall/CRC, London.

Section I

Andrieu, C., De Freitas, N., Doucet, A., *et al.* (2003) An introduction to MCMC for machine learning. *Machine Learning* **50**, 5–43.

Ben-Gal, I., Shani, A., Gohr, A., *et al.* (2005) Identification of transcription factor binding sites with variable-order Bayesian networks. *Bioinformatics* **21** (11), 2657–2666.

Burgess, C.J.C. (1998) A tutorial on support vector machines for pattern recognition. *Data Mining Knowledge Disc.* **2**, 121–167.

Chen, T., Lu, C. and Li, W. (2004) Prediction of splice sites with dependency graphs and their expanded bayesian networks. *Bioinformatics* **21** (4), 471–482.

De Raedt, L. and Kersting, K. (2004) Probabilistic inductive logic programming. *Proc. 15th Int. Conf. ALT*, Padova, pp. 19–36.

Dietterich, T. (1995) Overfitting and undercomputing in machine learning. *ACM Comput. Surveys* **27** (3), 326–327.

Eddy, S.R. (2004) What is a hidden Markov model? *Nat. Biotechnol.* **22** (10), 1315–1316.

Garthwaite, P. H., Jolliffe, I. T. and Jones, B. (1995) *Statistical inference*, 1st edn. Prentice Hall, Upper Saddle River, NJ.

Hodgman, T.C. (1988) A new superfamily of replicative proteins. *Nature* **333**, 22–23, 578.

Imoto, S., Higuchi, T., Goto, T., *et al.* (2003) Combining microarrays and biological knowledge for estimating gene networks via bayesian networks. *Bioinf. Comput. Biol.* **2** (1), 77–98.

Khan, Z., Balch, T. and Dellaert, F. (2004) An MCMC-based particle filter for tracking multiple interacting targets. *Proc. ECCV*, Prague, pp. 279–290.

Levitsky, V.G., Ignatieva, E.V., Ananko, E.A., *et al.* (2007) Effective transcription factor binding site prediction using a combination of optimization, a genetic algorithm and discriminant analysis to capture distant interactions. *BMC Bioinformatics* **8**, 481.

Miklós, I. (2003) MCMC genome rearrangement. *Bioinformatics* **19** (2), 130–137.

Minsky, M. and Papert, S. (1969) *Perceptrons*. MIT Press, Cambridge, MA.

Muggleton, S. and De Raedt, L. (1994) Inductive logic programming: Theory and methods, *Journal of Logic Programming*, **19**, 629–679.

Paschke, A. and Schröder, M. (2007) Inductive logic programming for bioinformatics in Prova. *Very Large Data Base (VLDB) Endowment*, Vienna.

Pedersen, J. and Hein, J. (2003) Gene finding with a hidden Markov model of genome structure and evolution. *Bioinformatics* **19** (2), 219–227.

Rabiner, L.R. (1989) A tutorial on hidden Markov models and selected applications in speech recognition. *Proc. IEEE* **77** (2), 257–286.

Rojas, R. (1996) *Neural networks.* Springer-Verlag, Berlin.

Rosenblatt, F. (1958) The perceptron: a probabilistic model for information storage and organization in the brain. *Psychol. Rev.* **65** (6), 386–408.

Semikhodskii, A. (2007) *Dealing with DNA evidence: A legal guide*. Routledge, Oxford.

Zou, M. and Conzen, S.D. (2005) A new dynamic Bayesian network (DBN) approach for identifying gene regulatory networks from time course microarray data. *Bioinformatics* **21** (1), 71–79.

Section K

Bansal, M., Gatta, G.D. and di Bernardo, D. (2006) Inference of gene regulatory networks and compound mode of action from time course gene expression profiles. *Bioinformatics* **22** (7), 815–822.

Brazma, A., Hingamp, P., Quackenbush, J., *et al.* (2001) Minimum information about a microarray experiment (MIAME) – toward standards for microarray data. *Nat. Genet.* **29**, 365–371.

Cline, M.S., Smoot, M., Cerami, E., *et al.* (2007) Integration of biological networks and gene expression data using Cytoscape. *Nat. Protoc.* **2** (10), 2366–2382.

Craigon, D.J., James, N., Okyere, J., *et al.* (2004) NASCArrays: a repository for microarray data generated by NASC's transcriptomics service. *Nucleic Acids Res.* **32** (Database Issue), D575–D577.

Darby, A.C. and Hall, N. (2008) Fast forward genetics. *Nat. Biotechnol.* **26**, 1248–1249.

Dougherty, E.R., Akutsu, T., Cristea, P.D., *et al.* (2007) Genetic regulatory networks. *J. Bioinform. Syst. Biol.* (special issue), **2007**, 17321.

Hubbell, E., Liu, W.M. and Mei, R. (2002) Robust estimators for expression analysis. *Bioinformatics* **18** (12), 1585–1592.

Ideker, T., Thorsson, V., Ranish, J.A, *et al.* (2001) Integrated genomic and proteomic analyses of a systematically perturbed metabolic network. *Science* **292** (5518), 929–934.

Irizarry, R.A., Warren, D., Spencer, F., *et al.* (2005) Multiple-laboratory comparison of microarray platforms. *Nat. Methods* **2**, 345–350.

Li, C. and Wong, W.H. (2001) Model-based analysis of oligonucleotide arrays: expression index computation and outlier detection. *Proc. Natl. Acad. Sci. USA* **98** (1), 31–36.

Lim, W.K., Wang, K., Lefebvre, C., *et al.* (2007) Comparative analysis of microarray normalization procedures: effects on reverse engineering gene networks. *Bioinformatics* **23** (13), 282–288.

Spieth, C., Streichert, F., Speer, N., *et al.* (2004) Optimizing topology and parameters of gene regulatory networks from time series experiments. *Lecture Notes Computer Sci.* **3102**, 461–470.

Stekel, D. (2003) *Microarray bioinformatics*. CUP, Cambridge.

The Wellcome Trust Case Control Consortium (2007) Genome-wide association study of 14,000 cases of seven common diseases and 3,000 shared controls. *Nature* **447**, 661–678.

Wu, Z., Irizarry, R.A., Gentleman, R., *et al.* (2004) A model-based background adjustment for oligonucleotide expression arrays. *J. Am. Stat. Assoc.* **99**, 909–917.

Section L

Cline, M.S., Smoot, M., Cerami, E., *et al.* (2007) Integration of biological networks and gene expression data using Cytoscape. *Nat. Protoc.* **2**, 2366–2382.

Cramer, C.J. (1961) Essentials of computational chemistry: Theories and models. 2nd ed. (ed Christopher J. Cramer). Wiley, Chichester.

Cui, J., Li, P., Li, G., *et al.* (2008) AtPID: *Arabidopsis thaliana* protein interactome database – an integrative platform for plant systems biology. *Nucleic Acids Res.* **36**, D999–D1008.

Dunkley, T.J.P., Watson, R., Griffin, J.L., *et al.* (2004) Localization of organelle proteins by isotope tagging. *Mol. Cell. Proteomics* **3**, 1128–1134.

Forler, D., Köcher, T., Rode, M., *et al.* (2002) An efficient protein complex purification method for functional proteomics in higher eukaryotes. *Nat. Biotechnol.* **21**, 89–92.

Fujii, K., Kondo, T., Yamada, M., *et al.* (2006) Toward a comprehensive proteome database. *Proteomics* **6**, 4856–4876.

Garwood, K., McLaughlin, T., Garwood, C., *et al.* (2004) PEDRo: A database for storing, searching and disseminating experimental proteomics data. *BMC Genomics* **5**, 68.

Gavin, A.C., Bösche, M., Krause, R., *et al.* (2002) Functional organization of the yeast proteome by systematic analysis of protein complexes. *Nature* **415**, 141–147.

Hames, D. and Hooper, N. (2005) *Instant Notes: Biochemistry*, 3rd edn. Taylor & Francis, London.

Höltje, H-D., Sippl, W., Rognan D. (2008) *Molecular modeling: Basic principles and applications*, 3rd edn. Wiley-VCH, Weinheim.

Jensen, F. (2007) *Introduction to computational chemistry*, 2nd edn. (ed F. Jensen). John Wiley & Sons, Chichester.

Leach, A.R. (2001) *Molecular modelling: Principles and applications,* 2nd edn. (ed A.R. Leach). Prentice Hall, Harlow.

Lesk, A. (2002) *Introduction to bioinformatics*. OUP, Oxford.

Laskowski, R.A., Luscombe, N.M., Swindells, M.B., *et al.* (1996) Protein clefts in molecular recognition and function. *Protein Sci.* **5**, 2438–2452.

Oh, J.M.C., Brichory, E., Puravs, R., *et al.* (2001) A database of protein expression in lung cancer. *Proteomics* **1**, 1303–1319.

Sander, C. and Schneider, R. (1991) Database of homology-derived protein structures and the structural meaning of sequence alignment. *Proteins* **9**, 56–68.

Section N

Bard, J., Kaufman, M., Dubreuil, C., *et al.* (1998) An internet-accessible database of mouse developmental anatomy based on a systematic nomenclature. *Mech. Dev.* **74**, 111–120.

Bodenreider, O. and Stevens, R. (2006) Bio-ontologies: current trends and future directions. *Brief. Bioinform.* **7**, 256–274.

Brune, R.M., Bard, J.B.L., Dubreuil, C., *et al.* (1999) A three-dimensional model of the mouse at embryonic day 9. *Dev. Biol.* **216**, 457–468.

Ringwald, M., Baldock, R., Bard, J., *et al.* (1994) A database for mouse development. *Science* **265**, 2033–2034.

Ringwald, M., Davis, G.L., Smith, A.G., *et al.* (1997) The mouse gene expression database GXD. *Cell Dev. Biol.* **8**, 489–497.

Schwarz, A.J., Danckaert, A., Reese, T., *et al.* (2006) A stereotaxic MRI template set for the rat brain with tissue class distribution maps and co-registered anatomical atlas: Application to pharmacological MRI. *NeuroImage* **32**, 538–550.

Toga, A.W., Santori, E.M., Hazani, R., *et al.* (1995) A 3D digital map of rat brain. *Brain Res. Bull.* **38** (1), 77–85.

Section O

Calvano, S.E., Xio, W., Richards, D.R., *et al.* (2005) A network-based analysis of systemic inflammation in humans. *Nature* **437**, 1032–1037.

Degtyarenko, K., de Matos, P., Ennis, M., *et al.* (2008) ChEBI: a database and ontology for chemical entities of biological interest. *Nucleic Acids Res.* **36**, D344–D350.

Duarte, N.C., Becker, S.A., Jamshidi, N., *et al.* (2007) Global reconstruction of the human metabolic network based on genomic and bibliomic data. *Proc. Natl Acad. Sci. USA* **104**, 1777–1782.

Fairlamb, A.H. (1996) Pathways to drug discovery. *The Biochemist* (Feb/Mar) **18**, 11–16.

Fell, D. (1997) *Understanding the control of metabolism.* Portland Press, London.

Hocker, C.G. (1994) Applying bifurcation analysis to enzyme kinetics. *Methods Enzymol.* **240**, 781–816.

Joe, Y., Miwa, T., Kida, T., *et al.* (2002) *Method for producing transformed plant having increased glutamic acid content*, US Patent 0100074 A1.

Kanehisa, M., Araki, M., Goto, S., *et al.* (2008) KEGG for linking genes to life and the environment. *Nucleic Acids Res.* **36**, D480–D484.

Klamt, S., Saez-Rodriguez, J. and Gilles, E. (2007) Structural and functional analysis of cellular networks with cellnetanalyzer. *BMC Syst. Biol.* **1**, 2.

OMIM:300661 http://www.ncbi.nlm.nih.gov/entrez/dispomim.cgi?id=300661

Palsson, B.O. (2006) *Systems biology: Properties of reconstructed networks.* CUP, Cambridge.

Schwarz, R., Musch, P., von Kamp, A., *et al.* (2005) YANA – a software tool for analysing flux modes, gene-expression and enzyme activities. *BMC Bioinformatics* **6**, 135.

Selkov Jr, E., Grechkin, Y., Mikhailova, N., *et al.* (1998) MPW: the metabolic pathways database. *Nucleic Acids Res.* **26**, 43–45.

Torres, N.V. and Voit, E.O. (2002) *Pathway analysis and optimization in metabolic engineering.* CUP, Cambridge.

Von Kamp, A. and Schuster, S. (2006) Metatool 5.0: fast and flexible elementary modes analysis. *Bioinformatics* **22**, 1930–1931.

Windass, J., Worsey, M.J., Pioli, E.M., *et al.* (1980) Improved conversion of methanol to single-cell protein by *Methylophilus methylotrophus*. *Nature* **287**, 396–401.

Section P

Bourn, D.W., Maleckar, M.M., Rodriguez, B., *et al.* (2006) Mechanistic enquiry into the effect of increased pacing rate on the upper limit of vulnerability. *Philos. Transact. A Math. Phys. Eng. Sci.* **364**, 1333–1348.

Bradley, C.P., Pullan, A.J. and Hunter, P.J. (1997) Geometric modeling of the human torso using cubic hermite elements. *Ann. Biomed. Eng.* **25** (1), 96–111.

Byrne, H.M., Alarcon, T., Owen, M.R., *et al.* (2006) Modelling aspects of cancer dynamics: A review. *Philos. Transact. A Math. Phys. Eng. Sci.* **364**, 1563–1578.

Byrne, H.M., Owen, M.R., Alarcon, T., *et al.* (2006) Modelling the response of vascular tumours to chemotherapy: A multiscale approach. *Mathematical Models and Methods in Applied Science* **16** (7S), 1219–1241.

Ch'en, F.F.T., Vaughan-Jones, R.D., Clarke, K., *et al.* (1998) Modelling myocardial ischaemia and reperfusion. *Prog. Biophys. Mol. Biol.* **69**, 515–538.

Crampin, E.J., Halstead, M., Hunter, P., *et al.* (2004) Computational physiology and the physiome project. *Exp. Physiol.* **89** (1), 1–26.

FDA (1997) *Eightieth meeting of the cardiovascular and renal drugs advisory committee, Center for Drug Evaluation and Research. FDA. Silver Spring, USA.*

Finch-Savage, W.E., Rowse, H.R. and Dent, K.C. (2005) Development of combined inhibition and hydrothermal threshold models to simulate maize (*Zea mays*) and chickpea (*Cicer arietinum*) seed germination in variable environments. *New Scientist* **165**, 825–838.

Highfield, R. (2006) *The virtual medical man.* (31-10-2006) Telegraph Online – accessed 31/8/2007.

Jönsson, H., Heisler, M.G., Shapiro, B.E., *et al.* (2006), An auxin-driven polarized transport model for phyllotaxis, *Proc. Natl. Acad. Sci.* **103** (5), 1633–1638.

Kohl, P., Noble, D., Winslow, R.L., *et al.* (2000) Computational modelling of biological systems: tools and visions. *Philos. Transact. A Math. Phys. Eng. Sci.* **358**, 579–610.

Kurth, W. (1996) Some new formalisms for modelling the interactions between plant and architecture, competition and carbon allocation. *4th Workshop on Individual-based Structural and Functional Models in Ecology*, Wallenfels.

Lindenmayer, A. (1968) Mathematical models for cellular interactions in development. *J. Theor. Biol.* **18** (3), 280–299.

McDougall, S., Dallon, J., Sherratt, J., *et al.* (2006) Fibroblast migration and collagen deposition during dermal wound healing: mathematical modelling and clinical implications. *Philos. Transact. A Math. Phys. Eng. Sci.* **364**, 1385–1405.

Mjolsness, E. and Meyerowitz, E. (2007) *The Computable Plant.* http://computableplant.ics.uci.edu/ Accessed 28-11-2007.

Mjolsness, E. (2006) Stochastic process semantics for dynamical grammar syntax: an overview. *Proc. Ninth International Symposium on Artificial Intelligence and Mathematics*, Fort Lauderdale, Florida, USA.

Mjolsness, E. (2006) The growth and development of some recent plant models: A viewpoint. *J. Plant Growth Regul.* **25**, 270–277.

Nash, N.P. and Hunter, P.J. (2000) Computational mechanics of the heart. *J. Elasticity* **61**, 113–141.

Noble, D. (2002) Modeling the heart – from genes to cells to the whole organ. *Science* **295**, 1678–1682.

Noble, D. (2002) Modelling the heart: insights, failures and progress. *BioEssays* **24**, 1155–1163.

Noble, D. (2006) *The music of life.* Oxford University Press, Oxford, UK.

Prusinkiewicz, P. and Lindenmayer, A. (1990) *The algorithmic beauty of plants* Springer-Verlag, London.

Prusinkiewicz, P., Mündermann, L., Karwowski, R., *et al.* (2001) The use of positional information in the modeling of plants. Proceedings of SIGGRAPH, LA, California, USA. 289–300.

Rolland-Lagan, A.-G., Bangham, J.A. and Coen, E. (2003) Growth dynamics underlying petal shape and symmetry. *Nature* **422**, 161–163.

Tawhai, M.H., Burrowes, K.S. and Hoffman, E.A. (2006) Computational models of structure function relationships in the pulmonary circulation and their validation. *Exp. Physiol.* **91** (2), 285–293.

University of Nottingham. Centre for Plant Integrative Biology (2007) http://www.cpib.eu/ Accessed 27-9-2007.

Warwick, H.R.I. (2007) *Simulation of seed germination and seedling emergence.* http://www2.warwick.ac.uk/fac/sci/whri/research/seedscience/simulation/

Wiese, A., Christ, M.M., Virnich, O., *et al.* (2007) Spatio-temporal leaf growth patterns of *Arabidopsis thaliana* and evidence for sugar control of the diel leaf growth cycle. *New Phytologist* **174**, 752–761.

Winslow, R.L., Tanskanen, A., Chen, M., *et al.* (2006) Multiscale modeling of calcium signalling in the cardiac dyad. *Ann. N. Y. Acad. Sci.* **1080**, 362–375.

Youm, J.B., Kim, N., Han, J., *et al.* (2006) A mathematical model of pacemaker activity recorded from mouse small intestine. *Philos. Transact. A Math. Phys. Eng. Sci.* **364**, 1135–1154.

Section Q

Canny, J. (1986) A computational approach to edge detection. *IEEE Trans. Pattern Anal. Mach. Intell.* **8** (6), 679–698.

Dryden, I.L. and Mardia, K.V. (1998) *Statistical shape analysis.* John Wiley, New York, NY.

Farid, H. and Popescu, A.C. (2001) Blind removal of lens distortion. *J. Opt. Soc. Am. A Opt. Image Sci. Vis.* **18** (9), 2072–2078.

Frost, A., French, A.P., Tillett, R.D., *et al.* (2004) A vision guided robot for tracking a live, loosely constrained pig. *Comp. Electron. Agric.* **44**, 93–106.

Garini, Y., Vermolen, B.J. and Young, I.T. (2005) From micro to nano: recent advances in high-resolution microscopy. *Curr. Opin. Biotechnol.* **16**, 3–12.

Gonzalez, R.C. and Woods, R.E. (2008) *Digital image processing*, 3rd edn. Prentice Hall, Upper Saddle River, NJ.

Isard, M. and Blake, A. (1998) Condensation – conditional density propagation for visual tracking. *Int. J. Comput. Vision* **29** (1), 5–28.

Jung, H. and Cho, H. (2002) An automatic block and spot indexing with k-nearest neighbors graph for microarray image analysis. *Bioinformatics* **18**(Suppl. 2), S141–S151.

Kalman, R.E. (1960) A new approach to linear filtering and prediction problems. *J. Basic Eng.* **82** (D), 35–45.

Kapelan Bio-Imaging. LabImage 1D. (2007) http://www.labimage.net/ Accessed 12-7-2007.

Non-linear dynamics. Bioinformatics and electrophoresis software for proteomics and genomics research. (2007) http://www.nonlinear.com/ Accessed 12-7-2007.

North, A.J. (2006) Seeing is believing? A beginner's guide to practical pitfalls in image acquisition. *J. Cell Biol.* **172** (1), 9–18.

Pawley, J. (2000) The 39 steps: A cautionary tale of quantitative 3-D fluorescence microscopy. *Biotechniques* **28** (5), 884–888.

Silk Scientific, I. Un-Scan-It gel. (2007) http://www.silkscientific.com/Accessed 12-7-2007.

Stekel, D. *Microarray bioinformatics* (2003) Cambridge University Press, Cambridge, UK.

Swaminathan, R. and Nayar, S.K. (2000) Non-metric calibration of wide angle lenses. *IEEE Trans. Patt. Anal. Machine Intell.* **22** (10), 1172–1178.

Swiss Institute of Bioinformatics. ExPASy Proteomics Server. (2007) http://expasy.org/Accessed 3-7-2007.

Tillett, N.D., Hague, T. and Miles, S.J. (2002) Inter-row vision guidance for mechanical weed control in sugar beet. *Comput. Electron. Agric.* **33** (3), 163–177.

Tsai, R.Y. (1986) An efficient and accurate camera calibration technique for 3D machine vision. *Proc. IEEE Conf. Comput. Vision. Patt. Recog.* Miami Beach, FL, pp. 364–374.

van der Weele, C.M., Jiang, H.S., Palaniappan, K.K., *et al.* (2003) A new algorithm for computational image analysis of deformable motion at high spatial and temporal resolution applied to root growth. Roughly uniform elongation in the meristem and also, after an abrupt acceleration, in the elongation zone. *Plant Physiol.* **132**, 1138–1148.

Viola, P. and Jones, M. (2001) Rapid object detection using a boosted cascade of simple features. *Int. Conf. Patt. Recog.* **1**, 511–518.

Wallace, W., Schaefer, L.H. and Swedlow, J.R. (2001) A workingperson's guide to deconvolution. *Biotechniques* **31** (5), 1076–1097.

Wu, J., Tillett, R.D., McFarlane, N., *et al.* (2004) Extracting the three-dimensional shape of live pigs using stereo photogrammetry. *Comput. Electron. Agric.* **44**, 203–222.

Yang, Y.H., Buckley, M.J. and Speed, T.P. (2001) Analysis of cDNA microarray images. *Brief. Bioinform.* **2** (4), 341–349.

Yoo, T.S. (2004) *Insight into images: Principles and practice for segmentation, registration, and image analysis.* AK Peters, Wellesey, MA.

Zucker, R.M. and Price, O. (2001) Evaluation of confocal microscopy system performance. *Cytometry* **44**, 273–294.

Section R

Feldman, R. and Sanger, J. (2007) *The text mining handbook.* CUP, New York.

Humphreys, K., Demetriou, G. and Gaizauskas, R. (2000) Bioinformatics applications of information extraction from scientific journal articles. *J. Inform. Sci.* **26**, 75–85.

INDEX